PHYSICS OF THE EARTH

Gemini XI photograph of the Gulf of Aden and the Red Sea by NASA astronauts Charles Conrad and Richard F. Gordon. This is one of the areas of particular interest in the theory of sea-floor spreading. A line of earthquake epicenters (see Figs. 4.1 and 7.4) extends from the ridge system in the Indian Ocean up the middle of the Gulf of Aden and into the Red Sea and is presumed to mark the axis of a new ridge, along which mantle material is rising and pushing Africa and Asia apart. Photograph courtesy of the National Aeronautics and Space Administration, Washington, D.C.

Physics of the Earth

FRANK D. STACEY

Reader in Physics
University of Queensland
Australia

John Wiley & Sons, Inc.

New York · London · Sydney · Toronto

Library of Congress Catalogue Card Number: 70-81330

SBN 471 81955 7

Printed in the United States of America

PREFACE

*... Obviously there are no well qualified students of
Earth, and all of us, in different degrees, dig our own
small specialised holes and sit in them."*

BULLARD, 1960, p. 92.

The first purpose of this book is to bring the fundamental problems in solid-earth geophysics to the attention of graduate and advanced undergraduate students of physics. The physical study of the Earth has a natural fascination, but for the physicist who has no previous contact with the Earth sciences there is a wide range of new concepts. Mathematical developments are here simplified or relegated to appendices, so that they do not obscure the physical arguments; I hope that this has made most of the text digestible to students of geology and related sciences. I have also aimed to extend its usefulness by a careful selection of references from the now vast literature.

Most geophysical effects are not accessible to scientific manipulation and they are often complex. The materials concerned are neither pure nor homogeneous and exact analysis is often out of the question. In these circumstances vague suggestions become hypotheses and hypotheses are called theories. Improbable theories abound and are difficult to dispose of convincingly, if only because someone can find another, possibly irrelevant factor which has not previously been considered. Geologists are aware of this situation and allow for it. A physicist meeting it for the first time may be disillusioned and so needs to be warned. In many problems rigorous analysis does not take us very far; numerous loose ends are tied together only by intriguing speculations. Our task is to sift the plausible from the improbable and to devise tests to distinguish them. Order-of-magnitude arguments play an important part, and a feeling for the magnitudes of geophysical quantities is necessary to a sense of relevance. In particular, calculation of the energies involved in various processes often allows us to narrow the range of possible ultimate causes.

The field is so wide that a selection of topics, chosen with some bias of personal interest, is inevitable. My selection is based partly upon an attempt to predict the topics which will increase in relative importance. Paleomagnetism is presented as a central theme. I am impressed by the revolution in geophysical thinking which it has caused by establishing continental drift as an acceptable hypothesis after decades of general disfavor. The discussions of the geomagnetic field, tectonics, or mechanics of the crust and mantle, thermal history, and even gravity are all influenced more or less profoundly by the conclusions derived from rock magnetism. The impact of recent geochemical research, notably on isotopic abundances and meteorite compositions, also merits specific recognition. Some basic observations are introduced in Chapters 1 and 8, although they are properly the domain of geochemistry and a comprehensive discussion is beyond the scope of this book. Some emphasis is given to the application of solid state physics to geophysical problems. This approach will become increasingly useful as our knowledge of the Earth improves. The relevant solid state topics are not all readily available in simple presentations which make their geophysical applications obvious. In these cases I have diverged from the strictly geophysical discussion to deal with the relevant solid state problems.

Some topics are mentioned only in passing, others not at all. Instrumental techniques and problems of exploration geophysics are not considered, although there are several points of contact. Texts on applied geophysics are available at several levels (Nettleton, 1940; Heiland, 1946; Jakosky, 1950; Dobrin, 1960; Parasnis, 1962, 1966; Griffiths and King, 1965; Grant and West, 1965; Society of Exploration Geophysicists, 1967). For a unified treatment at an advanced level, the reader should refer to Grant and West. Many geophysical discussions appeal to geochemical and geological evidence and, in these respects, useful companion volumes are *Principles of Geochemistry* (B. Mason, 1958) and *Principles of Physical Geology* (Holmes, 1965).

The principal casualty in the omission of extended mathematical developments is the theory of elastic waves, as applied to layered media, but several monographs cover this field comprehensively (Ewing et al., 1957; Brekhovskikh, 1960; Cagniard, 1962; Kolsky, 1963). In some respects the material presented here overlaps with earlier texts (Gutenberg, 1959; Howell, 1959; Jacobs et al., 1959), although both the approach and the coverage are substantially different.

I thank publishers and authors for their response to my requests for permission to use their data and figures; the sources are acknowledged in captions.

Colleagues who have read the draft manuscript, and drawn my attention to defects, are J. Cleary, W. Compston, H. J. Dorman, H. Doyle, D. H. Green, E. Irving, B. Isacks, W. H. K. Lee, J. F. Lovering, M. W. McElhinny, W. D. Parkinson, M. S. Paterson, J. R. Richards, W. I. Riley, G. J. Tuck, and J. P. Webb, whose interest and help are very much appreciated.

Frank D. Stacey

Brisbane
1969

CONTENTS

Chapter 1

THE SOLAR SYSTEM

"...the solar system possesses several oddities. ..."

GAMOW, 1963, p. 63.

1.1 The Planets

Our knowledge of the solar system has made major advances in the past two decades. For many years theories of the origin of the planets have been built around the observed regularity in planetary orbits. More recently the emphasis has been on chemical considerations, with appeals to the densities of the planets and isotopic abundances in the Earth and in meteorites. It is now apparent that the Earth had a common origin with the other planets, about 4.5×10^9 years ago, in a cloud of gas and dust surrounding the then youthful Sun. The abundances of the elements in the cloud were approximately what would be expected from theories of nuclear synthesis, so that the formation of the solar system required no special conditions; rather we suppose that there are millions of similar systems of planets, even in our own galaxy.

Densities and orbital radii of the planets are listed in Table 1.1. Reliable values of density are much more recent than the orbital data, because precise measurements of planetary diameters are notoriously difficult to make and they enter as the third power in density estimates. However, the remaining uncertainties leave no doubt about the significant differences in composition, even between the basically similar inner four (terrestrial) planets. These differences are considered in Section 1.5.

The approximate geometrical progression of orbital radii of the planets is known as Bode's law or sometimes as the Titius-Bode law (Roy, 1967). In its original form this law gives the orbital radius R_n of the nth planet (counted outwards) as

$$R_n = a + b \cdot 2^n \tag{1.1}$$

1

TABLE 1.1: PLANETARY ORBITS AND DENSITIES

n	Planet	Orbit Radius (R_n) Relative to Earth's Orbit	$\dfrac{R_n}{R_{n-1}}$	Mass Relative to Earth	Radius Relative to Earth	Density (gm cm^{-3})	Estimated Density at Zero Pressure
1	Mercury	0.387	—	0.055_3	0.3820	5.47	5.3
2	Venus	0.723	1.86	0.815_5	0.9506	5.24	3.9
3	Earth	1.000	1.38	1.00000	1.000	5.517	4.04
	Moon			0.01230	0.273	3.33	3.3
	Earth + Moon			1.01230		5.44	3.96
4	Mars	1.524	1.52	0.107	0.530	4.0	3.8
5	Asteroids	Mean about 2.7	1.77	—	—	3.9	3.9[a]
6	Jupiter	5.203	1.92	317.9	10.97	1.35	Largely
7	Saturn	9.539	1.83	95.1	9.03	0.71	} Gaseous
8	Uranus	19.18	2.00	14.6	3.72	1.56	
9	Neptune	30.06	1.56	$16._1$	3.8_3	1.58	
	Pluto	40. (eccentric)	—	0.09?	0.5?	4 ?	

NOTE: There are numerous tabulations in the literature, with slight discrepancies. Important improvements in the data for terrestrial planets are by Ash et al. (1967) and for Neptune by Taylor (1968).
[a] Average for total mass of meteorite falls.

a and b being appropriate constants. In reviewing theories of the origin of the solar system, Ter Haar and Cameron (1963) pointed out that a better fit was obtained with a simple geometrical progression

$$R_n = R_0\, m^n \tag{1.2}$$

in which $m = 1.89$. Bode's law played an important role in the discovery of the asteroids or minor planets, which are concentrated in the region of the "missing planet" between Mars and Jupiter. In Table 1.1 and Fig. 1.1 the

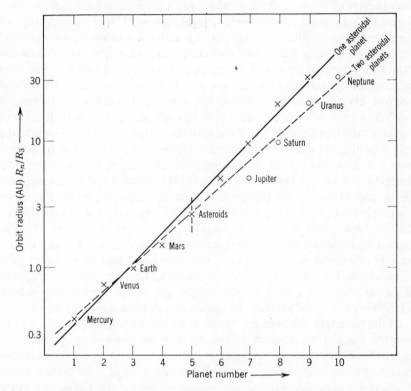

Figure 1.1: Radii of the planetary orbits, plotted to show the approximate geometrical progression (Bode's law).

asteroids have been counted together as a single planet, to satisfy Bode's law. This is justified by identifying them with meteorites, whose physical and chemical properties indicate an origin in one or more substantial planetary bodies, perhaps as large as the Moon. However, the approximate nature of Bode's law is indicated by the fact that it fits the sequence of orbital radii equally well if we assume not one but two missing planets, as in Fig. 1.1. Most recent discussions of the law are influenced by von Weizsäcker's (1944) model of a turbulent solar nebula, in which the planets accumulated at the boundaries between vortices, whose size increased regularly outwards from the Sun.*

Another important aspect of the orbital motions of the solar system is the distribution of angular momentum. With three apparently related exceptions, all of the planets revolve about the Sun and the satellites about the planets in the same direction and the orbits are approximately circular and coplanar. The exceptions are Pluto, whose orbit is inclined at 17° to the ecliptic (Earth's orbital plane) and is so elliptical that it crosses the orbit of Neptune, and the two satellites of Neptune, Triton, which has a reversed orbital motion about Neptune, and Nereid, whose orbit is highly elliptical. It is possible that all three were once satellites of Neptune, but suffered a dramatic gravitational interaction, throwing Pluto into an independent solar orbit. For this reason Pluto is not counted as one of the numbered planets in Table 1.1. The axial rotations of the Sun and planets are also in the same sense as the orbital motions, except for Uranus, which has a reversed rotation, and Mercury and Venus,† whose rotations have been virtually stopped by tidal friction of the solar tides (see Section 2.4). Thus with the axial rotation of Uranus as the only difficulty, we can envisage the development of the Sun and its planets from a disk-shaped nebula of gas and dust, all rotating in the same sense. Hartman and Larson (1967) and Alfvén (1967) pointed out that, except for Venus and Mercury, the planets and asteroids have approximately equal rotation periods. One of the detailed problems is then to explain how the Sun itself acquires nearly 99.9 % of the mass of the solar system but only 2 % of its angular momentum. Alfvén (1954) emphasized the necessity for an outward transfer of angular momentum from the contracting Sun to the outer parts of the cloud. He assumed that the protosun had a very extensive magnetic field, which exerted a drag on the ionized gas of the surrounding nebula, accelerating the outer parts of the nebula and slowing down the Sun.‡

Of the planetary satellites the Moon is by far the largest in relation to its primary, the Earth; so much so that we may be justified in regarding the Earth-Moon system as a double planet, rather than as a planet plus satellite. For comparison of density and composition with those of other planets it may therefore be more appropriate to consider values for the Earth and Moon together, rather than the Earth alone, and this possibility is allowed for in Table 1.1. However, the Moon is similar in both size and density to the inner

---

* Reviews of the subject are given by Spencer Jones (1956) and Ter Haar and Cameron (1963).

† Radar observations of Venus (Dyce et al., 1967) indicate a very slow reverse rotation. The dense atmosphere of Venus prevents optical observation of the solid surface. The rotations of Venus and Mercury are now explained in terms of tidal resonances (Goldreich and Peale, 1966; Bellomo et al., 1967).

‡ See also Hoyle (1960).

satellites of Jupiter and Saturn. In particular Io, a satellite of Jupiter, is very close in size to the Moon but has a mean density of about 4.06, which is even greater than the mean lunar density, 3.33. In composition Io is almost certainly very similar to the terrestrial planets. A comparison of the masses and diameters of planetary satellites (which are conveniently tabulated by Blanco and McCuskey, 1961, Table VII pp. 288–289) shows that the satellites of Jupiter and Saturn follow a pattern of decreasing density outwards, similar to the outward decrease in uncompressed densities of the planets. It is probable that the same process of segregation of the elements occurred in the Jovian and Saturnine satellites as occurred in the solar system as a whole.

The obvious division of the planets into two groups can be made equally well on the basis of size or density. The four major planets, Jupiter to Neptune, are more remote from the Sun and larger but less dense than the four smaller terrestrial (Earth-like) planets, Mercury to Mars, with which must be grouped also the asteroids. The terrestrial planets and meteorites are composed mainly of nonvolatiles, especially iron and silicon in varying states of oxidation, whereas the major planets have densities which are so low that they must be composed largely of light, volatile materials, especially hydrogen, with the nonvolatiles in much smaller but unknown amounts.

1.2 Meteorites and Their Compositions

Meteorites are iron and stone bodies, which arrive on the Earth in small numbers and quite randomly, apparently from within the solar system. The study of these objects has become a science with its own name, *meteoritics*, and has been comprehensively reviewed by Mason (1962), Wood (1965) and Anders (1964). Meteorites should not be confused with meteors, which are the transient luminous trails in the sky produced by small particles, known as meteoroids, many of which can be identified from their orbits as fragments of comets. There is no association between the arrivals of meteorites and the occurrences of meteor showers, which are observed when the Earth passes through bands of orbiting comet debris, although it must be expected that a few of the sporadic meteors have the same origin as the meteorites (Jacchia, 1963).

The total number of observed *falls* from which the recovery of meteorites has been documented is about 700, although the number of *finds*, which were not seen to fall but are certainly meteorites, rather more than doubles the total number available for study. It is therefore not surprising that only one fall has been observed with sufficient scientific control (by photographing the trail from two well separated points) to allow a reliable calculation of its orbit. This occurred at Pribram, Czechoslovakia, in 1959 and the fall was one of the familiar type known as chondrites. The orbit calculated by Ceplecha

(1961) is reproduced in Fig. 1.2.* Although this is only a single observation, it strengthens the usual conclusion, derived from estimates of meteorite velocities, that at least the majority of meteorites have come from the asteroidal belt on orbits which are sufficiently elliptical to cross the Earth's orbit. Asteroids or asteroidal fragments smaller than 1 km across are not detectable astronomically but are presumed to be numerous from the fact that the distribution of sizes in the observable range shows a strong increase in numbers with decreasing size (Blanco and McCuskey, 1961, pp. 262–268).

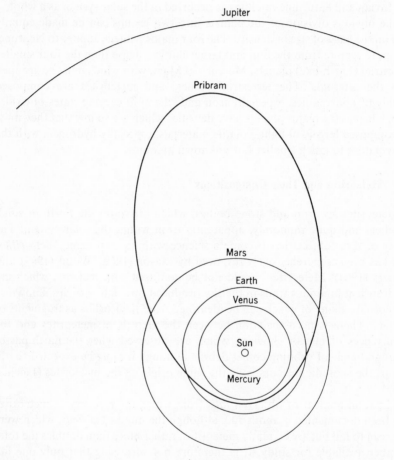

Figure 1.2: Orbit computed by Ceplecha (1961) for the meteorite which fell at Pribram, Czechoslovakia, in April 1959. Figure reproduced, by permission, from Mason (1962).

* Similar orbits have been deduced from reports of visual observations of the flights of meteorite fireballs; see, for example, Krinov (1960).

A lower limit to the size range is imposed by the Poynting-Robertson effect (Section 1.4).

Öpik (1966) has drawn attention to a mechanical difficulty which arises in the conventional theory of meteorites as asteroidal collision fragments. Asteroids orbiting in a common direction within the main asteroidal belt do not have velocity differences great enough to project impact fragments into Earth-crossing orbits. However, the asteroidal origin remains acceptable if the source of meteorites is restricted to the asteroids in elliptical orbits and Wetherill (1968) has pointed out that afternoon falls of chondritic meteorites are twice as frequent as morning falls. This requires the chondrites to be orbiting substantially faster than the Earth when intercepted and their orbit aphelia must be near to Jupiter.

Although meteorites are, for convenience, classified into several structural and chemical types, there are no clear lines of demarcation and slightly different systems of classification have been in favor at different times. The four groups listed by Mason (1962, p. 52) are chondrites, achondrites, stony-irons, and irons. A convenient checklist of their properties is given by Kaula (1968, pp. 380–382).

The irons are predominantly metallic iron, with nickel in solid solution averaging about 11%. Smaller amounts of sulphide and graphite are found and there are occasional inclusions of silicate. Two metal phases occur, the body-centered cubic α form (kamacite) with about 5.5% nickel, and the face-centered cubic γ form (taenite) with variable nickel content, generally exceeding 27%. Normally both phases occur in close association, evidently as a phase separation from solid solution after solidification as a single phase from the melt. The metal crystals are commonly very large, up to a meter or so across, indicative of extremely slow cooling. Quantitative evidence of the cooling rate has been obtained from the variations in composition across the kamacite-taenite phase boundaries. The phase separation usually forms a characteristic pattern, which is rendered obvious by etching, and is known as the Widmanstätten structure. A good example is shown in Fig. 1.3. The mutual solubility of iron and nickel decreases with falling temperature, but the rate of atomic diffusion also decreases. Thus, as cooling proceeds, the compositions of kamacite and taenite cannot maintain mutual equilibrium, except at the phase boundary itself, and a diffusion zone is formed. The kamacite (nickel-poor phase) is depleted in nickel near to the boundary and the taenite (nickel-rich phase) is further enriched. The width of the diffusion zone has been used to estimate cooling rates by Short and Anderson (1965) and Goldstein and Short (1967). The rates of cooling were variable from 0.4°C to 40°C per 10^6 years through the critical range from 650 to 350°C, in which the diffusion was effective. These slow rates are in keeping with the large crystal sizes and coarse Widmanstätten structures of iron meteorites and

Figure 1.3: Widmanstätten structure of the metal phase in the Glorietta Mountain pallasite (stony-iron meteorite). The scale line is 1 cm. Photograph courtesy of J. F. Lovering.

suggest burial in parent bodies 100 km or more in radius. However, the wide range of cooling rates appears not to be compatible with a single parent, but rather with several of different sizes.

Chondrites are the most common of the four types of meteorite, comprising 85% of observed falls, although iron meteorites are generally larger and the mass of meteoritic matter in space is probably less than 85% chondritic. The characteristic feature of chondrites is the occurrence of *chondrules*, spherical grains of silicate with diameters of about 1 mm, which are distributed through the matrix of silicates and nickel-iron (Fig. 1.4.). Chondrules do not occur in any of the terrestrial rocks accessible to us and they indicate that a large fraction of meteoritic material underwent a special process, of which we have no direct evidence on the Earth. The principal theories of chondrule formation have been reviewed by Anders (1964), but none avoid what appear to be improbable stages in meteorite evolution.

Wood (1963b) argued persuasively that the chondrules condensed as independent objects in the solar nebula and that they were subsequently incorporated in the accretion of the meteorite parent bodies. Evidence in favor of this hypothesis includes the enrichment of the isotope Xe^{129}, observed by Merrihue (1963), in chondrules relative to the groundmass in the chondrite Bruderheim. Xe^{129} is the key isotope in discussions of the very early history of the solar system (Section 8.5). It is the decay product of I^{129}, whose

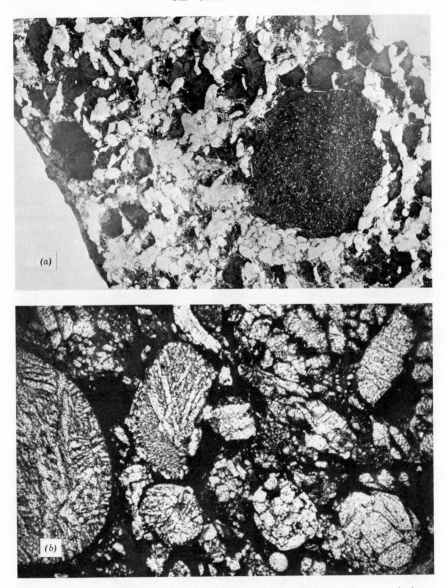

Figure 1.4: (a) Polished section of the Bencubbin meteorite. The bulk of the meteorite has a well-developed crystalline structure, but a large chondritic inclusion appears as the dark area on the right-hand side of this photograph. On the left hand side is a carbonaceous chondritic inclusion. (b) Enlarged section of the chondrite fragment, showing the structure of the chondritic spherules, just apparent in (a). The significance of the occurrence of the three meteorite types in a single specimen is discussed by Lovering (1962). Photographs courtesy of J. F. Lovering.

17 million year half-life is so short that its incorporation in a meteorite, and the subsequent enrichment of the meteorite in Xe^{129}, implies that the meteorite both formed and cooled within a few hundreds of millions of years of the cessation of nuclear synthesis. Greater Xe^{129} enrichment of the chondrules therefore implies that the chondrules formed first. Most other authors (especially Ringwood, 1966b) have favored chondrule formation by rapid crystallization from a melt, after accretion of one or several parent bodies of substantial size, perhaps, but not necessarily, as large as the Moon. Dodd and Teleky (1967) found evidence of alignment of olivine crystals within chondrules, such as is observed in terrestrial rocks which have flowed as partially crystallized magma. They concluded that this observation is compatible only with a volcanic origin of chondrules. The subsequent mixing of chondrules with silicate groundmass presents no problem, but it is difficult to envisage a chondritic parent body reaching the stage of volcanism while still retaining near to its surface the free metal which is an integral constituent of the chondrites.

Much less frequent than chondrites are the achondrites, which are essentially similar to terrestrial rocks, being crystalline silicates with virtually no metal phase. The stony-irons hardly merit separate classification, being merely about halfway through the continuous range between stones and irons. However, there is an important class of chondrites which deserves special mention, the rare carbonaceous chondrites (for a review of their properties see Du Fresne and Anders, 1963). Their rarity in meteorite collections is probably due more to their extreme friability and lack of strikingly meteoritic features than to an absolute rarity in space. As the name implies, they contain several percent of carbon and carbon compounds. Metal phases are virtually absent, the iron occurring in silicate, oxide, and sulfide phases. Substantial amounts of volatiles, especially water, are present, so that these chondrites can never have been strongly heated. Of all of the materials available for laboratory examination, the subclass of carbonaceous chondrites known as type I must be nearest to the primitive dust from which the terrestrial planets were formed. Types II and III are more like ordinary chondrites.

No listing of meteorite types would be complete without mention of the tektites, although they cannot strictly be classed with the other groups. Tektites are rounded pieces of silica-rich glass which have been found in tens of thousands in limited areas, notably the Philippines and South East Asia, Southern Australia, and Czechoslovakia. Their shapes clearly indicate rapid flight through the atmosphere, although none has been observed to fall. For each geographical group the ages determined from potassium and argon isotopes agree and coincide with the age of the geological formations in which they are found, although their compositions are unrelated to their environments. The range of ages (time since fusion) of the several groups is 0.3 to

35×10^6 years,* many orders of magnitude less than the meteorite ages, which exceed 4×10^9 years. Metallic (nickel-iron) spherules have been found in some tektites. The favored, but not universally accepted theory is that each geographical group was formed as a fused splash by a major meteorite impact on the Moon, so that the tektites are fused lunar surface material with traces of meteoritic matter. O'Keefe (1966) has reviewed the tektite problem with emphasis on a variation of the lunar origin. The other serious possibility is that tektites were produced by meteorite impacts with the Earth.

1.3   Cosmic Ray Exposures of Meteorites

The "age" of a meteorite normally means the time elapsed since it was formed as solid material. More specifically, this is the *solidification age*, determined by the methods used for dating rocks which are discussed in Chapter 8. The principal methods, based on the decay of uranium to lead, rubidium to strontium, and potassium to argon, are all applicable to stony meteorites, which contain the parent elements of these decay schemes. Iron meteorites contain negligible amounts of the parents and can be dated only from lead isotope ratios by presuming that the ultimate source material is the same as for the stones (and for the Earth). The numerous age determinations indicate that the meteorites underwent a major process of chemical evolution 4.5×10^9 years ago, simultaneously with a similar process in the Earth.

In considering meteorites we are also interested in the *cosmic ray exposure ages*, the intervals of time which have elapsed since the meteorites were broken down into meter-sized pieces and exposed to bombardment in space by cosmic radiation. It is only the outer 1 m or so of each independent body that is exposed to cosmic radiation, so that each fragmentation event exposes fresh material. From the abundances of certain short-lived cosmogenic (cosmic ray produced) nuclides (Ar^{39}, C^{14}, Cl^{36}) a *terrestrial age* can also be estimated. This is the time since a meteorite arrived on the earth and cosmic radiation was thus "switched off."

Extremely energetic cosmic ray protons cause violent disruption (spallation) of the atomic nucleii in exposed meteorites. Anders (1962, 1963), who reviewed the whole subject of meteorite ages, used the following example to illustrate the spallation process:

$$Fe^{56} + H^1 \rightarrow Cl^{36} + H^3 + 2He^4 + He^3 + 3H^1 + 4n \qquad (1.3)$$

Many products can arise from numerous similar reactions and in principle the total cosmic ray exposure can be estimated from the content of a particular

---

* Tektites fall into distinct age groups, which show that they were produced by four, or at most five events in the past 35 million years (see tabulation of ages by Kaula, 1968). It must, however, be admitted that there may have been earlier tektite-producing events, but that the tektites produced have been lost geologically.

product if its half-life exceeds the estimated exposure age and if its concentration in the unexposed meteorite can be neglected or reasonably estimated. Uncertainties in cosmic ray intensities and shielding by burial in a large meteorite are largely avoided by comparing concentrations of two cosmogenic nuclides, one stable and the other radioactive and having a half-life short compared with the cosmic ray exposure times. Further selection to include only pairs of nuclides whose production cross sections have similar dependences upon cosmic ray energy, and which are sufficiently close in mass number to be produced from the same target nuclides, leaves two isobaric pairs (H^3—He^3 and Cl^{36}—Ar^{36}) and two isotopic pairs (Ar^{38}—Ar^{39} and K^{44}—K^{41}) as species of greatest interest. The first three of these pairs also avoid an uncertainty in initial composition which arises in the case of non-volatile spallation products. Spallation products were produced in the solar nebula before planetary accretion and the nonvolatiles were therefore incorporated in the meteorites at the time of their formation. Under the conditions thus imposed, the cosmic ray exposure age is given by

$$t = \frac{S}{R} \cdot \frac{\sigma_R}{\sigma_S} \cdot \frac{t_{1/2}}{\ln 2} \tag{1.4}$$

where S, R are the concentrations of the stable and active nuclides, σ_S, σ_R are their production cross sections from laboratory data, and $t_{1/2}$ is the half-life of the active nuclide (Anders, 1962). In the cases of the isobaric pairs the stable nuclide is produced by decay of the active one as well as directly, so that the equation becomes

$$t = \frac{S}{R} \cdot \frac{\sigma_R}{\sigma_S + \sigma_R} \cdot \frac{t_{1/2}}{\ln 2} \tag{1.5}$$

The recently developed method of dating geological materials by counting fission tracks (Fleischer and Price, 1964; Fleischer, Price, and Walker, 1965; see also Section 8.3) has an interesting application to cosmic ray exposure ages. Fission tracks in meteorites are produced not only by the spontaneous fission of U^{238}, as in rocks, but also by neutron-induced fission of U^{235}. "Excess" tracks, due to U^{235}, can be estimated by comparing the solidification age with the expected number of tracks from U^{238}, and the excess represents the integrated exposure of the sample to cosmic ray neutrons or heavy cosmic ray primaries. In particular, Fleischer, Naesser, Price, and Walker (1965) found a complete absence of tracks attributable to cosmic rays in tektites, allowing them to conclude that the tektites cannot have existed independently in space for a period as long as 300 years.

If the meteorites all had the same cosmic ray exposure age, we would have to conclude that they were produced simultaneously by disruption of a single parent body or by the collision of two parents. However, the concentrations of spallation products are highly variable and indicate a more complex history. Nearly half of the exposure ages for stony meteorites are closely grouped around 23×10^6 years, but others cover the range from 2.8×10^6 to 500×10^6 years (Anders, 1962, Table III), with a tendency to minor groupings of different types also at 4.5×10^6 and 10×10^6 years. Some of the lower estimates are probably invalidated by diffusion losses because the same meteorites have small potassium-argon ages. Iron meteorites have generally had much greater exposures, up to a maximum of 1500×10^6 years with groupings at 550×10^6 and 900×10^6 years but not at 23×10^6 years (Anders, 1962, Table IV). There is a wide scatter and, in many cases, very imperfect agreement between the alternative methods. Nevertheless, it is evident that the meteorites suffered a multiplicity of fragmentation events which were very much more recent than their solidifications.

The measurements do not preclude the possibility of a primary fragmentation of one or two parent bodies, because the large number of asteroids requires that we assume many more recent collisions of small bodies, although few major collisions of large ones. However, if collisions are occurring at a rate such that the interval between them, for any one fragment, is short compared with the time since the suppose primary fragmentation, then the cosmic ray exposures will indicate the more recent fragmentations. The fact that stony meteorites are more easily broken up, and are on average significantly smaller than the irons, is consistent with their shorter cosmic ray exposures. Furthermore, the exposed surfaces of stones are eroded more quickly by impact of small dust particles than are the surfaces of irons, an additional reason why stony material with very prolonged cosmic ray exposures may not be available for measurement (Fisher, 1966). We are therefore faced with a sampling difficulty and must regard the evidence as inconclusive. A more important reason for entertaining the hypothesis that the asteroids are collision fragments of planetismals which have never accreted to form a planet arises from the difficulty of devising a satisfactory mechanism for the break-up of more than a very modest (i.e., sublunar-sized) planet.

The grouping of exposure ages suggests that the irons and stones were produced by different events and therefore that the bodies from which they came were chemically different. However, measurements of isotopic ratios of lead and strontium, which have been used to date the meteorites (Section 8.4), show that the meteoritic parents differentiated from a common source and, if they are of asteroidal origin, then the asteroids had a history of chemical differentiation before the break-up of those sampled as meteorites. It is this differentiation process which is dated at 4.5×10^9 years ago.

1.4   The Poynting-Robertson Effect

Solar radiation has an important influence on the orbits of small particles whose ratio of surface area to mass is large. Its effects on the meteor streams have been studied in detail and a historical and physical discussion is given by Lovell (1954). Particles up to about 10 cm diameter are affected on a time scale of 10^9 years.

It is convenient to distinguish three effects of solar radiation pressure, although they are not really independent. First, there is a simple outward force from the Sun. For particles with diameters of a few thousand Angstroms or less this force may exceed the gravitational attraction of the Sun and blow them out of the solar system. This problem is complicated by the fact that the critical particle size is comparable to the wavelength of the radiation and the effective optical cross section is not the simple physical cross section. We are concerned here with much larger particles. Second, the solar radiation received by a particle is Doppler-shifted to cause an increase in radiation pressure if the particle is approaching the Sun and a decrease if it is receding; elliptical orbits are thus reduced to nearly circular orbits. Third, the angular momentum of an orbiting particle is progressively destroyed by the fact that it receives solar radiation, which has only a radial momentum from the Sun (neglecting the solar rotation), and reradiates this energy with a forward momentum corresponding to its own motion about the Sun. This is the essential feature of the Poynting-Robertson effect, which is most conveniently analyzed as a problem in relativity.

We consider the special case of a spherical particle of mass m and diameter d in a circular orbit at radius r. Its orbital velocity is

$$v = \left(\frac{GM}{r}\right)^{1/2} \tag{1.6}$$

M is the mass of the Sun and G the gravitational constant, so that the total orbital energy is

$$E = -\frac{GMm}{r} + \frac{1}{2}mv^2 = -\frac{GMm}{2r} \tag{1.7}$$

It is convenient to consider separately the processes of absorption and re-radiation of the energy.

In time dt the particle receives energy $d\epsilon$ as solar radiation, and this causes an increase in mass

$$dm = \frac{d\epsilon}{c^2} \tag{1.8}$$

c being the velocity of light. But since this radiation traveled radially from the Sun it carried no orbital angular momentum and the total angular momentum of the particle is conserved, so that

$$m\, d(vr) = -vr\, dm = -\frac{v}{c^2} r\, d\epsilon \qquad (1.9)$$

The particle then reradiates the energy $d\epsilon$, but it does so isotropically in its own frame of reference and this process involves no reaction on the particle. The orbital velocity is therefore conserved in the radiation process and since the mass dm is lost, a net loss of angular momentum by $vr\,dm$ occurs. This angular momentum is carried away by the radiation, which, when viewed in the stationary reference frame of the Sun, is seen to be Doppler-shifted; the energy and momentum projected forward from the particle exceed the energy and momentum radiated backward.

The rate of loss of orbital angular momentum may be equated to a retarding torque L:

$$L = m\frac{d(vr)}{dt} = -\frac{v}{c^2} r\frac{d\epsilon}{dt} \qquad (1.10)$$

so that

$$\frac{dE}{dt} = L\frac{v}{r} = -\frac{v^2}{c^2}\frac{d\epsilon}{dt} \qquad (1.11)$$

Now $d\epsilon/dt$ is the rate at which the particle receives solar radiation and is given by

$$\frac{d\epsilon}{dt} = S\left(\frac{r_E}{r}\right)^2 A \qquad (1.12)$$

where S is the solar constant, the energy flux through unit area at the distance r_E of the Earth's orbit, 1.39×10^6 ergs cm^{-2} sec^{-1} (1400 wm^{-2}), and $A = (\pi/4)\, d^2$ is the cross-sectional area of the particle. Thus by differentiating Eq. (1.7) and equating to Eq. (1.11) with the substitution of Eq. (1.12) we obtain

$$\frac{GMm}{2r^2}\frac{dr}{dt} = -\frac{v^2}{c^2} S\left(\frac{r_E}{r}\right)^2 A \qquad (1.13)$$

and since v is given in terms of r by Eq. (1.6), we obtain the differential equation for r:

$$r\frac{dr}{dt} = -\frac{2Sr_E^2 A}{mc^2} \qquad (1.14)$$

Integrating from the initial condition, $r = r_0$ at $t = 0$:

$$\frac{r_0^2 - r^2}{r_E^2} = \frac{4SA}{mc^2} t \qquad (1.15)$$

which, for a spherical particle of density ρ, becomes

$$\left(\frac{r_0}{r_E}\right)^2 - \left(\frac{r}{r_E}\right)^2 = \frac{6S}{d\rho c^2} t \tag{1.16}$$

where (r/r_E) is the radius of a particle orbit, expressed in astronomical units (AU).

We are interested in the time taken by particles, of diameter d, originating in the asteroidal belt at $2.7 \, r_E$, to reach the Earth's orbit, r_E. Assuming a particle density of 4 gm/cm^3 and d in centimeters, this is

$$t = 8.6 \times 10^7 \, d \, \text{years} \tag{1.17}$$

A more complete analysis (Lovell, 1954, pp. 402–409) shows that a particle in an elliptical orbit is first reduced to a nearly circular orbit, just inside its initial perihelion distance. Since this process also depends upon the Doppler shift of radiation due to motion of the particle relative to the Sun, the time required is similar to that for the spiraling effect.

The Poynting-Robertson effect thus ensures that any small particles in the common meteoroid range (less than 1 cm diameter), which originated in the asteroidal belt about 10^8 years ago, would have passed the Earth's orbit and spiraled into the Sun. McKinley (1961, pp. 169–171) has pointed out that very few meteors appear to be due to particles having the density of stone or rock. They are envisaged as loose, dusty aggregates, similar to the supposed structures of comets and quite different from the meteorites. The relative rarity of very small meteorites is consistent with the conclusion that they must be products of recent asteroidal collisions. Further, we can see that if a primary asteroidal fragmentation had occurred very early in the history of the solar system, say 4×10^9 years ago, then all primary fragments smaller than 50 cm would have spiraled into the Sun and the terrestrial collection would be strongly biased toward the shorter cosmic ray exposures of more recent, secondary fragmentations. Thus the currently available exposure age data do not permit us to decide whether the meteorites originated in one or two fairly large or many smaller parent bodies. Although the complexity of the chemical evidence appears to demand at least four parent bodies, semi-independent physical evidence is very desirable.

1.5 Compositions of the Terrestrial Planets

In spite of their uncertain mechanical histories, meteorites have had a profound influence upon our ideas about the composition, internal structure, and history of the Earth. They provide us with samples of the compositions of the terrestrial planets which are far more representative of the planets as as whole than are the rocks to which we have access near the surface of the Earth. Chemical considerations now dominate the discussion of the nature and origin of the Earth. Important reviews, although with somewhat divergent views, are given by Urey (1952, 1957, 1963) and Ringwood (1966a); MacDonald (1963a) has reviewed the physical aspects of studies of internal constitutions of the terrestrial planets.

Meteorite compositions are in satisfactory accord with spectroscopic estimates of the solar abundances of nonvolatile elements. The dominance of Si, Mg, and Fe strongly indicates that all of the terrestrial planets are composed essentially of magnesian silicates and iron, either as metal or oxide. The average density of the Earth (Table 1.1) and its internal structure deduced from seismology (Chapter 4) agree well with the presumption that the Earth has a liquid iron core, of uncompressed density $\rho_0 = 7$ gm cm^{-3}, and a solid silicate mantle with $\rho_0 = 3.3$ gm cm^{-3}. The proportions are estimated from the measured radius of the core. In the same way we can interpret the average densities of the other terrestrial planets in terms of iron cores and silicate mantles in different proportions, although we have no direct estimates of the core sizes. Mercury, whose uncompressed density is substantially greater than that of the Earth, must have a larger core in proportion to its total volume. At the other extreme, the low density of the Moon leads us to conclude that it has virtually no core.

Venus is very similar to the Earth in both size and density and is presumed to be essentially similar internally, but if we assign to Mars an iron core of radius calculated to give the observed mean density with a silicate mantle of $\rho_0 = 3.3$ gm cm^{-3}, then we run into difficulty. From the motions of its satellites the mass of Mars is well determined. The moment of inertia is also estimated, assuming approximate hydrostatic equilibrium for the surface, as in the calculations given for the Earth in Section 2.1 (Wilkins, 1967; Runcorn, 1967b)* and is found to correspond much more nearly to uniform density than in the case of the Earth. Allowing for uncertainty of the radius, the core can be no more than 10% of the total mass, and is probably substantially less; moreover, a much higher density must be assigned to the mantle than in the case of the Earth. This observation gives strong support to Ringwood's (1966a, 1966c) theory of planetary evolution, which leads to different oxidation states for the terrestrial planets, so that in Mars virtually all of the iron has remained oxidized and therefore has not separated from the silicates. According to Ringwood the overall Fe/Si ratio in Mars is approximately the same as that in the Earth, the iron occurring as oxide, which has a density of 5.2 gm cm^{-3}. When added to silicate of $\rho_0 = 3.3$ gm cm^{-3}, it brings the uncompressed silicate density up to $\rho_0 = 3.7$ gm cm^{-3}, to which only a small core need be added to give the observed Martian density. In view of the high

---

* Disagreement between the observed surface ellipticity and that expected from the dynamical ellipticity raises doubt about the assumption of equilibrium, but this is not sufficient to invalidate the conclusion drawn here.

iron oxide content of its mantle, it is not surprising that the surface of the
"red planet" should appear rusty. Ringwood also favors a more oxidized
state for Venus than for the Earth, although the difference must be much less
than in the case of Mars. When space technology is sufficiently advanced to
soft-land seismometers on Venus and Mars (assuming that they have Venus-
quakes and, more doubtfully, Marsquakes) we may be able to estimate directly
their core sizes and thus the oxidation states.

In Table 1.2 the more important constituents of the Earth are listed with
estimates of their proportions, as derived from a number of chemical and
physical considerations. Ringwood's estimates of the proportions obtained
by reduction of type I carbonaceous chondritic material are also listed for
comparison. The scientific value of these chondrites is very great; their composition is evidently very close indeed to the primeval cosmic dust from which
the Earth accreted.

TABLE 1.2: CHEMICAL COMPOSITION OF THE EARTH Percentages by mass of
major constituents according to estimates by Mason (1966) and Ringwood
(1966a) compared with Ringwood's calculation of the composition derived
by reduction of type I carbonaceous chondrites. Nonsignificant figures are
retained to make totals add to 100%.

		Mason	Ringwood	Reduction Calculation
Mantle	SiO_2	32.51	31.16	29.84
	MgO	21.06	25.86	26.29
	FeO	8.59	0.31 ⎫ 5.86	6.38
	Fe_2O_3		5.55 ⎭	
	Al_2O_3	2.04	2.44	2.69
	CaO	1.57	2.16	2.57
	Na_2O	0.76	0.39	1.23
	Others	1.07	1.16	
Core	Fe		23.6	25.87
	Ni	32.4	3.4	1.66
	Si		4.0	3.47

In Table 1.2 the occurrence of elemental silicon in the core is assumed,
following the arguments of Ringwood (1959, 1966a).* This would have the
effect of bringing the core density into accord with the evidence discussed in
Section 4.5, but it is not the only possibility. The fact that elemental silicon
is found in the metal phases of enstatite (highly reduced) chondrites shows
that it can enter the metal phase under suitably intense reducing conditions
in planetary formation. The assumption that it is a constituent of the Earth's

---

* See also MacDonald and Knopoff (1958) and Balchan and Cowan (1966).

core imposes two important requirements in the early chemical history of the Earth. First, carbon played an important role in the reduction process because hydrogen would not reduce silica. Second, the mantle, which contains some magnetite, is not in chemical equilibrium with the core because iron oxide and silicon react, producing metallic iron and silica; the materials of the core and mantle could not have formed an intimate mixture at any stage. The separation of at least a major part of the core must therefore have occurred simultaneously with the reduction process which produced the metal and, since there must have been a mechanism for the escape of carbon monoxide, the reduction was probably part of the accretion itself. The occurrence of silicon in the core and magnetite in the mantle thus requires that the reducing conditions vary during the accretion of the Earth. It is possible that the reaction $Fe_3O_4 + 2Si \rightarrow 3Fe + 2SiO_2$ has subsequently occurred across the core-mantle boundary, depositing iron in the core and leaving at the bottom of the mantle a silica-rich layer which is responsible for complications in the seismic velocity profile of this region. Alder (1966) has given an alternative explanation for the core density which avoids these consequences. He argued that at the temperature and pressure of the core-mantle boundary MgO is soluble in iron to the extent of about 10% and is a more likely core constituent than Si. If this argument is correct, the core and mantle may not be in such serious disequilibrium as Ringwood requires. A complication in seismic velocities of the bottom of the mantle could then arise from depletion of MgO relative to SiO_2. Another possible core constituent is sulfur,* which is associated with the metal phases (as troilite, FeS) in iron meteorites and chondrites.

All of the important, long-lived radioactive elements, uranium, thorium, rubidium, and potassium, are strongly oxyphile and remained with the silicate during the process of reduction and melting which produced the iron meteorites and metal phases of the chondrites. Similarly, the iron core of the Earth can contain no significant radioactive material, so that radioactive heating cannot be invoked as a source of power to maintain the geomagnetic dynamo (Sections 5.4 and 9.4). In the Earth the fractionation of these elements is even more marked because they have large ionic radii and are expelled from the close-packed spinel structures to which silicates transform at the pressures of the lower mantle. The concentration of radioactive elements in continental rocks is the most striking evidence of the chemical differentiation of the mantle whereby the continents were formed. The age zones of the continents (Section 8.3) indicates that the differentiation was not a sudden process at an early stage in the Earth's history, but has occurred progressively, or possibly spasmodically, since.

* O. L. Anderson has drawn the author's attention to the fact that this is an old idea and that it appears to be demanded by the cosmic abundance of sulfur.

Chapter 2

ROTATION AND THE FIGURE OF THE EARTH

"... consider the wobble induced by a hula dancer of mass m on the geographic north pole ... "

MUNK AND MACDONALD, 1960b, p. 55.

2.1 Figure of the Earth

The centrifugal effect of the Earth's rotation causes an equatorial bulge, which is the principal departure of the Earth from spherical shape. If the whole Earth were covered with a shallow sea, then, apart from minor disturbances due to wind, etc., the surface would assume the shape determined by hydrodynamic equilibrium of the water subjected to gravitation and rotation; the sea level equipotential surface is the geoid or figure of the Earth.* Tidal effects are superimposed upon the mean geoid by gravitational gradients of the Moon and Sun, but are very small in comparison with the rotational ellipticity with which we are concerned here. Crustal features, continents and mountain ranges, are significant departures of the actual surface of the Earth from the geoid, but mass compensation at depth (the principle of isostasy discussed in Section 3.3) reduces the influence of surface features on the geoid.

The form of the geoid has been determined from astrogeodetic surveys over several extended continental survey arcs, the vertical, or direction of the local gravity vector, being determined at each point of observation by reference to stars. The process is described in detail by Bomford (1962) and is indicated in Fig. 2.1. Values of the ellipticity of the geoid deduced in this way from surveys completed between 1900 and 1960 are within the range 1/297 to 1/298.3 and deductions from gravity observations (Section 3.1) cover the same range.

* For a physical idea of the geoid in continental areas, picture canals cut through the continents and connected with the oceans, so that the water level is the geoidal surface.

These estimates are derived from measurements limited to the continental areas of the Earth's surface. A more accurate value which properly represents the Earth as a whole is now available from analyses of satellite orbits (Section 3.2), which give 1/298.25 with considerable confidence, so that this figure is now preferred. It is, however, important to note that the geodetic estimates are not in conflict with the satellite value, although they are biased to one side of it.

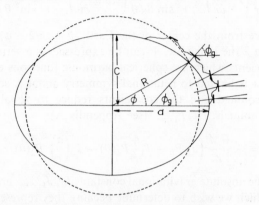

Figure 2.1: A comparison of the equilibrium geoid (solid line) with a sphere of equal volume (dashed line). The ellipticity of the geoid is exaggerated by a factor of about 50. The radius of the sphere is R $= (a^2c)^{1/2}$, a and c being the major and minor semi-axes of the geoidal ellipsoid. ϕ is the geocentric latitude at any point and ϕ_g is the geographic latitude or angle between the normal to the geoidal equipotential surface at the point of observation and the equatorial plane. This angle is determined by reference to stars at each point around an astrogeodetic survey arc, so that the local orientation of the geoidal surface is measured.

The geoid is a surface of constant geopotential U_0, and the total geopotential at any point is a sum of gravitational potential V and rotational potential:

$$U = V - \tfrac{1}{2}\omega^2(x^2 + y^2) = V - \tfrac{1}{2}\omega^2 r^2 \cos^2 \phi \qquad (2.1)$$

where ω is the angular velocity of the Earth's rotation, which is taken to be about the z axis, and (x, y) or (r, ϕ) are coordinates of a point on the surface of the Earth. An additional potential term due to pressure must be added for Eq. 2.1 to apply to points in the interior. Gravitational acceleration at the surface is normal to the geoid, and is given by

$$g = -\operatorname{grad} U \qquad (2.2)$$

The problem of calculating the form of the geoid thus reduces to obtaining an expression for V. If the distribution of mass within the earth were known, V could be obtained by direct integration, but we have to proceed in the reverse direction and derive information about the Earth's interior from the form of the geoid.

We may approach the problem in a general way through Laplace's equation, which V must satisfy at all points external to the Earth and therefore, in the limit, on the surface itself:

$$\nabla^2 V = \frac{1}{r^2}\frac{\partial}{\partial r}\left(r^2\frac{\partial V}{\partial r}\right) + \frac{1}{r^2 \sin\theta}\frac{\partial}{\partial\theta}\left(\sin\theta\frac{\partial V}{\partial\theta}\right) + \frac{1}{r^2 \sin^2\theta}\frac{\partial^2 V}{\partial\lambda^2} = 0 \quad (2.3)$$

r being distance from the center of the Earth, $\theta = (\pi/2 - \phi)$ the geocentric colatitude, and λ the longitude. V can be expressed as a series of powers of $1/r$ with coefficients which are spherical harmonic functions of θ and λ. Our present interest is limited to rotational symmetry about z, so that variation with λ is discounted and the coefficients reduce to zonal harmonics or Legendre polynomials, $P_0, P_1, \ldots$ (see Appendix A):

$$V = -\frac{GM}{r}\left(J_0 P_0 - J_1\frac{a}{r}P_1(\theta) - J_2\left(\frac{a}{r}\right)^2 P_2(\theta)\cdots\right) \quad (2.4)$$

in which G is the absolute gravitational constant, $J_0, J_1, \ldots$ are dimensionless coefficients, which we wish to determine because they represent the distribution of mass within the earth, and a is the equatorial radius of the Earth.

J_0 is known to be unity from the fact that at great distances all other terms become insignificant and we are, in effect, considering the potential due to a point mass M:

$$V = -\frac{GM}{r} \quad (2.5)$$

By taking the origin of the coordinate system to be the center of mass of the Earth we make J_1 identically zero. Our principal interest is in the J_2 term, which is the one required to give the observed oblate ellipsoidal form of the geoid. The departure of the geoid from an ellipsoid is represented by higher harmonics with amplitudes smaller by factors of order 1000; they are disregarded here but will be considered in Section 3.2 in connection with the undulations of the geoid revealed by satellite orbits.

We now have

$$V = -\frac{GM}{r} + \frac{GMa^2 J_2}{2r^3}(3\sin^2\phi - 1) \quad (2.6)$$

This is the starting point for two approaches to the problem of the geoid. In Section 3.1 we consider the gravity on the equilibrium geoid as the gradient of the geopotential. Here we are concerned with the internal distribution of mass.

Having accepted the two terms in Eq. 2.6 as sufficient for the present purpose, we can express J_2 in terms of the principal moments of inertia of the Earth. The geometry of the problem is given in Fig. 2.2.

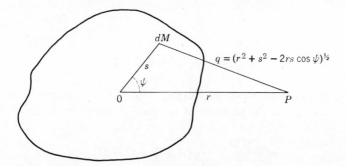

Figure 2.2: Geometry for the integration of gravitational potential to obtain MacCullagh's formula. The potential is calculated at a point P external to the mass M and distant r from its center of mass, O; r is a constant in the integration, the variables being s and ψ, the coordinates of the mass element with respect to O and the line OP.

The gravitational potential at P due to the mass element is

$$dV = -G\frac{dM}{q} = -\frac{G\,dM}{r[1 + s^2/r^2 - 2(s/r)\cos\psi]^{1/2}} \tag{2.7}$$

This may be expanded in powers of $1/r$ to $1/r^3$ by noting that

$$\left(1 + \frac{s^2}{r^2} - 2\frac{s}{r}\cos\psi\right)^{-1/2} = \left(1 + \frac{s}{r}\cos\psi - \frac{1}{2}\frac{s^2}{r^2} + \frac{3}{2}\frac{s^2}{r^2}\cos^2\psi + \cdots\right)$$

$$= \left(1 + \frac{s}{r}\cos\psi + \frac{s^2}{r^2} - \frac{3}{2}\frac{s^2}{r^2}\sin^2\psi + \cdots\right) \tag{2.8}$$

To this order the total potential can be expressed as a series of integrals obtained by substituting Eq. 2.8 in Eq. 2.7:

$$V = -\frac{G}{r}\int dM - \frac{G}{r^2}\int s\cos\psi\,dM - \frac{G}{r^3}\int s^2\,dM$$

$$+ \frac{3}{2}\frac{G}{r^3}\int s^2\sin^2\psi\,dM \tag{2.9}$$

The first integral is the potential of the centered mass, $-GM/r$. The second is identically zero because the center of mass was chosen as the origin. By assigning to the elementary mass coordinates (x, y, z), the axes being arbitrary, we can write the third integral as

$$-\frac{G}{r^3} \int (x^2 + y^2 + z^2)\, dM = -\frac{G}{2r^3}\left[\int (y^2 + z^2)\, dM + \int (x^2 + z^2)\, dM \right.$$

$$\left. + \int (x^2 + y^2)\, dM\right]$$

$$= -\frac{G}{2r^3}(A + B + C) \tag{2.10}$$

where A, B, C are moments of inertia about the x, y, z axes. The fourth integral in Eq. 2.9 is the moment of inertia I of the mass M about OP, so that

$$V = -\frac{GM}{r} - \frac{G}{2r^3}(A + B + C - 3I) - \cdots \tag{2.11}$$

This is the general form of MacCullagh's formula. We may write

$$I = Al^2 + Bm^2 + Cn^2 \tag{2.12}$$

where l, m, n are direction cosines of OP with respect to x, y, z. Then with rotational symmetry of the Earth about z,

$$A = B \tag{2.13}$$

$$n^2 = \sin^2 \phi = 1 - l^2 - m^2 \tag{2.14}$$

so that

$$V = -\frac{GM}{r} + \frac{G}{2r^3}(C - A)(3 \sin^2 \phi - 1) \tag{2.15}$$

Thus the coefficient J_2 of the gravitational potential is

$$J_2 = \frac{C - A}{Ma^2} \tag{2.16}$$

The total geopotential is therefore

$$U = -\frac{GM}{r} + \frac{G}{2r^3}(C - A)(3 \sin^2 \phi - 1) - \tfrac{1}{2}r^2\omega^2 \cos^2 \phi \tag{2.17}$$

and the geoid has been defined as the surface of constant potential U_0.

At the equator ($r = a$, $\phi = 0$) and poles ($r = c$, $\phi = \pi/2$) we obtain

$$U_0 = -\frac{GM}{a} - \frac{G}{2a^3}(C - A) - \tfrac{1}{2}a^2\omega^2 \tag{2.18}$$

$$U_0 = -\frac{GM}{c} + \frac{G}{c^3}(C - A) \tag{2.19}$$

from which

$$a - c = \frac{C - A}{M}\left(\frac{a}{c^2} + \frac{c}{2a^2}\right) + \frac{1}{2}\frac{ca^3\omega^2}{GM} \qquad (2.20)$$

and since $a \simeq c$, the ellipticity ϵ is slight:

$$\epsilon = \frac{a - c}{a} = \frac{3}{2}\frac{C - A}{Ma^2} + \frac{1}{2}\frac{\omega^2 a^3}{GM} \qquad (2.21)$$

This gives the ellipticity correct to first order and, since $\epsilon \simeq 10^{-3}$, the error in neglecting terms of order ϵ^2 is comparable in magnitude to the neglected higher harmonics in the gravitational potential. To this accuracy the equation of the surface is

$$r = a(1 - \epsilon \sin^2\phi) \qquad (2.22)$$

Equation 2.21 allowed the absolute difference $(C - A)$ between the moments of inertia of the Earth about polar and equatorial axes to be calculated in terms of the geodetically determined ellipticity ϵ. However, since $(C - A)/Ma^2$ has been determined from satellite orbits with precision more than an order of magnitude greater than that obtained geodetically, the ellipticity of the geoid estimated from satellite data is now used in geodesy. The second term in Eq. 2.21 is well determined from the ratio of centrifugal acceleration to gravitational acceleration at the equator (see Eq. 3.8):

$$m = \frac{\omega^2 a^3}{GM} = \frac{\omega^2 a}{GM/a^2} = 3.4678 \times 10^{-3} \qquad (2.23)$$

Using the satellite value of J_2 obtained by Kaula (1965) and now adopted as the standard for geodetic reference (International Union of Geodesy and Geophysics, 1967),

$$J_2 = 1.08270 \times 10^{-3} \qquad (2.24)$$

we obtain for the ellipticity by Eq. 2.21:

$$\epsilon = \tfrac{3}{2}J_2 + \tfrac{1}{2}m = 3.3579 \times 10^{-3} \qquad (2.25)$$

Retaining second-order terms in the theory, we obtain a better value:

$$\epsilon = 3.35280 \times 10^{-3} = \frac{1}{298.26} \qquad (2.25a)$$

It is useful at this point to anticipate a result obtained in Section 2.2. The gravitational attractions of the Sun and Moon, acting on the equatorial bulge of the Earth, exert torques which cause a precession of the axis of rotation and from the rate of precession the dynamical ellipticity of the Earth, H, is estimated:

$$H = \frac{C - A}{C} = 3.2732 \times 10^{-3} = \frac{1}{305.51} \tag{2.26}$$

From Eqs. 2.16, 2.24, and 2.26 we obtain the axial moment of inertia of the Earth:

$$C = \frac{J_2}{H} Ma^2 = 0.33078 Ma^2 \tag{2.27}$$

The value of C thus obtained is a vital boundary condition on calculations of the radial density profile within the Earth, so that reasonably certain estimates can be made of the densities of the various internal layers discovered by seismology (Section 4.3). An inspection of Eq. 2.27 shows that the moment of inertia is smaller than for a uniform sphere, for which the numerical constant is 0.4, and therefore that there is a strong concentration of mass toward the center of the Earth.

Assuming a knowledge of the density profile from seismology (Chapter 4) and hydrostatic equilibrium at all levels, we can bring the problem of the figure of the Earth back to its starting point and calculate the equilibrium figure. Comparison of the geoid and dynamical ellipticities (Eqs. 2.25 and 2.26) shows that the inner layers are less elliptical than the surface, because if the ellipticities of all layers of constant density were equal, we would have

$$\frac{C - A}{C} = \frac{a^2 - (1/2)(a^2 + c^2)}{a^2} \approx \frac{a - c}{a} \tag{2.28}$$

The observed difference is expected because the internal surfaces of constant potential enclose material of higher average density than the Earth as a whole, so that the rotational contribution to the potential is smaller, relative to the gravitational contribution.

The hydrostatic theory is given to first order by Jeffreys (1962), following an elegant approximation procedure due to R. Radau, which shows that, to this order, the surface ellipticity, can be expressed in terms of the moment of inertia:

$$\epsilon_H = \frac{(5/2) m}{1 + (25/4)[1 - (3/2)(C/Ma^2)]^2} \tag{2.29}$$

in which m is given by Eq. 2.2 and C/Ma^2 by 2.27. This is insufficient, however, to establish the significance of the difference between the hydrostatic and observed ellipticities. Corrections have been applied numerically (Jeffreys, 1963) from the density profile, which is sufficiently well known to give the corrections to Eq. 2.29 to second order, although not adequate for a direct calculation to this order. These calculations give $\epsilon_H = 1/299.7$. The difference between ϵ and ϵ_H is not in doubt; the Earth is more elliptical than the hydrostatic theory would suggest by about 0.5 %.

As Goldreich and Toomre (1969) have pointed out, the excess ellipticity is not remarkable. When the second-order harmonics of gravitational potential are considered alone, and the equilibrium ellipticity is subtracted, the Earth appears as a triaxial ellipsoid with principal moments of inertia $A' > B' > C'$, such that $(C' - A')/(B' - A') \approx 2$. This is the axial ratio to be expected with highest probability in a randomly evolving spheroid. By causing such a body to rotate, a rotational bulge is superimposed. If the body is imperfectly elastic, the bulge assumes the equilibrium (hydrostatic) value, superimposed upon the intrinsic ellipticity, and the body turns so that its axis of greatest intrinsic (nonequilibrium) moment of inertia coincides with the axis of rotation. It now appears unlikely that the excess ellipticity represents a delay in the response of the equatorial bulge to slowing of the Earth's rotation, as first suggested by Munk and MacDonald (1960a).

The departure from hydrostatic conditions implies the existence of shear stresses within the mantle, either statically or dynamically maintained. The calculated magnitude of the stress depends upon the assumed depth of the density variations which are responsible. The excess ellipticity may be attributed to the core-mantle boundary, where there is a strong density contrast, but it is probably erroneous to separate it from the higher harmonics of the gravitational potential, whose source cannot be that deep without invoking improbably high long-term strength to the lower mantle. It therefore appears probable that the density variations occur within the lower mantle, where the stresses required to support them are of order 100 kg cm^{-2} (10^8 dyne cm^{-2}). Undulations in a vague boundary between the upper and lower mantles, which are probably slightly different chemically (Press, 1968), could also be responsible.

2.2 Precession of the Equinoxes

Equation 2.15 shows that, in addition to the principal term in r^{-1}, the gravitational potential of the Earth has a smaller term in r^{-3}, which arises from the ellipticity and has an angular dependence. Thus there is, in addition to the central gravitational force $-m(\partial V/\partial r)$ exerted on mass m at (r, ϕ), a torque $-m(\partial V/\partial \phi)$. The reaction is therefore an equal and opposite torque exerted by the mass on the Earth. The torques exerted by the Moon and the Sun on the Earth's equatorial bulge are responsible for the precession of the equinox.

The magnitude of the precession deserves some emphasis because it is commonly supposed that the axis of the Earth's rotation has a fixed direction in space. It is inclined at about 23.5° to the pole of the ecliptic (normal to the

Earth's orbital plane) and precesses slowly about it, so that the orientation of the north pole will diverge from the pole star, reaching a maximum deviation of 47° before returning 25,800 years later. Navigation by the stars would be much more complex if the precession were not so slow. Nevertheless, it is sufficiently rapid to be measured astronomically with considerable precision, the mean rate being about 50″2 per year (19″6 pole movement per year).

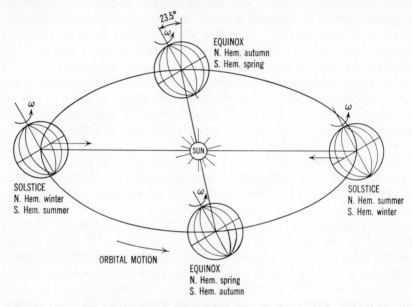

Figure 2.3: Cause of the precession of the equinox. Gravitational action of the Sun on the equatorial bulge causes a torque in the same sense at both solstices but no torque at the equinoxes.

A rigorous solution to the problem of precession requires the application of Euler's equations (see, for example, Gray, 1959) but since the angular rate of precession is very small compared with the rate of rotation (by a factor of nearly 10^7) a simpler approach, in which precession is treated as a perturbation of the rotation, is sufficient for the present discussion. We consider the Sun, of mass $M_\odot$ at (R, ϕ) with respect to the center of the Earth, as in Fig. 2.4, so that the gravitational potential at the center of the Sun due to the Earth is given by Eq. 2.15 and the torque exerted on the Earth is therefore

$$L = M_\odot \frac{\partial V}{\partial \phi} = \frac{3GM_\odot}{R^3}(C - A)\sin\phi\cos\phi \qquad (2.30)$$

The component of the Earth's rotational angular momentum in the direction of R, the line joining the centers of the Earth and Sun, is $C\omega\sin\phi$, so that the instantaneous angular rate of the precession is given by the ratio of the torque to the component of angular momentum along the Earth-Sun axis:

$$\omega_{p_s} = -\frac{L}{C\omega \sin \phi} = -\frac{3GM_\odot}{R^3} \cdot \frac{C-A}{C\omega} \cdot \cos \phi \qquad (2.31)$$

The mean precessional rate is therefore

$$\overline{\omega_{p_s}} = -3\frac{G}{\omega}\frac{C-A}{C} \cdot \frac{M_\odot}{R^3} \cos \phi \qquad (2.32)$$

which gives, by a trigonometric integration,

$$\overline{\omega_{p_s}} = -\frac{3}{2}\frac{G}{\omega}\frac{C-A}{C} \cdot \frac{M_\odot}{R^3} \cos \theta \qquad (2.33)$$

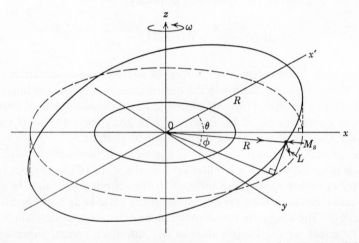

Figure 2.4: Geometry of the precessional torque. O is the center of the Earth and Oxy the equatorial plane. The Earth's orbit about the Sun is equivalent for this purpose to rotation of the Sun about the Earth at radius R in a plane $Ox'y$ which makes an angle θ with the equatorial plane. The projection of the "solar orbit" on to the equatorial plane is shown as a dashed ellipse. ϕ is the instantaneous geocentric latitude of the Sun.

The axis of the rotation precesses about the pole of the ecliptic in a sense opposite to the rotation itself, as in the familiar case of precession of a spinning top.

Equation 2.33 gives the contribution to the rate of precession due to the solar torque only. The lunar torque is obtained similarly:

$$\overline{\omega_{p_L}} = -\frac{3}{2}\frac{G}{\omega}\frac{C-A}{C} \cdot \frac{M_{\leftmoon}}{r^3} \cos \theta' \qquad (2.34)$$

Since the plane of the lunar orbit is close to the ecliptic, θ' is not very different from θ (23.5°). The proximity of the Moon more than compensates for its smaller mass, the effect being that of a gravitational gradient inversely proportional to the cube of distance, and $\overline{\omega_{pL}}$ is slightly more than twice as great as $\overline{\omega_{ps}}$. The total mean rate of precession is the observed quantity

$$\overline{\omega_p} = \overline{\omega_{ps}} + \overline{\omega_{pL}} = 50''.2 \text{ year}^{-1} \tag{2.35}$$

A complete treatment of the precessional motion shows that there is superimposed upon the precessional rotation a small nutation or "nodding" of the pole toward and away from the pole of the ecliptic. In fact there are several nutations arising from ellipticities and misalignments of the orbits and also minor effects of other planets, all of which can be adequtely accounted for. Since all of the quantities in Eqs. 2.33 and 2.34, except the moments of inertia, are observable, the dynamical ellipticity

$$H = \frac{C - A}{C} = 0.0032732 = \frac{1}{305.51} \tag{2.36}$$

is determined. Until recent observations of space probe acceleration by the Moon, the greatest uncertainty in the determination of H was the lunar mass $M_{\mathbb{D}}$. A satellite estimate (1/81.303 times the Earth's mass) was used to obtain the value quoted in Eq. 2.36. Equation 2.36 gives the value of H used in Section 2.1 to obtain a value of the moment of inertia C.

The preceding analysis applies to a simplified situation, in which the Moon is assumed to be in a circular orbit in the plane of the ecliptic. In fact the orbit is both eccentric and inclined to the ecliptic plane and the eccentricity and inclination both vary by virtue of the three-body Sun-Earth-Moon interactions. The long-term trends in the lunar orbit are considered in Section 2.4. The effect on the Earth's motion is to produce several superimposed periodicities with periods in the range 10^4 to 10^5 years, which are of interest in connection with climatic variations (Section 6.8).

2.3   The Chandler Wobble

Independently of its gravitational interaction with external bodies* the Earth undergoes a free, Eulerian precession, commonly called the free nutation in geophysical literature. To distinguish this from the forced motions due to interactions with other bodies, the term wobble, or, in the name of its discoverer, Chandler wobble is preferred. The wobble results from rotation of the Earth about an axis which departs slightly from its axis of greatest

---

* Except for Mach's principle, whereby the stationary axes to which absolute rotation refers are determined by the whole mass of the universe.

moment of inertia. The total angular momentum remains constant in magnitude and direction, but the Earth wobbles so that, relative to the surface features, the pole of rotation describes a circle about the geometrical axis of greatest moment of inertia. The spin is almost fixed in absolute orientation, so that the wobble is apparent as a cyclic variation of latitude with a period estimated to be 430 to 435 days and variable amplitude averaging 0.14 second of arc. It is superimposed upon a 12-month (seasonal) variation of similar

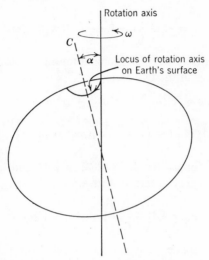

Figure 2.5: The Chandler wobble. A cyclic variation of latitude is observed as the Earth moves bodily so that the axis of rotation, which is virtually constant in space, describes a cone of semi-angle α about the axis of maximum moment of inertia.

amplitude. Melchior (1957) has reviewed the observations. The free precession of rigid bodies is given in terms of Euler's equations by standard texts on dynamics (see, for example, Gray, 1959) and a detailed discussion of the wobble of the Earth has been given by Munk and MacDonald (1960b); a simplified treatment follows here.

We consider initially a rigid Earth with principal moments of inertia $C, A, A(C > A)$, rotating with angular velocity ω about an axis at a very small angle α to the C axis. The total energy of rotation is the sum of energies of the components of rotation about the three principal axes:

$$E_T = \tfrac{1}{2}(Cm_3^2 + Am_2^2 + Am_1^2)\omega^2 \qquad (2.37)$$

where m_1, m_2, m_3 are direction cosines of the rotational axis with respect to principal axes, and

$$m_1^2 + m_2^2 + m_3^2 = 1 \qquad (2.38)$$

$$m_1^2 + m_2^2 = \alpha^2 \qquad (2.39)$$

The energy of rotation with the same angular momentum about the C axis, i.e., with $\alpha = 0$ and no wobble, is

$$E_0 = \tfrac{1}{2}C\omega_0{}^2 \qquad (2.40)$$

so that we can write the wobble energy as

$$E_w = E_T - E_0 = \tfrac{1}{2}C\omega^2\left[m_3{}^2 + \frac{A}{C}(m_1{}^2 + m_2{}^2)\right] - \tfrac{1}{2}C\omega_0{}^2$$

$$= \tfrac{1}{2}C\omega^2\left[1 - \left(\frac{C-A}{C}\right)(m_1{}^2 + m_2{}^2)\right] - \tfrac{1}{2}C\omega_0{}^2 \qquad (2.41)$$

Since the components of angular momentum add vectorially we can equate the angular momenta in the two states:

$$C\omega_0 = (C^2 m_3{}^2 + A^2 m_2{}^2 + A^2 m_1{}^2)^{1/2}\omega$$

$$= \left[1 - \left(\frac{C^2 - A^2}{A^2}\right)(m_1{}^2 + m_2{}^2)\right]^{1/2} C\omega \qquad (2.42)$$

We thus substitute for ω_0 from Eq. 2.42 in 2.41 and use also Eq. 2.39 to obtain the wobble energy in terms of its amplitude α:

$$E_w = \tfrac{1}{2}A\left(\frac{C-A}{C}\right)\omega^2\alpha^2 = \tfrac{1}{2}AH\omega^2\alpha^2 \qquad (2.43)$$

The total energy is greater than for symmetrical rotation about the C axis and the excess energy tends to restore the state of symmetrical rotation, thereby exerting, in effect, a gyroscopic torque

$$L = -\frac{dE_w}{d\alpha} = -AH\omega^2\alpha \qquad (2.44)$$

Treating it in the same way as the external torque in forced precession, we obtain the angular velocity of the free precessional motion at the Chandler frequency:

$$\omega_c = \frac{L}{A\omega\alpha} = -H\omega \qquad (2.45)$$

The period of the wobble is $C/(C - A)$ days, viewed in stationary axes, or $A/(C - A)$ days when referred to Earth axes. The precessional rotation is opposite to the spin, as observed, but the period calculated for a rigid Earth is 305 days compared with the observed period of about 430 days. A 305-day variation of latitude was sought for many years before the discovery in 1891 by S. C. Chandler of the 430-day period, which was subsequently shown to be the free Eulerian period. The difference is due to elastic yielding of the

Earth, which allows a partial accommodation of the equatorial bulge to the instantaneous axis of rotation and thus reduces the free precessional torque. As Lord Kelvin pointed out, by making assumptions about the interior of the Earth, the observed Chandler period allows a mean effective rigidity to be estimated. He showed that, for an assumed homogeneous Earth, the rigidity exceeded that of steel. However, a poorly determined single parameter for the rigidity of the Earth as a whole is of limited interest as long as it agrees, as it does, with the internal structure deduced from seismology. The general equation for the wobble period of an elastic Earth is given by Kaula (1968, p. 190).

Geophysical interest in the wobble centers on the mechanisms for excitation and damping, which are still problematical. They are discussed further in Sections 2.5 and 7.4.

2.4   Tidal Friction and the History of the Earth-Moon System

The astronomical and geophysical consequences of tides are very far-reaching. The subject owes much to the pioneering work of G. H. Darwin, whose 1898 monograph is still fascinating reading and has recently been reprinted (Darwin, 1962).

Figure 2.6 illustrates the principle by which the dissipation of tidal energy causes a slowing of the Earth's rotation. Both lunar and solar tides are responsible, but the solar tide accounts for less than one quarter of the total dissipation at the present time and still less in the past, so that for a simplified first-order treatment we may suppose the Earth-Moon to be a closed system, conserving angular momentum. If there were no dissipation of tidal energy,

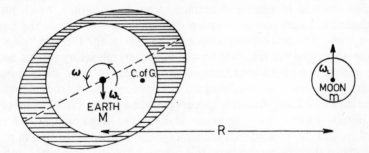

Figure 2.6: Origin of the tidal torque. The gravitational gradient of the Moon raises a tidal bulge in the Earth and in particular in the sea, but dissipative processes cause a lag in the tidal response. The gravitational potential gradient of the tidal ellipticity thus exerts an orbital accelerating force on the Moon and slows the Earth's rotation. The Earth spins on its axis with angular velocity ω and the Earth and Moon rotate about their common centre of gravity with angular velocity ω_L. The moon maintains a constant orientation with respect to the Earth.

the tidal bulge would be aligned exactly with the Moon and there would thus be no retarding torque; this would be the case for a perfectly elastic Earth and nonviscous ocean and core. Direct observation of the tidal lag has only just become possible from perturbations of satellite orbits (Newton, 1968), but the energy dissipation by tidal currents in the sea is calculable and a recent estimate (Miller, 1966) gives $1.5 \times 10^{19} \pm 50\%$ ergs sec^{-1}, which does, however, fall short of the dissipation indicated by astronomical data.

Observations of star transits give a direct measure of variations in the Earth's rotation, but the steady retardation due to tidal friction cannot be distinguished from substantial irregular changes due to fluctuations in core motion (Section 2.5). However, angular momentum in the Earth-Moon system is conserved* so that the slowing of the Earth's rotation is accompanied by an orbital acceleration of the Moon. This causes the Moon to swing out into an increasingly wide orbit, so that its angular velocity decreases. The present rate of change of the lunar orbital angular velocity, ω_L, is (Munk and MacDonald, 1960b):

$$\frac{d\omega_L}{dt} = -22''.4 \text{ century}^{-2} = -1.09 \times 10^{-23} \text{ rad sec}^{-2} \qquad (2.46)$$

That this rate applies not merely to the past two centuries of observations but is the average value for the past 370 millions years is indicated by Runcorn's (1964) analysis of counts of daily growth rings in Devonian corals by J. W. Wells and C. T. Scrutton, who found that there were 397 ± 7 days per year and 30.59 ± 0.13 days per month in the Devonian period. Since the length of the year is virtually unaffected by tidal friction, the absolute lengths of the Devonian day and month are determined from these figures, which justify our calculation based on conservation of angular momentum in the Earth-Moon system. The accelerating torque of the Sun on the thermal "tide" in the atmosphere and the retarding torque of the interplanetary plasma acting on the geomagnetic field have featured prominently in some discussions of the Earth's rotation, but since they do not conserve the rotational angular momentum of the Earth-Moon it appears that these torques must be an order of magnitude smaller than the lunar tidal torque. Munk and MacDonald (1960b, pp. 222–227) found that the atmospheric tide leads the Sun in phase and estimated that the atmospheric tide injects 2.2×10^{18} ergs sec^{-1} into the Earth's rotation. This energy is lost by the Earth's orbital motion about the Sun. It is much smaller than the energy lost to lunar tides, as calculated below, but approximately cancels the solar tidal interaction with the oceans and solid part of the Earth.

---

* Neglecting the much smaller solar tidal friction.

Equating the centripetal force of lunar rotation about the center of gravity of the Earth (mass M) and Moon (mass m) to the gravitational attraction at separation R,

$$m\omega_L{}^2\left(R\,\frac{M}{M+m}\right) = \frac{GMm}{R^2}$$

we obtain Kepler's third law:

$$\omega_L{}^2 R^3 = G(M+m) \tag{2.47}$$

and, by differentiation, the rate of the lunar recession:

$$\frac{dR}{dt} = -\frac{2}{3}\,\frac{R}{\omega_L}\,\frac{d\omega_L}{dt} = 1.04 \times 10^{-7}\ \text{cm sec}^{-1}$$

$$= 3.3\ \text{cm year}^{-1} \tag{2.48}$$

A simple linear extrapolation backward in time to the origin of the Earth 4.5×10^9 years ago gives a total change of 1.5×10^5 km in the orbital radius of the Moon, nearly half of the current value, 3.84×10^5 km. Since tidal friction is a strong function of the distance, this extrapolation merely shows that the friction has played an important role in the history of the Earth and Moon and requires more careful examination.

We neglect here the effect of the solar tide, which is much smaller than that of the Moon and is virtually constant so that the Earth and Moon are treated as an isolated system rotating in the manner of Fig. 2.6. Then conservation of the total angular momentum of the Earth about its axis and of the Earth and Moon about their common center of gravity gives

$$C\omega + \frac{Mm}{M+m}\,R^2\omega_L = \text{constant} = (1+K)C\omega_0 \tag{2.49}$$

where $K = 4.884$ is the present ratio of the orbital angular momentum to the Earth's rotational angular momentum and ω_0 is the present rotational angular velocity. The axial angular momentum of the Moon is neglected. Differentiating Eq. 2.49 with respect to t and substituting for dR/dt from Eq. 2.48, we obtain

$$\frac{d\omega}{dt} = \frac{1}{3}\,\frac{m}{M+m}\,\frac{MR^2}{C}\,\frac{d\omega_L}{dt} = -4.85 \times 10^{-22}\ \text{rad sec}^{-2} \tag{2.50}$$

Newton (1968) has made a direct estimate of the tidal retardation of the Earth's rotation from a satellite determination of the phase lag of the tidal bulge. The range of values is 16.8 to 21.4 parts in 10^{11} per year, i.e. 3.9 to 4.95×10^{-22} rad sec^{-2}.

From Eq. 2.50 the lunar tidal torque is

$$C\frac{d\omega}{dt} = -3.9 \times 10^{23} \text{ dyne cm} \tag{2.51}$$

We may also calculate the energy dissipation. The total energy is a sum of three terms due to axial rotation of the Earth, rotation of Earth and Moon about the center of gravity, and mutual potential energy:

$$E = \tfrac{1}{2}C\omega^2 + \tfrac{1}{2}R^2\omega_L{}^2\left(\frac{mM}{M+m}\right) - \frac{GMm}{R} \tag{2.52}$$

which, by Eq. 2.47, reduces to

$$E = \tfrac{1}{2}C\omega^2 - \frac{1}{2}\frac{GMm}{R} \tag{2.53}$$

Thus

$$\frac{dE}{dt} = C\omega\frac{d\omega}{dt} + \frac{1}{2}\frac{GMm}{R^2}\frac{dR}{dt} \tag{2.54}$$

Successive substitutions in the second term by Eqs. 2.48, 2.50, and 2.47 give

$$\frac{dE}{dt} = C(\omega - \omega_L)\frac{d\omega}{dt} = -2.74 \times 10^{19} \text{ ergs sec}^{-1} \tag{2.55}$$

The present numerical value obtained in Eq. 2.55 is significantly greater than the estimate of tidal dissipation in the sea, although there is considerable uncertainty in the latter. Internal friction in the solid Earth can certainly be neglected (Section 7.4) and the core is therefore implicated by default. The possibility that the core dissipates tidal energy at a rate of about 10^{19} ergs sec^{-1} is of considerable interest in the theory of the origin of the geomagnetic field (Chapter 5) as well as the geothermal flux from the deep interior (Chapter 9).

By tracing the history of the Earth-Moon system backward in time, the Moon is found to have been much closer to the Earth in the remote past than it is now. If the mechanisms of tidal dissipation were linear and if the tide was an equilibrium tide, calculation of the mechanical history of the Earth and Moon would be a tractable proposition, but in fact the marine tidal dissipation is known to be principally a result of turbulence (Jeffreys, 1962) and the tides themselves are very variable, so that the assumption of an equilibrium tide is at best a poor approximation. This means that we can have only a rough idea of the variation of dissipation with R, ω_L, and ω and can only guess at the time scale of changes in the lunar orbit. However, from a consideration of the foregoing equations we can obtain an idea of the magnitude of the change.

Combining Eqs. 2.47 and 2.49 to eliminate R, we relate ω_L and ω:

$$\omega_L[(1 + K)C\omega_0 - C\omega]^3 = \frac{G^2M^3m^3}{M + m} \tag{2.56}$$

Consider now the condition $\omega_L = \omega$, that is the Moon revolving about the Earth synchronously with the rotation of the Earth but with total angular momentum equal to the present value. Putting $\omega_L = \omega$ in Eq. 2.56 and dividing by $C^3\omega_0{}^4$ to make the equation conveniently dimensionless, we obtain

$$\frac{\omega}{\omega_0}\left(1 + K - \frac{\omega}{\omega_0}\right)^3 = \frac{G^2M^3m^3}{C^3\omega_0{}^4(M + m)} \tag{2.57}$$

which is, with insertion of numerical values,

$$\frac{\omega}{\omega_0}\left(5.88 - \frac{\omega}{\omega_0}\right)^3 = 4.24 \tag{2.58}$$

The two real solutions are $\omega/\omega_0 = 4.95, 0.0211$, corresponding to rotational periods of 4.85 hours and 47.4 days and orbital distances of 14,000 km (2.2 Earth radii) and 645,000 km. These two orbital states are asymptotic limits for the system since there is no tidal friction when the rotations are synchronous. The Moon is moving away from the close synchronous orbit toward the distant one.

This calculation is in several respects too simple; solar tidal friction, inclination and ellipticity of the lunar orbit and the figures of the Earth and Moon have been neglected. The intriguing results of a more complete calculation appeared first in a paper by H. Gerstenkorn (reviewed by Alfvén, 1965) and have led to several more papers on the subject (MacDonald, 1964a, 1967; Kaula, 1964; Goldreich, 1966; Gerstenkorn 1967a, 1967b, Singer, 1968). In tracing the lunar orbit back in time, the inclination increases until, at a stage when the Earth and Moon were very close together, the Moon was in a nearly polar orbit, and, earlier still, in a retrograde orbit. Tidal friction on the Moon in a retrograde orbit causes it to approach the Earth, so that proceeding further backward in time beyond the close approach and reversal of the orbit, we see the Moon receding on an orbit of increasing ellipticity which eventually becomes parabolic. The Gerstenkorn history of the Moon is thus an initial capture by the Earth from a solar orbit into a highly elliptical, retrograde terrestrial orbit with small perigee, followed by a very close approach to the Earth, during which tidal friction flipped the orbit over into one of rotation in the same sense as the Earth. The subsequent tidal friction has caused the recession from the Earth to the present orbit.

In calculating the time scale of the capture event, it is usually assumed that the tidal amplitude is proportional to the tidal potential, independently of its period, as for the equilibrium tide. This is perfectly valid for the mantle, of which the free resonances are all much shorter than the shortest tidal periods, but not for the oceans, which are responsible for more than half of the present dissipation. Further difficulty arises from the fact that dissipation by turbulence in shallow seas is approximately proportional to the cube of tidal velocity (compared with the square for linear dissipation). We can therefore derive no more than an order of magnitude estimate of the time scale from the present observed rate of dissipation using a simple linear theory, according to which the rate of energy dissipation is

$$\frac{dE}{dt} = A\left(\frac{\omega - \omega_L}{R^3}\right)^2 \tag{2.59}$$

where A is a constant of proportionality, which is determined from Eq. 2.55. On this basis MacDonald (1964) found the closest approach of the Moon to have been 1.78×10^9 years ago.

There is necessarily much uncertainty in the details of such a calculation. In particular, the theory depends critically on the closeness of the approach of the Moon and Earth. Inside the Roche limit of about 2.89 Earth radii the Moon is gravitationally unstable—it would literally be torn apart by the gravitational gradient of the Earth. The effect on the Earth of a lunar capture process would also have been quite dramatic. In Section 7.4 the present tidal dissipation in the mantle is estimated to be 6×10^{17} ergs sec^{-1}. With the Moon at 3 Earth radii, 20 times closer than at present, the tidal amplitude would be increased by the factor $(20)^3$; with the naive assumption that the mechanical properties of the mantle remain the same under these conditions as at present, the energy dissipation would be $(20)^6 = 6 \times 10^7$ times the present value. This is 4×10^{25} ergs sec^{-1}, a prodigious, if short-lived, heating and mechanical stirring of the Earth; recognizable geological features could hardly have survived, although if the mantle were fluidized by partial melting, the tidal dissipation would have been greatly reduced. Cooper, Richards, and Stacey (1967) have suggested that mantle lead isotopes were homogenized at about the right time and they therefore favored the capture theory.

Many geophysicists are naturally shy of the violent consequences of the capture hypothesis and see no geological evidence for a catastrophe on the scale it demands. But what are the alternative postulates for the origin of the Moon? Fission of the Earth is hardly less violent; it may be chemically attractive, but it is dynamically unacceptable and is now discounted. If the Moon had separated from the Earth, it would either have broken away completely or returned, but it could not have gone into orbit. A more favored alternative is that the Moon accreted in orbit; if the tidal friction calculations are correct, it could not have done so when the Earth itself was formed, 4.5×10^9 years ago, because its orbital lifetime appears to be less than

3×10^9 years. Since it is now supposed that the meteorites and the Earth accreted no more than a few hundred million years after the cessation of nuclear synthesis (see Section 8.5), a delay of 2×10^9 years in the accretion of the Moon appears improbable. We must therefore consider seriously the suggestion that the tidal friction calculations are misleading. The greater part of the present tidal dissipation occurs in shallow seas, which could be relatively recent additions to the Earth. If in pre-Cambrian times, more than 6×10^8 years ago, continental shelves were the actual margins of the Earth's land masses, then there would have been virtually no shallow seas and very much less tidal dissipation.* There are two reasons for supposing that this may have been so. Continental material has gradually accumulated by differentiation from the mantle (Section 8.3) and the ocean water may be at least partly a product of mantle outgassing. Thus, although the Moon was much closer in the remote past than it is now, reduction of the pre-Cambrian marine tidal friction by a factor of 3 relative to a simple extrapolation from the present dissipation appears possible and allows the Moon to have formed as a double planet with the Earth.

The origin of the Moon therefore remains an unsolved problem, but it does not appear to be beyond the scope of scientific determination. There is a further observation whose implications are not understood but which will have to be accounted for by an acceptable theory. Öpik (1961) examined the ellipticities of lunar craters and, presuming them to have been circular when formed and to have distorted with the decrease in ellipticity of the Moon as it receded from the Earth, he estimated that none of the craters was formed when the Moon was closer than 6 Earth radii. However, tidal friction and orbital evolution would have been extremely rapid if the Moon was ever that close, and if Öpik's interpretation is correct it is compatible with the capture theory.

In the very remote future, the Earth and Moon will approach a state of coherent rotation at the 47-day period, presenting constant faces to one another. (The fact that the Moon already presents a constant face to the Earth is almost certainly due to tidal friction in the Moon.) However, the solar tide will continue a slow transfer of the rotational angular momentum of the Earth-Moon pair to orbital motion about the Sun so that the Moon will again approach the Earth, eventually coalescing with it.

2.5   Fluctuations in Rotation and the Excitation of the Wobble

Superimposed upon the slow decrease in angular velocity of the Earth's rotation there are irregular fluctuations amounting to a few parts in 10^8. External influences are insignificant; the phenomenon is due to internal

---

* Pannella, MacClintock and Thompson (1968) have found evidence from fossil growth rings that tidal friction has varied greatly, even in the past 5×10^8 years.

redistribution of angular momentum. The spectrum of fluctuations has a cutoff at about 0.1 cycle per year (Munk and MacDonald, 1960b), which is therefore indicative of a time constant of about $1/(2\pi \times 0.1) = 1.6$ years. The magnitude of the changes and the time scale clearly point to motions of the fluid core and Bullard (1949) showed that electromagnetic coupling of the mantle to variable core motion was capable of explaining the length-of-day (l o d) fluctuations. It is natural to assign the time constant to the coupling mechanism and in the following discussion we pursue the consequences of doing so. Mechanically we can equally well suppose that the coupling is very tight and that the l o d spectrum is a feature of the core motions, but in this case it is difficult to devise a method of sufficiently tight mechanical coupling.

Electrical conduction in the lower mantle is due to the semiconducting properties of silicates at elevated temperatures (Section 5.3). It results in an electromagnetic coupling of the mantle to the core, in which the geomagnetic field is generated. An essential feature of the theory of the origin of the field (Section 5.4) is that the fluid core is not rotating coherently. The mantle coupling is therefore a dynamic balance between the influences of different parts of the core. Transfer to the mantle of irregularities in convective core motion or in the field thus provide a natural explanation for the l o d fluctuations. The adequacy of the electromagnetic coupling is the subject of papers by Rochester (1960) and Roden (1963). Here the effect is treated empirically in terms of a single coupling constant.

Assume for simplicity that the core is rotating coherently but at a different angular rate from the mantle and that equilibrium is restored by the coupling between the two. The moments of inertia of the core and mantle are I_c and I_m and instantaneously the angular velocities differ by $\Delta\omega_c$ and $\Delta\omega_m$ from the common value ω at equilibrium. Then conservation of angular momentum gives

$$I_m \, \Delta\omega_m + I_c \, \Delta\omega_c = 0 \tag{2.60}$$

so that the angular velocity difference between the core and mantle is

$$\Delta\omega = \Delta\omega_m - \Delta\omega_c = \Delta\omega_m\left(1 + \frac{I_m}{I_c}\right) \tag{2.61}$$

This is the angular rate at which the geomagnetic field lines are drawn through the conducting mantle. Calculation of the consequent torque is a complex problem depending upon the conductivity of the mantle and its penetration by the field, but without knowing the details we can see that the induced currents are proportional to $\Delta\omega$ and ohmic dissipation is proportional to $(\Delta\omega)^2$. We define a coupling coefficient K_R in terms of this dissipation:

$$\frac{dE_R}{dt} = -K_R(\Delta\omega)^2 = -K_R(\Delta\omega_m)^2\left(1 + \frac{I_m}{I_c}\right)^2 \tag{2.62}$$

E_R is the total rotational energy:

$$E_R = \tfrac{1}{2}I_m(\omega + \Delta\omega_m)^2 + \tfrac{1}{2}I_c(\omega + \Delta\omega_c)^2 \qquad (2.63)$$

which gives, with Eq. 2.60

$$\frac{dE_R}{dt} = I_m\left(1 + \frac{I_m}{I_c}\right)\Delta\omega_m \frac{d}{dt}(\Delta\omega_m) \qquad (2.64)$$

dE_R/dt is necessarily negative because $\Delta\omega_m$ and $(d/dt)(\Delta\omega_m)$ have opposite signs. $\Delta\omega_m$ increases if it is negative and decreases if it is positive. Equating (2.62) and (2.64) we obtain

$$\frac{d(\Delta\omega_m)}{\Delta\omega_m} = -K_R\left(\frac{1}{I_m} + \frac{1}{I_c}\right) dt \qquad (2.65)$$

$\Delta\omega_m$ thus decays exponentially with a time constant

$$\tau_R = \frac{1}{K_R(1/I_m + 1/I_c)} \approx 1.6 \text{ years} \qquad (2.66)$$

It should be noted that we expect the coefficient K_R to be constant only for the particular geometry of field and induced currents appropriate to differential axial rotation of the core and mantle. However, we expect it to be comparable in magnitude to the coefficient for coupling of the wobble to the core. We can therefore examine in a simple way the possibility that core-mantle coupling is adequate to excite or damp the wobble, bearing in mind that the true situation is much more complex than that analyzed, as we have three superimposed motions—rotation, precession, and wobble—all with irregularities and not strictly independent.

The internal motions within the core are here neglected; it is treated as a rigid sphere. Munk and MacDonald (1960b) considered the dynamics of a core and mantle with variable coupling; with zero coupling the core does not follow the wobble at all and there is no damping, and with perfectly tight coupling the core follows exactly, again without damping. For any intermediate degree of coupling the core follows the wobble with reduced amplitude and a phase lag with consequent dissipation of the wobble energy. They found that with optimum coupling for maximum dissipation the wobble would be damped with a time constant as short as 100 days. For the observed time constant we may take the value 12.4 years (with a wide range of uncertainty) quoted by Munk and MacDonald (1960b, p. 174) from the work of P. Fellgett. If this is to be attributed to core-mantle coupling we must therefore assume either that the coupling is much weaker than the optimum and that the core hardly follows the wobble, or that the coupling is much stronger and that the core follows almost exactly. Rochester and Smylie

(1965) examined the requirements of geomagnetic field strength and mantle conductivity for damping by core mantle coupling and concluded that they were inadequate. For the purpose of this calculation we may therefore assume that the coupling is weak, this being particularly simple because we can assume the mantle to be wobbling about a nonwobbling core without serious error.*

In the notation of this section, the energy of a wobble of angular amplitude α is, by Eq. 2.43,

$$E_W = \tfrac{1}{2} I_m H \omega^2 \alpha^2 \tag{2.67}$$

The relative angular velocity of the mantle over the core is

$$\Delta\omega' = \omega_c \alpha \tag{2.68}$$

where ω_c is the Chandler angular frequency. Although assumed here to be so, the axis of instantaneous rotation is not strictly constant but wobbles with amplitude $(\omega_c/\omega)\alpha$, to conserve angular momentum. Now we can define a wobble coupling coefficient K_W, which would be sufficient to explain the wobble:

$$\frac{dE_W}{dt} = -K_W(\Delta\omega')^2 = -K_W \omega_c^2 \alpha^2 \tag{2.69}$$

and compare this with the differential of Eq. 2.67:

$$\frac{dE_W}{dt} = I_m H \omega^2 \alpha \frac{d\alpha}{dt} \tag{2.70}$$

from which we obtain

$$\frac{1}{\alpha}\frac{d\alpha}{dt} = -\frac{K_W \omega_c^2}{I_m H \omega^2} \tag{2.71}$$

This is an exponential decay with a time constant

$$\tau_W = \frac{I_m H}{K_W}\frac{\omega^2}{\omega_c^2} = 12.4 \text{ years} \tag{2.72}$$

Equations 2.66 and 2.72 allow us to compare the values of K_W and K_R required if the spectra of both the l o d fluctuations and the wobble are to be explained in terms of core-mantle coupling:

$$\frac{K_W}{K_R} = \frac{\tau_R}{\tau_W}\left(\frac{\omega}{\omega_c}\right)^2 H\left(1 + \frac{I_m}{I_c}\right) = 800 \tag{2.73}$$

It therefore appears that coupling is deficient by a large factor to explain the wobble, the conclusion reached by Rochester and Smylie (1965).

---

* It is, of course, understood that the core must follow the wobble slightly if there is any coupling.

The suggestion that the wobble is excited by earthquakes has recently been renewed by Mansinha and Smylie (1967), although this mechanism had been discounted by Munk and MacDonald (1960b, pp. 163–164). Earthquakes may be regarded as random impulses which produce instantaneous small changes in the moments of inertia of the Earth. A simple order-of-magnitude calculation, based on the dislocation theory of earthquakes (Section 4.2), suffices to show what is possible. An essential point of this theory is that the stresses within the Earth are dynamically balanced both before and after an earthquake; the stress release therefore also forms a dynamically balanced system of forces. A consequence of this is that the center of gravity of the Earth cannot be moved by an earthquake or, in mathematical terms,

$$\int x \, dm = 0 \tag{2.74}$$

where x is the displacement in a particular direction of mass dm and the integration extends over the whole focal volume of the earthquake (or of the Earth). Thus, although in many earthquakes there may appear to be a predominance of uplift over downthrust of the crust or vice versa, when viewed on a large enough scale the displacements must be balanced.

Now consider a simple model of an earthquake which causes a mutual displacement of two masses, m_1, m_2, whose distances, r_1, r_2, from a selected axis are changed by Δr_1, and Δr_2, respectively. By Eq. 2.74

$$m_1 \, \Delta r_1 + m_2 \, \Delta r_2 = 0 \tag{2.75}$$

and the change in moment of inertia of the Earth about the chosen axis is

$$\Delta I = m_1[(r_1 + \Delta r_1)^2 - r_1{}^2] + m_2[(r_2 + \Delta r_2)^2 - r_2{}^2] \tag{2.76}$$

Using Eq. 2.75 and neglecting terms of second order in small quantities,

$$\Delta I = 2m_1 \, \Delta r_1 (r_1 - r_2) \tag{2.77}$$

We can estimate the magnitude of ΔI for a very large earthquake by taking the dimensions of a dislocation model of the 1964 Alaskan earthquake (Fig. 4.12) to be representative of the occasional very large shocks. Then $\Delta r_1 = 22m$, $(r_1 - r_2) = 200 \, km$ (the scale of the stress field normal to the fault plane must be comparable to the smaller dimension of the fault face) and, allowing for grading of the displacement to zero at the edges of the fault plane and assuming a fault length of 800 km, the displaced masses may be treated as blocks not exceeding (800 km × 200 km × 200 km) in volume and having a density of 3 gm cm^{-3}, so that $m_1 = 10^{23}$ gm and, by Eq. 2.77,

$$\Delta I = 9 \times 10^{33} \text{ gm cm}^2 \tag{2.78}$$

To estimate the effect of such displacements on the wobble, ΔI is compared with the fractional difference in moments of inertia of the Earth about polar and equatorial axes, $(C - A) = 2.6 \times 10^{42}$ gm cm^2. Favorably oriented displacements could thus excite a wobble of angular amplitude

$$\Delta\alpha = \frac{\Delta I}{C - A} = 3.5 \times 10^{-9} \text{ rad} \tag{2.79}$$

which is very much smaller than the average observed amplitude, 0.14 sec = 7×10^{-7} rad. A calculation based on the movements of blocks disallows the contributions to ΔI from parts of the displacement field well removed from the fault plane. It therefore underestimates the effect, but not by a big factor; Eq. 2.78 is quite close to the estimate by Mansinha and Smylie (1967) based on a dislocation model with an extended displacement field.

Earthquake energy release is apparently dominated by relatively few very large shocks (Section 4.1, especially Eq. 4.9); by dislocation theory changes in moment of inertia vary approximately as (energy)^2 and are therefore even more strongly dominated by the largest shocks, so that even if a sequence of earthquakes were synchronized, they would be ineffective in exciting the wobble. It is therefore paradoxical that Mansinha and Smylie (1968) claim to have found evidence of a correlation between earthquakes and small shifts in the rotational axis. The possibility of a common causal connection must be considered. However, an immediate difficulty to the supposition that the main cause of the wobble is in the mantle is that the damping is unexplained. It cannot be a mantle effect as this demands an improbably low mechanical Q (Section 7.4).

At first sight atmospheric movements appear to provide a more plausible mechanism of wobble generation. Although the available mass is limited, the displacements are comparable to the dimensions of the Earth. Equation 2.77 is not directly applicable to this case, but the mass Δm which must be moved by a distance equal to the radius of the Earth a to produce a change of moment of inertia $\Delta I = 10^{36}$ gm cm^2 is

$$\Delta m = \frac{\Delta I}{a^2} = 2.5 \times 10^{18} \text{ gm} \tag{2.80}$$

which is only 5×10^{-4} times the mass of the atmosphere. Munk and Hassan (1961) examined monthly means of the moments and products of inertia of the atmosphere, from all available barometric pressure data, for the period 1873 to 1950. They concluded that the 12-month (annual) wobble was excited atmospherically, but that at the 14-month period the power in the atmospheric fluctuations was one or two orders of magnitude too small to excite the Chandler wobble.

Torques exerted on the magnetosphere by fluctuations and irregularities of particle emissions from the sun (the solar wind) are also too small, by many orders of magnitude.

Apart from the possibility that there is a completely unsuspected mechanism, it is evident that some important factor has been neglected in one of the mechanisms considered. It is a matter of personal conjecture which this is. To the author, core-mantle coupling appears most promising, partly because its complexity has been underestimated. Large-scale core motions, excited by the precession and responsible for the geomagnetic field (Section 5.4), must react back on the mantle and an analysis in terms of a rigid core is too simple. The core motion is evidently turbulent, so that linear equations which allow the precession and wobble to be considered separately are probably inapplicable.

Chapter 3

THE GRAVITY FIELD

"... we are entitled to regard ourselves as general practitioners and to restrict ourselves to the kinds of peculiarities that occur in physics...."

JEFFREYS AND JEFFREYS, 1962, p. 5.

3.1 Gravity as Gradient of the Geopotential

Gravitational acceleration on the geoid can be related directly to the ellipticity, so that one can be determined from the other without appealing to independent evidence of the moments of inertia of the Earth. The first-order theory suffices for many purposes and is given here; derivation of the second-order terms is given by Jeffreys (1962). Satellites have provided more precise values of the low-order harmonics in gravitational potential than can possibly be obtained from surface gravity data, so that gravity surveys can now be referred to an ellipsoid derived from the satellite work and internationally adopted as standard (International Union of Geodesy and Geophysics, 1967).

Gravity g is obtained by differentiating the total geopotential, as given by Eqs. 2.1 and 2.6:

$$U = -\frac{GM}{r} + \frac{GMa^2}{2r^3} J_2(3\sin^2\phi - 1) - \tfrac{1}{2}\omega^2 r^2 \cos^2\phi \qquad (3.1)$$

Since

$$g = -\operatorname{grad} U$$

we have

$$g = -\left[\left(\frac{\partial U}{\partial r}\right)^2 + \left(\frac{1}{r}\frac{\partial U}{\partial \phi}\right)^2\right]^{1/2} \qquad (3.2)$$

But the normal to the geoid departs from the radial direction only by a small angle $(\phi_g - \phi)$ which is of order ϵ, as in Fig. 2.1, so that to the first order in small quantities the second term in Eq. 3.2 is negligible, whence

$$-g = \frac{\partial U}{\partial r} = \frac{GM}{r^2} - \frac{3}{2}\frac{GMa^2}{r^4}J_2(3\sin^2\phi - 1) - \omega^2 r(1 - \sin^2\phi) \quad (3.3)$$

Now we may substitute the value of r on the geoid, at arbitrary latitude ϕ:

$$r = a(1 - \epsilon \sin^2\phi) \quad (3.4)$$

where a is the equatorial radius and ϵ is the ellipticity given by Eq. 2.25:

$$\epsilon = \tfrac{3}{2}J_2 + \tfrac{1}{2}m \quad (3.5)$$

We substitute for r from (3.4) in (3.3) and use the binomial expansion

$$(1 - \epsilon\sin^2\phi)^{-n} = (1 + n\epsilon\sin^2\phi \ldots) \quad (3.6)$$

This allows products of small quantities to be neglected, and since the second and third terms in Eq. 3.3 are themselves of order ϵ times the first term, the expansion is applied only to the first term:

$$-g = \frac{GM}{a^2}(1 + 2\epsilon\sin^2\phi) - \frac{3}{2}\cdot\frac{GM}{a^2}J_2(3\sin^2\phi - 1) - \omega^2 a(1 - \sin^2\phi)$$

$$(3.7)$$

The equatorial gravity is therefore

$$-g_e = \frac{GM}{a^2}(1 + \tfrac{3}{2}J_2 - m) \quad (3.8)$$

since, by (2.24),

$$m = \frac{\omega^2 a^3}{GM}$$

Then, again to the first order in ϵ, the gravity at latitude ϕ is given in terms of the equatorial value g_e, from Eqs. 3.7 and 3.8:

$$g = g_e[1 - (\tfrac{9}{2}J_2 - 2\epsilon - m)\sin^2\phi] \quad (3.9)$$

which by Eq. 3.5 takes two more useful forms:

$$g = g_e[1 + (2m - \tfrac{3}{2}J_2)\sin^2\phi] \quad (3.10)$$

or

$$g = g_e[1 + (\tfrac{5}{2}m - \epsilon)\sin^2\phi] \quad (3.11)$$

The basic result, Eq. 3.11, is known as Clairaut's theorem. Historically its value has been to provide an estimate of ellipticity independently of astro-geodetic surveys. By retaining higher-order terms, the following result is obtained:

$$g = g_e\left[1 + \left(\frac{5}{2}m - \epsilon - \frac{17}{14}m\epsilon\right)\sin^2\phi + \left(\frac{\epsilon^2}{8} - \frac{5}{8}m\epsilon\right)\sin^2 2\phi + \cdots\right] \quad (3.12)$$

With recent values of the constants, this gives the geodetic standard to which gravity surveys should be referred:

$$g = (978.03090 + 5.18552\sin^2\phi - 0.00570\sin^2 2\phi)\ \text{cm sec}^{-2} \quad (3.13)$$

The international gravity formula adopted in 1930, with earlier values of the constants, will probably continue to be used for some time where gravity surveys have been based on it, making a change inconvenient:

$$g = 978.0490(1 + 0.0052884\sin^2\phi - 0.0000059\sin^2 2\phi)\ \text{cm sec}^{-2} \quad (3.14)$$

These equations refer to gravity on an ideal geoidal (sea-level) surface. Gravity survey data are referred to the standard latitude variations, Eq. 3.13 or 3.14, with corrections for the elevation of the Earth's surface, where measurements are actually made. Such corrections imply some knowledge of the crustal structure and therefore, to some extent, beg the question which a gravity survey attempts to answer. However, surveys show local departures from the standard of reference amounting to 30 times the third term of the reference formula and thus give clear evidence of density variations in the crust. The large-scale features of the gravity field, which reflect deep-seated variations, are best indicated by analyses of satellite orbits.

3.2   The Satellite Geoid

In Section 2.2 the torques exerted by the Sun and Moon on the Earth's equatorial bulge were shown to cause precession of the Earth. Equal and opposite torques are of course exerted by the Earth on the Sun and Moon; in the case of the Moon the influence on its orbit is appreciable. A precisely similar torque is exerted by the equatorial bulge on artificial satellites whose masses are too small to influence appreciably the motion of the Earth, but whose orbits provide the most precise evidence of the large-scale departures of the Earth from spherical symmetry.

Consider initially an axially symmetric Earth whose external gravitational potential is of simple ellipsoidal form, being represented by a second-order zonal harmonic as in Eq. 2.6:

$$V = -\frac{GM}{r} + \frac{GMa^2 J_2}{2r^3}(3\sin^2\phi - 1) \quad (3.15)$$

To first order, a satellite's motion is controlled by the central force, given by the first term in Eq. 3.15, and its orbit is therefore an ellipse with the center of the Earth at one focus. It is convenient to deal with the special case of a circular orbit of radius r. The effect of the second term can then be treated as a perturbation by calculating the torque L exerted as in Section 2.2. Writing the satellite mass as m,

$$L = m \frac{\partial V}{\partial \phi} = \frac{3GMa^2 J_2 m}{r^3} \sin \phi \cos \phi \qquad (3.16)$$

We may consider this torque to cause a precession of the satellite orbit, an effect known as regression of the nodes, the succession of points at which the orbit crosses the Earth's equatorial plane (as seen in stationary coordinates, i.e., not in the Earth's rotating coordinates). The torque acts on the angular momentum a of the satellite,

$$a = mr^2 \omega_s \qquad (3.17)$$

where ω_s is its orbital angular velocity given by Kepler's third law (as in Eq. 2.47):

$$\omega_s^2 r^3 = GM \qquad (3.18)$$

As in the analysis of Section 2.2 the instantaneous angular rate of the precessional motion of the satellite is

$$\omega_p = -\frac{L}{a \sin \phi} = -\frac{3GMa^2 J_2}{r^5 \omega_s} \cos \phi \qquad (3.19)$$

and, if the inclination of the plane of the orbit to the equatorial plane is i, the mean precessional rate is

$$\bar{\omega}_p = -\frac{3}{2} \frac{GMa^2 J_2}{r^5 \omega_s} \cos i \qquad (3.20)$$

Thus the angular change $\Delta\Omega$, per orbital revolution, in the position of a node is given by

$$\frac{\Delta\Omega}{2\pi} = \frac{\bar{\omega}_p}{\omega_s} = -\frac{3}{2} \frac{a^2}{r^2} J_2 \cos i \qquad (3.21)$$

where the simplification is effected by Eq. 3.18. The factor J_2 is thus determined, in principle, by very simple observations on regression of nodes.

In practice the situation is considerably more complicated.* First, orbits are elliptical and the ellipticity enters the equation for nodal regression.

---

* Apart from a more complicated gravitational field and orbit ellipticity, atmospheric drag and lunar and solar accelerations must also be allowed for.

However, there is also a steady motion of the perigee and apogee of the orbit which can be useful as a check. Of greater interest is the fact that the geoid has small departures from the simple ellipsoidal form. These are smaller by factors of 1000 or more than the ellipticity but are nevertheless observable by means of satellite orbits. Still regarding the Earth as an axially symmetric body, the gravitational potential can be represented as an infinite series of *zonal harmonics*:

$$V = -\frac{GM}{r} \left[1 - \sum_l \left(\frac{a}{r}\right)^l J_l P_l(\sin \phi) \right] \qquad (3.22)$$

The harmonics contribute to nodal regression by amounts depending upon the inclination i of a satellite orbit and the lower-order coefficients J_l can be estimated from the rates of regression for satellites with different orbital inclinations. Several zonal harmonics in the gravitational potential have been determined with reasonably accuracy because the rotation of the Earth ensures that, averaged over a sufficient time, the Earth appears to have rotational symmetry.

Departures from axial symmetry are equally interesting but are less easily determined because they cause shorter period perturbations in the satellite orbits, for example, an oscillation in the rate of nodal regression. Variations in gravitational potential with longitude are represented in terms of tesseral harmonics, the potential being written in the general spherical harmonic form (see Appendix A):

$$V = \frac{GM}{r} \left\{ 1 - \sum_{l=2}^{\infty} \left(\frac{a}{r}\right)^l \sum_{m=0}^{l} P_l^m(\sin \phi)[C_l^m \cos m\lambda + S_l^m \sin m\lambda] \right\} \qquad (3.23)$$

where $P_l^m(\sin \phi)$ is the associated Legendre polynomial. The general mathematical treatment of satellite orbits in a potential field of the form of Eq. 3.23 is very involved; it has been reviewed by A. H. Cook (1963a) and discussed more simply by Kaula (1965). Reliable determinations of even the low-order tesseral harmonics require extensive and accurate data on satellite orbits and heavy computer processing. Kozai (1966) has summarized the results available up to 1966 and has shown that, in spite of the difficulties, several independent determinations of the form of the geoid have converged to substantial agreement on the general features which are apparent in Fig. 3.1. The harmonic coefficients C_l^m and S_l^m in Eq. 3.23, which were used to plot this figure, are given in Table 3.1.

It is significant that the highs and lows of the geoid show no correlation with the crustal features of the Earth (continents and oceans). This is the most striking evidence for mass compensation at depth (isostasy) on a continental scale. If the continents were simply superimposed upon a perfectly layered ellipsoidal Earth, then there would be an exact correlation of the geoid with continental shapes and the differences between highs and lows would be about 10 times greater than they are. Approximate isostatic balance of features such as mountain ranges has been established for many years from gravity surveys and is considered in Section 3.3. Independence of the geoidal and continental features compels the inference that the features of the geoid are due either to density differences deep within the mantle—deeper than the "soft" layer in the upper mantle, which is presumed to allow isostatic balance to be maintained—or else to density differences maintained dynamically (by convection), in which case they are more likely to be in the upper mantle. Cook (1963b) speculated that the low-order harmonics could be undulations of the boundary between the solid mantle and fluid metal core, but pointed out that the higher harmonics could not originate so low.

Lee and MacDonald (1963) made a spherical harmonic analysis of all of the observed values of heat flow through the Earth's surface (Fig. 9.1) and concluded that there was a correlation between geoidal lows and high heat flow and vice versa. This might be expected in a simple regime of thermal convection; the greater heat flow occurring above the hotter, rising material, which, being less dense, causes a low in the geoid. However, the correlation is very imperfect and the coverage of the Earth by heat flow data is too incomplete to regard the correlation as anything more than an interesting suggestion which will bear further examination. Meanwhile there is no obvious relationship between the geoid and currently active tectonic regions and although a satisfactory theory of the convection process is still needed (see Section 7.3), it is more likely that the low-order harmonics of the geoid have an origin in the lower mantle.

3.3   Crustal Structure and the Principle of Isostasy

The distinctness of continents and oceans is made apparent by plotting the areas of the Earth's solid surface at different levels above and below sea level. Such a plot is known as a hypsometric curve and is given as a histogram in Fig. 3.2. This shows a bimodal distribution with a peak near to sea level, corresponding to continents, and another at about 5-km depth, which corresponds to the ocean basins. Mountain ranges and deep ocean trenches occupy very small fractions of the surface area, but the most significant feature of the hypsometric curve is the smallness of the area between elevations of -1 km and -3 km. In other words, the continents have sharp submarine margins, which separate them from the oceans. This is apparent also in the smooth curve in Fig. 3.2, which is the integral of the data in the histogram and is thus an averaged characteristic of the surface elevation.

Sea level is necessarily the reference level for Fig. 3.2, but it is not the demarcation between continental and oceanic structures. If sea level had happened to be, say, 1.5 km lower, it would have marked the true continental

TABLE 3.1: HARMONIC COEFFICIENTS OF THE EARTH'S GRAVITATIONAL POTENTIAL. (Data by Guier and Newton, 1965, normalized to the harmonic functions $p_l{}^m$ defined by Eq. A.13 in Appendix A. These are the data used to plot the geoid reproduced as Fig. 3.1. Units are 10^{-6}.)

	J_l	$C_l{}^m$ $S_l{}^m$							
l	m 0	1	2	3	4	5	6	7	8
2	1082.64	—	7.53 3.79						
3	2.68	6.90 0.80	4.56 −2.54	2.47 3.66					
4	−1.61	−2.39 −1.87	1.77 1.88	3.59 0.03	−0.89 0.81				
5	0.03	0.64 −0.78	1.26 −1.58	0.43 0.49	−2.29 −1.22	−0.16 −3.14			
6	0.71	0.00 0.51	−0.82 −0.79	2.70 0.26	−1.56 −2.59	−0.90 −2.60	0.07 −1.18		
7	0.59	0.69 0.51	2.51 0.33	2.16 −1.13	−0.75 0.00	−0.31 −1.02	−2.49 4.15	0.48 −0.79	
8	0.13	−0.86 −0.28	0.55 −0.22	−0.32 1.26	−0.40 0.22	0.47 −0.01	−0.13 3.88	0.99 −0.41	−0.85 0.55
9	−0.18								
10	0.09								

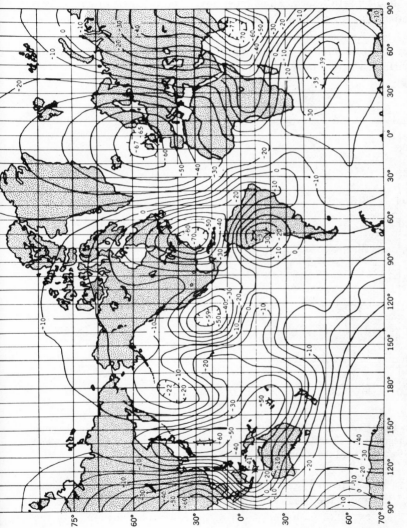

Figure 3.1: Satellite geoid. This represents departures in meters of the geoidal surface from an ellipsoid with ellipticity corresponding to the measured values of J_2, plus also the effect of the higher zonal harmonic J_4, which is associated with J_2 in the second-order theory of the ellipticity. Reproduced by permission from Guier and Newton (1965.)

margins reasonably well and we might have supposed that the sea level was in some way responsible for continental structure. However, even allowing for the significance of erosion processes in geomorphology, it is apparent that the general forms and distribution of continents cannot be results simply of forces at the Earth's surface but must have a deep-seated origin. This is considered further in Chapters 7 and 8. By including submerged areas of continents, which we may define roughly as the areas where the sea is less than 1 km deep, the total area of the continents is found to be nearly 40% of the Earth's surface, whereas only 29% is actually above sea level.

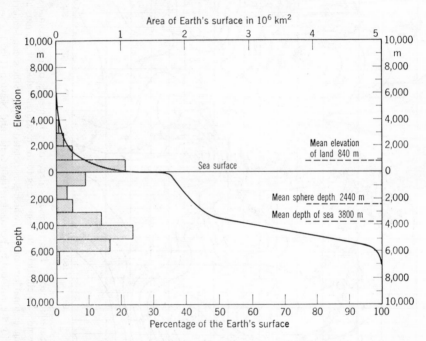

Figure 3.2: Histogram of areas of the Earth's solid surface in intervals of 1 km in elevation with the integral curve showing the area above any depth. After Sverdrup, Johnson, and Fleming (1942) who used the compilations of E. Kossinna.

Since, as we see from Fig. 3.2, the continental masses are distinct areas standing 5 km above the ocean floor and their individual areas are large, the fact that their existence is not apparent in the low-order harmonics of the satellite geoid requires mass compensation at depth to quite a high degree of adjustment. Note that satellite observations indicate mass compensation on a *continental scale*; they are not suitable for the determination of local gravity variations, which would be apparent as high harmonics of the geoid, and for which surface observations are required. Nevertheless, the principle of mass compensation has been known for over a century, since a geodetic survey across North India showed that the Himalayas caused much less deflection of the vertical than if they were simply a prominence upon a uniform Earth. This discovery was the birth of the theory of isostasy, according to which the total mass of rock (and sea where it occurs) in any vertical column of unit cross section is constant. The columns may be measured from a particular level of "compensation" below which the Earth can be assumed uniform. Thus variations in elevation of the crust are supported hydrostatically. The balance is nowhere perfect, but on a continental scale it is very nearly so.

The distribution of mass at depth cannot be determined from surface gravity data without additional evidence and, even with the approximate location by seismic methods of boundaries between layers of different densities, there is generally quite a wide range of admissible alternatives. Heiskanen and Meinesz (1958) and Garland (1965) have reviewed the subject and presented computed density profiles for different regions, in which the seismic evidence is taken into account. These incorporate in different proportions the two rival hypotheses of isostasy which grew directly from the original survey work in India. In 1854 J. H. Pratt suggested that the higher parts of the crust were elevated by virtue of their lower densities, as in Fig. 3.3a, and in 1855 G. B. Airy proposed the scheme represented in Fig. 3.3b. Airy visualised the crustal masses as logs (all of the same density) floating in water. A log appearing higher out of the water than its neighbors must extend correspondingly deeper. This accords with the seismic evidence that the continental crust (35 to 40 km average) is thicker than the oceanic crust (about 5 km) by an amount greatly exceeding the difference in surface elevations. The considered opinion of W. A. Heiskanen (see Heiskanen and Meinesz, 1958) is that, on average, 63% of the isostatic balance of the crust is achieved by the Airy principle of depth compensation and 37% by density differences, as envisaged by Pratt.

The observation that the form of the satellite geoid is not correlated with continents and oceans and that the departures of the geoid from an ellipsoid of revolution are slight compared with what one would expect if the continents were simply superimposed upon an ellipsoidally layered Earth, allows us to estimate the degree of exactness of the isostatic balance of the continents. It is convenient to discuss the effect of continents on the geoid in terms of the geometrically simple pair of continents represented in Fig. 3.4. First consider the effect on the satellite geoid of a pair of circular continents, radius r, height h, and density ρ_c, superimposed upon an otherwise spherically symmetrical earth, as in Fig. 3.4a. The resulting ellipticity of the satellite geoid is calculable from the difference between the moments of inertia of the Earth about axes 1 and 2, by Eq. 2.21, neglecting the second (rotational) term, since satellites see only the gravitational component of the geopotential. The

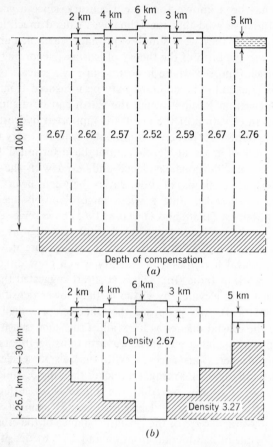

Figure 3.3: Isostatic compensation according to (*a*) J. H. Pratt and (*b*) G. B. Airy, with numerical values of density given by W. A. Heiskanen. Figures based on Heiskanen and Meinesz(1958).

model is otherwise spherically symmetrical; thus we need consider only the moments of inertia of the continents, which are, for $r \ll a$,

$$
\begin{aligned}
I_1 &\approx 2(\pi r^2 h \rho_c) \frac{r^2}{2} \\
I_2 &\approx 2 \int_a^{a+h} (\pi r^2 \rho_c) x^2 \, dx = 2\pi r^2 \rho_c a^2 h, \qquad (h \ll a)
\end{aligned}
\right\} \tag{3.24}
$$

so that approximately

$$
I_1 - I_2 = -2\pi r^2 \rho_c a^2 h \tag{3.25}
$$

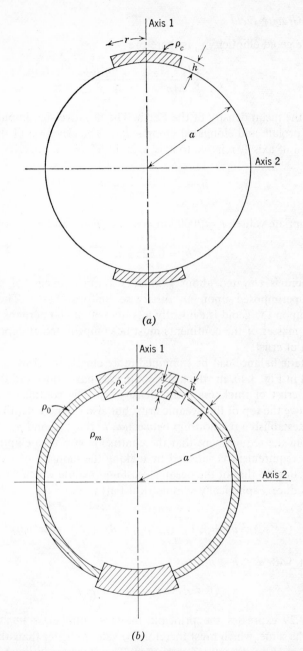

(a)

(b)

Figure 3.4: Simple model of a symmetrical pair of continents, illustrating their effect upon the satellite geoid. (*a*) Continents superimposed upon an otherwise spherically symmetric earth. (*b*) Continents of density ρ_c overlaying a mantle of density ρ_m and isostatically balanced with an oceanic crust of density ρ_0.

The satellite geoid ellipticity ϵ_g is therefore

$$\epsilon_g = \frac{3}{2} \frac{I_1 - I_2}{Ma^2} = -\frac{9}{4} \frac{r^2 h}{a^3} \frac{\rho_c}{\bar{\rho}} \tag{3.26}$$

where $\bar{\rho}$ is the mean density of the Earth. The negative sign implies that the ellipsoid is prolate and elongated along axis 1. The elevation of the geoid in the direction of axis 1, relative to axis 2, is therefore

$$h_g = \epsilon_g a = \frac{9}{4} \frac{r^2}{a^2} \frac{\rho_c}{\bar{\rho}} h \tag{3.27}$$

With appropriate values, $r = 2000$ km $= a/3$, $\rho_c/\bar{\rho} = \frac{1}{2}$ and $h = 5$ km, we have

$$h_g = 0.625 \text{ km}$$

i.e., the geoid is elevated about $\frac{1}{8}$ km per 1 km thickness of continental material superimposed upon an otherwise uniform Earth. The effect of continents upon the geoid is evidently less than 50 m and perhaps much less, so that the masses of the continents must be compensated at depth to better than 400 m of crust.

The isostatic balance can be examined more closely in terms of the model represented in Fig. 3.4*b*, in which the continents have roots and there is also an oceanic crust of thickness t and density ρ_0. The continents reach to a height h above the top of the oceanic crust but also extend a depth d below it, and we can establish a relationship between d, h, t, ρ_c, ρ_0 and ρ_m, the mantle density, from the requirement that the continents give zero ellipticity to the geoid. This requirement is satisfied by making the moment of inertia of the continents equal to that of the oceanic crust and mantle which would replace them to produce a spherically symmetrical Earth:

$$2 \int_{a-d}^{a+h} (\pi r^2 \rho_c) x^2 \, dx = 2 \int_{a-d}^{a-t} (\pi r^2 \rho_m) x^2 \, dx + 2 \int_{a-t}^{a} (\pi r^2 \rho_0) x^2 \, dx \tag{3.28}$$

from which, with d, t, $h \ll a$,

$$\rho_c(h + d) = \rho_m(d - t) + \rho_0 t \tag{3.29}$$

Equation 3.29 expresses the principle that the total mass in any vertical column is the same, which must therefore be valid to better than about 8 % of 5 km or 400 m of crust, as noted previously. This is the general case of isostatic balance, without preference for the Pratt or Airy principles. The balance would be achieved by Pratt's method if $d = t$, so that

$$\rho_c(h + t) = \rho_0 t \tag{3.30}$$

or by Airy's method if $\rho_c = \rho_0$, in which case

$$\rho_c(h + d - t) = \rho_m(d - t) \tag{3.31}$$

Departures from isostatic balance must be expected to occur on a small scale, i.e., over areas which are not much larger than the depth of the crust, up to perhaps 100 km. Alternatively, an isostatic anomaly could be maintained dynamically by crustal movements, driven by internal forces, as considered in Chapter 7. However, even active mountain ranges appear to be in approximate isostatic balance. Probably the most striking dynamic features of the crust are the island arcs, such as that represented in Fig. 4.2, which do have characteristic gravity anomalies.

The restoration of isostatic equilibrium in an area where crustal movements have occurred is a matter of considerable interest in connection with the rheological properties of the upper mantle (Chapter 7). The steady rise of a large area around the Gulf of Bothnia (Fig. 7.6) is regarded by some authors as a process of isostatic rebound from the depression of the area caused by heavy glaciation during the last ice age. It is supposed that, at the depth of compensation, the mantle behaves as a fluid with a viscosity of 10^{21} to 10^{22} poises. If Newtonian viscosity is an appropriate concept for the mantle, this is not an unreasonable figure, although the assumption that the Fenno-Scandian uplift is a glacial rebound has been doubted.

The exactness of isostatic balance on a continental scale, as deduced from the satellite geoid, is referred to the observed ellipticity of the Earth, but the ellipticity itself departs quite significantly from hydrostatic equilibrium (see Section 2.1). We assume that the upper few hundred kilometers of the mantle, which appear to allow isostatic adjustment, are hydrostatically balanced for all low order harmonics of the geopotential, including the ellipticity term, and that the nonequilibrium ellipticity has a deep-seated origin. If the upper mantle contributes negligibly to the departure from equilibrium, then the lower mantle must be correspondingly further from equilibrium. It must have a substantial strength, even on a time scale of millions of years. This conclusion is important to theories of the rheology of the mantle; in particular, it precludes the consideration of convection in the lower mantle (see Chapter 7).

3.4 Earth Tides

The earth tide is the deformation of the solid Earth by the gravitational fields of the Sun and Moon. The periodicity in gravitational potential at any point on the surface of the Earth is due essentially to the rotation of the Earth in the lunar and solar gravitational fields and so gives diurnal and semidiurnal periods of about 24 and 12 hours. Since these are more than 10 times longer than the slowest of the free oscillations of the Earth (Section 4.6), the tidal deformations can be treated as static deformations of the solid

Earth,* to forces which are well known from astronomical observations. The observation of tides thus gives some information about the overall elasticity of the Earth, although for a detailed picture we must appeal to seismology (Chapter 4). Melchior (1966) has provided a comprehensive review of the theory and observation of the solid tide, and there are several shorter reviews (Melchior, 1958, 1964; Tomaschek, 1957).

Consider initially a spherical, rigid Earth subjected to the gravitational field of the Moon (or Sun). The bodies are separated by distances much greater than the diameter of the Earth; thus the gravitational attractions to the Moon of the different parts of the Earth differ by very small fractions of the average attraction calculated by assuming the Earth to be concentrated at its center of mass. It is therefore sufficient to treat tides as small quantities so that squares and products are negligible. Since the average Earth-Moon force provides the orbital acceleration about the common center of mass, the tidal forces are merely the local deviations from this average. The modification to gravity which they cause is calculable from the geometry of Fig. 3.5.

The gravitational attraction of the Moon at A in Fig. 3.5 may be written as components in the direction of the local vertical $\mathbf{r}$ (g_{Ar}) and normal to it ($g_{A\psi}$):

$$g_{Ar} = \frac{Gm}{R'^2} \cos \psi' \qquad (3.32)$$

$$g_{A\psi} = -\frac{Gm}{R'^2} \sin \psi' \qquad (3.33)$$

The negative sign in Eq. 3.33 indicates that the positive attraction is in the direction of *decreasing* ψ. It is the differences between these values and the gravitational attraction by the Moon at the center of the Earth g_0 which is apparent as a tidal force. Resolving g_0 into the same two directions:

$$g_{0r} = \frac{Gm}{R^2} \cos \psi \qquad (3.34)$$

$$g_{0\psi} = -\frac{Gm}{R^2} \sin \psi \qquad (3.35)$$

* This is not true of marine tides, which seriously disturb observations of solid Earth tides except at inland continental stations. However, since the luni-solar orbits are elliptical and differ from the equatorial plane, there are, in addition to the diurnal and semidiurnal tides, monthly, semimonthly, and annual tides, for which the ocean is reasonably near to equilibrium. These long-period tides are therefore more useful for many observations.

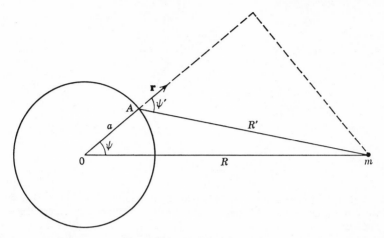

Figure 3.5: Geometry used to calculate the tidal force at a surface point A, due to the Moon, mass m and distance R from the center of the Earth, 0.

The resolved tidal gravity forces are thus

$$\Delta g_r = (g_{Ar} - g_{0r}) = \frac{Gm}{R^2} \left(\frac{R^2}{R'^2} \cos \psi' - \cos \psi \right) \qquad (3.36)$$

$$\Delta g_\psi = (g_{A\psi} - g_{0\psi}) = \frac{Gm}{R^2} \left(\sin \psi - \frac{R^2}{R'^2} \sin \psi' \right) \qquad (3.37)$$

These expressions simplify through the trigonometrical relationships:

$$R \sin \psi = R' \sin \psi' \qquad (3.38)$$

$$R \cos \psi = R' \cos \psi' + a \qquad (3.39)$$

$$R'^2 = R^2 + a^2 - 2aR \cos \psi \qquad (3.40)$$

Since $a \ll R$, we limit the analysis here to terms of first order in a/R. Then:

$$\Delta g_r = \frac{Gma}{R^3} (3 \cos^2 \psi - 1) \qquad (3.41)$$

$$\Delta g_\psi = -\frac{3}{2} \frac{Gma}{R^3} \sin 2\psi \qquad (3.42)$$

The vector sum of these forces is represented in Fig. 3.6.

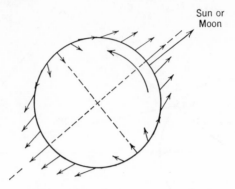

Figure 3.6: Tidal forces due to the Moon (or Sun). Apart from a phase lag due to dissipative processes (Section 2.4), the pattern is symmetrical about the Earth-Moon (or Sun) line.

It is commonly convenient to represent the tidal forces in terms of the tidal potential W_2, of which Δg_r and Δg_ψ are derivatives:

$$W_2 = -\frac{Gma^2}{2R^3}(3\cos^2\psi - 1) \tag{3.43}$$

$$\Delta g_r = -\frac{\partial W_2}{\partial a} \tag{3.44}$$

$$\Delta g_\psi = -\frac{1}{a}\frac{\partial W_2}{\partial \psi} \tag{3.45}$$

Δg_r and Δg_ψ, as calculated above, refer to a hypothetical rigid Earth and are not directly observable but deform the Earth. However, we can see the order of magnitude of the tidal disturbance to gravity by comparing Δg_r and Δg_ψ with the gravity g of the undisturbed Earth (Mass M):

$$g = -\frac{GM}{a^2} \tag{3.46}$$

The fractional change in the intensity of the gravitational field for a rigid Earth is

$$\frac{\Delta g}{g} = \frac{\Delta g_r}{g} = -\frac{m}{M}\left(\frac{a}{R}\right)^3(3\cos^2\psi - 1) \tag{3.47}$$

and the deflection of the vertical is:

$$\alpha = \tan^{-1}\left(\frac{\Delta g_\psi}{g}\right) = \tan^{-1}\left[\frac{3}{2}\frac{m}{M}\left(\frac{a}{R}\right)^3\sin 2\psi\right] \tag{3.48}$$

For the lunar tides $(m/M)(a/R)^3 = 5.7 \times 10^{-8}$ and for the solar tides it is 0.46 times this value.

The fact that the solid Earth is deformed by the tidal forces modifies the observed tidal potential. For a hypothetical, uniform Earth of known elasticity the modification is readily calculated because the deformation takes the same form as the disturbing potential W_2, which is a second-order zonal harmonic. For any magnitude of deformation the potential is calculable in terms of the moments of inertia by MacCullagh's formula (see Eqs. 2.11 and 2.15). However, for the real Earth both density and elastic constants vary with depth and the calculation is very heavy. Comparison of calculations with observation is made in terms of numerical values of two dimensionless parameters h and k^*,† which were introduced by A. E. H. Love and are referred to as Love's numbers, and a third, l, due to T. Shida, defined as follows:

h is the ratio of the height of the body tide to the height of the equilibrium (static) marine tide.

k^* is the ratio of the additional potential produced by the deformation to the deforming potential.

l is the ratio of horizontal displacement of the crust to that of the equilibrium fluid tide.

Observations of various tidal effects give different combinations of these dimensionless parameters. The simplest case is the equilibrium ocean tide. As seen by the ocean, the disturbing potential is W_2 plus the potential due to the solid Earth deformation k^*W_2, so that the amplitude of tidal response is $(1 + k^*)W_2/g$. However, this tide is observed relative to the solid Earth, which is itself deformed by hW_2/g. The observable tide is therefore

$$(1 + k^* - h)\frac{W_2}{g} = \frac{\gamma W_2}{g} \tag{3.49}$$

Since W_2 is known, γ is determined from an observation of tides. Similarly there are three contributions to the observed variations in the strength of the gravitational field, the disturbing potential W_2, the tidal displacement of the observing site, and the potential of the deformed mass of the Earth. This gives another combination of Love's numbers:

$$\Delta g = -(1 + h - \tfrac{3}{2}k^*)\frac{\partial W_2}{\partial r} = -\delta\frac{\partial W_2}{\partial r} \tag{3.50}$$

---

† Gutenberg (1959) used k^* instead of k to avoid confusion with bulk modulus and this practice is followed here.

Further combinations of Love's and Shida's numbers are obtained from different observations; values for h, k^*, and l can be determined from equations such as (3.49) and (3.50). Serious discrepancies still exist between different sets of data but tentative preferred values are

$$h = 0.58$$

$$k^* = 0.29$$

$$l = 0.05$$

Tidal observations are related to two other problems in geophysics: the latitude variation (Section 2.3) and the rigidity of the Earth's core. The relationship between the theoretical period of the Chandler wobble for a rigid Earth, $\tau_R = 305$ days, and the observed period $\tau_0 = 437$ days, is (Munk and MacDonald, 1960b)

$$\tau_0 = \tau_R \frac{k_f^*}{k_f^* - k^*} \tag{3.51}$$

where k_f^* is the value of k^* for a hypothetical fluid Earth with the density profile of the actual Earth.

Takeuchi (1950) has calculated Love's numbers for the realistic Earth models of K. E. Bullen (see Section 4.5), from which he was able to show that, to be compatible with tidal observations, the rigidity of the Earth's core must be less than 10^{10} dynes cm^{-2}. Fluidity of the core is indicated by the fact that shear waves are not propagated in the Earth below a depth of 2900 km (the core-mantle boundary), but it is of interest to confirm that there is a genuine absence of rigidity and not merely strong shear wave attenuation.

Chapter 4

SEISMOLOGY AND THE INTERNAL STRUCTURE OF THE EARTH

"Unwary readers should take warning that ordinary language undergoes modification to a high pressure form when applied to the interior of the Earth ..."

BIRCH, 1952, p. 234.

4.1 Seismicity of the Earth

By seismicity* we mean the geography of earthquakes, particularly their relationships to surface features and their magnitudes (or energies). The general geographical distribution of earthquakes was established in early compilations of F. de Montessus de Ballore; the word seismicity is now associated particularly with the classic work of Gutenberg and Richter (1954), upon which the discussion in this section is largely based.

The worldwide distribution of earthquakes is apparent in Fig. 4.1. The most intense activity is around the circum-Pacific belt; according to the tabulations of Gutenberg and Richter (1954), 75.4% of the energy release by shallow earthquakes during the period 1904 to 1952 occurred there. A further 22.9% was released around the trans-Asiatic or Alpide belt, which extends from Indonesia through the Himalayas to the Mediterranean, leaving less than 2% for the rest of the world. The energy release in the intermediate-(70–300 km) and deep-focus (>300 km) earthquakes is even more strongly concentrated in the circum-Pacific belt. However, the pattern of energy release is dominated by a few large earthquakes and the belts of relatively minor seismicity are also important indicators of Earth structure. Clearly apparent on Fig. 4.1 are the ocean ridges, best-known of which is the Mid-Atlantic ridge, which are now recognized as centers of a pattern of ocean-floor

* Seismicity ≡ seismic activity.

65

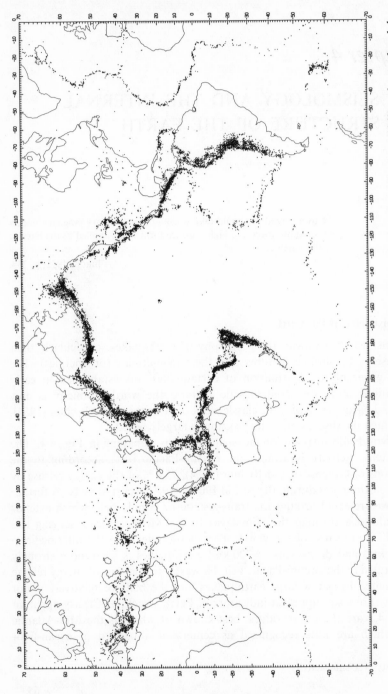

Figure 4.1: Epicenters of 29,000 earthquakes 1961-1967, depths 0-700 km, plotted by Barazangi and Dorman (1969) and reproduced, by permission, from their manuscript.

spreading (see Section 7.1). The ridge system is shown in Fig. 7.4 and the significance of these features as surface expressions of deep-seated mantle motions is considered in Section 7.1.

The most intense seismic activity occurs along arcuate geographic features, which, in the most striking cases, show a common pattern of deep ocean trench, volcanic island arc, and an associated steep gravity anomaly, with earthquake foci distributed about a plane dipping at an angle of about 45°; in cases where the nearest continent is close, the plane dips under the continent. Figure 4.2 is a plot of earthquake foci under the Tonga arc. As numerous

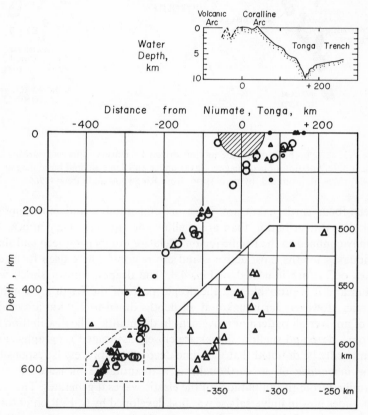

Figure 4.2: Dipping plane of earthquake foci under the Tonga Island arc. Reproduced, by permission, from a manuscript by Sykes et al. (1969).

authors have noted, the system of ocean ridges and ocean trench-island arc structures is very suggestive of a mechanism of ocean floor spreading by an underlying pattern of mantle convection. A particularly interesting variation of this hypothesis is illustrated in Fig. 4.3. The problem of mantle motions is discussed further in Chapter 7.

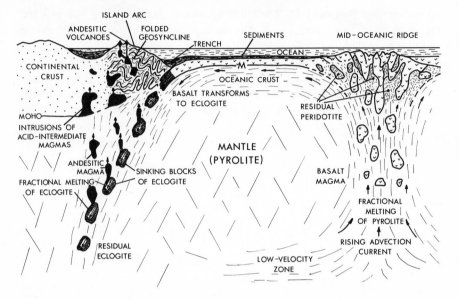

Figure 4.3: A hypothesis of mantle motion driven by density differences arising during chemical differentiation of acid magma (lava) under an island arc and basic magma under an ocean ridge. Reproduced, by permission, from Ringwood and Green (1966).

Deep-focus earthquakes are fewer in total number and more limited in geographical distribution than are shallow earthquakes. They appear to be virtually confined to the Pacific region. Relative numbers of deep and shallow earthquakes for the most active island arc regions, where deep focus earthquakes occur, are indicated in Fig. 4.4. The deepest known shock was at 720 km and it is probable that the deeper parts of the Earth are completely aseismic. However, earthquakes at all depths down to 720 km are evidently part of an over-all pattern of movement in the mantle. The distribution is not random—deep and shallow shocks are closely associated geographically and it can hardly be doubted that the same deep-seated process is responsible.

The *magnitude* of an earthquake is a quantitative measure of its size, determined from the amplitudes of the elastic waves it generates. The scale of magnitudes now in universal use was first developed by C. F. Richter for local earthquakes in California and subsequently improved and generalized to earthquakes at any distance. It is known as the Richter magnitude scale; the historical development of its present form has been well summarized in Richter's (1958) own book and the whole problem of magnitudes has been reviewed by Båth (1966b). As Richter points out, the success of the magnitude classification is due to the logarithmic scale, which gives a fine subdivision while allowing an enormous range of sizes between the largest and the smallest measurable earthquakes to be represented.

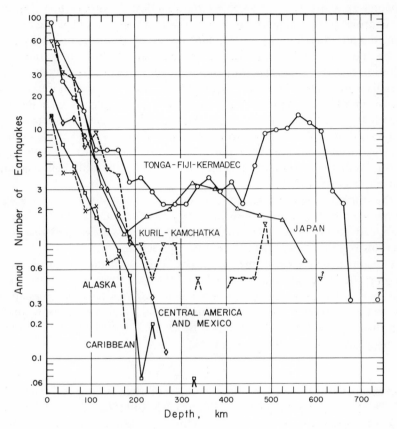

Figure 4.4: Variations in earthquake frequency with depth for six regions. Figure reproduced, by permission, from Isacks et al. (1968).

Following Båth (1966b), magnitude M may be defined by the equation

$$M = \log_{10}\left(\frac{a}{T}\right) + f(\Delta, h) + C \tag{4.1}$$

where a is the amplitude of the ground motion (in microns) for a particular type of wave [strictly surface waves; if body waves (Section 4.3) are used, a different magnitude estimate m is obtained], T is the dominant wave period (in seconds), Δ is the distance, measured as the angle subtended at the center of the Earth, between the earthquake and the seismometer, and h is the depth of the focus or origin of the earthquake. $f(\Delta, h)$ is a term found from a study of many recordings, which accounts for the diminution of wave amplitude with distance, due principally to geometrical spreading of the wave, as studied by Carpenter (1966), but also partly to anelastic attenuation (as discussed in

Chapter 7), and C is a station correction to adjust the observations for local peculiarities of seismometer siting, etc. The function $f(\Delta, h)$ has the effect of reducing all observations to a standard epicentral distance, originally taken to be 100 km, at which the wave amplitudes are directly comparable; the selection of a very small standard amplitude to correspond to magnitude zero then fixes the scale. The largest earthquake recorded in modern times was in Assam in 1952 and had a Richter magnitude of 8.7. The Alaska shock of March 1964 had a magnitude of 8.4. Modern, highly sensitive seismometers can record local shocks down to magnitude -2 or -3 and with reasonable control, i.e., a good spread of reporting stations, magnitudes are determined to 0.1, so that the classification is indeed a sensitive one, although uncertainties may amount to ± 0.5.

The original magnitude scale was developed from a study of surface wave amplitudes; body waves are now in more general use and a slightly different scale leads to a body wave magnitude m, but M and m are directly related, so that this is a problem of detail and not of principle.

Fundamental interest in the magnitude scale is greatly enhanced by the direct relationship between magnitude and the total elastic wave energy of an earthquake. Here the significance of the factor a/T in Eq. 4.1 becomes apparent, as this ratio is a measure of the actual ground strain in a seismic wave. We can thus represent the relationship between total energy and magnitude in terms of an obvious empirical equation

$$\log_{10} E = A + BM \tag{4.2}$$

Noting that wave energy per unit volume of rock is proportional to the square of strain, if earthquakes of different magnitudes produced wave trains of similar forms we would have $B = 2$. However, both the spectra and lengths of wave trains are functions of earthquake magnitude and the appropriate values of A and B must be determined by comparing magnitudes with integrated wave energies for earthquakes with a range of magnitudes. Values of the constants have varied considerably since the first estimates but recently have converged to better agreement; with the values preferred by Båth (1966a), Eq. 4.2 is, for E in ergs,

$$\log_{10} E = 12.24 + 1.44M \tag{4.3}$$

or

$$\log_{10} E = 6.5 + 2.3m \tag{4.3A}$$

Substitution in Eq. 4.3 of $M = 8.7$ for the largest earthquakes gives the prodigious energy of 5×10^{24} ergs. This is 0.05% of the annual energy dissipation by flow of heat from the entire Earth, which is taken to be 10^{28} ergs year^{-1}, equivalent to about 1.5 μcal cm^{-2} sec^{-1}. The processes responsible for generating the strain energies of large earthquakes demand so large an energy source that they must constitute a basic feature of the properties of the mantle and cannot be merely incidental phenomena.

Gutenberg and Richter (1954) also estimated the average annual energy release by all earthquakes from the observed variation of earthquake frequency with magnitude. They used an earlier version of Eq. 4.3, which gave a value for the energy that was much too high, but the calculation is readily repeated using the revised equation. By plotting on a logarithmic scale the number ΔN of shallow shocks in magnitude intervals of 0.1 for $4 < M < 8$, they observed the relationship

$$\log_{10}(\Delta N) = -0.48 + 0.90(8 - M) \tag{4.4}$$

in substantial accord with laboratory observations by Scholz(1968) on microfracturing in strained rock. This is more conveniently represented as

$$\log_{10}\left(\frac{1}{10}\frac{dN}{dM}\right) = 6.72 - 0.90M \tag{4.5}$$

which is equivalent to

$$\frac{dN}{dM} = 5.25 \times 10^7 \exp(-2.07M) \tag{4.6}$$

where dN is the number of shocks per year in the magnitude range dM. Equation 4.3 is rewritten in a similar form

$$E = 1.74 \times 10^{12} \exp(3.32M) \tag{4.7}$$

which combines with Eq. 4.6 to give the total energy in ergs per year, $d\varepsilon$, for the dN earthquakes in the magnitude range dM:

$$d\varepsilon = EdN = 9.13 \times 10^{19} \exp(1.25M)dM \tag{4.8}$$

Equation 4.8 integrates to give the energy in any magnitude interval M_1 to M_2:

$$\varepsilon_{12} = 7.35 \times 10^{19}[\exp(1.25M_2) - \exp(1.25M_1)] \tag{4.9}$$

It is apparent from Eq. 4.9 that the smaller magnitudes contribute negligibly to the total energy release in spite of their much larger numbers. If we take the upper magnitude limit to be $M_2 = 8.7$, then the total energy is 4×10^{24} ergs, or, allowing a 15% addition for intermediate- and deep-focus earthquakes, 4.6×10^{24} ergs; this may be a slight overestimate since the statistical result (4.4) cannot strictly be applied to the very largest shocks, but it is apparent that there is as much energy in the very occasional magnitude 8.7 earthquake as the annual average for all earthquakes.

The strength of ground movement during an earthquake is measured by strong motion instruments or accelerometers where these are available, but the use of reports on the effects on humans and buildings gives much more complete coverage of populated areas. To allow comparisons to be made on a semiquantitative basis, several scales of local earthquake intensity have been devised. The one most generally used was due originally to G. Mercalli and has subsequently been modified by a number of authors; the 1956 version is given by Richter (1958). It classifies effects on a scale of 12 distinguishable intensities, so that contours of equal intensity, or isoseismals, may be drawn on a map of the epicentral region of an earthquake to indicate its extent and the area of greatest intensity. A further elaboration by S. V. Medvedev, W. Sponheuer, and V. Karnik has been presented as the M.S.K. intensity scale. Since the object of the intensity scale is to make a quantitative assessment in terms of qualitative observations, the scale intensity, I, was related approximately to ground acceleration, a, by Gutenberg and Richter (1956):

$$\log_{10} a = \frac{I}{3} - \frac{1}{2}$$

I being an integer in the range 1 to 12 expressing the local severity of structural damage and a in cm sec^{-2}.

The magnitude and intensity scales are independent and assess different aspects of an earthquake; in particular, the Mercalli intensity represents a rather qualitative assessment and each earthquake has a range of intensities, depending upon where it is observed. Nevertheless, there is a correlation between magnitude and the maximum intensity I_{max} for shallow shocks, which appears to be quite well represented by the empirical relationship (Karnik, 1961)

$$M = 0.67I_{max} + 1.7 \log_{10} h - 1.4 \tag{4.10}$$

where h is the depth of focus in kilometers.

4.2   The Seismic Focal Mechanism

Elucidation of the causes and mechanisms of earthquakes is one of the basic problems in geophysics. The more difficult and obscure part of the problem, the physical mechanism and underlying cause, is discussed in Chapter 7. In this section we consider only the process of strain release, for which the basic theory has been generally accepted.

The focus of an earthquake is the point within the Earth from which the elastic waves radiate. The point source is, of course, an idealized concept and in considering the earthquake mechanism we are dealing with a focal volume, but still refer to the focus as the point from which the first wave radiates. The term epicenter refers to the point on the Earth's surface immediately above the focus. A large, shallow earthquake is normally accompanied by substantial deformation of the ground over hundreds of kilometers and this is indicative of the volume of rock from which elastic strain is released. Comparison of extensive geodetic observations in California before and after the San Francisco earthquake of 1906* led Reid (1911) to present the elastic rebound theory of earthquakes; the principle had been discussed previously by a number of seismologists without general agreement, but following Reid's exposition it became the central idea in virtually all theories of the earthquake mechanism. Reid summarized his theory in five statements:

1. The fracture of the rock, which causes a tectonic earthquake, is the result of elastic strains, greater than the strength of the rock can withstand, produced by the relative displacements of neighboring portions of the earth's crust.

2. These relative displacements are not produced suddenly at the time of the fracture, but attain their maximum amounts gradually during a more or less long period of time.

3. The only mass movements that occur at the time of the earthquake are the sudden elastic rebounds of the sides of the fracture towards positions of no elastic strain; and these movements extend to distances of only a few miles from the fracture.

4. The earthquake vibrations originate in the surface of fracture; the surface from which they start has at first a very small area, which may quickly become very large, but at a rate not greater than the velocity of compressional elastic waves in the rock.†

5. The energy liberated at the time of an earthquake was, immediately before the rupture, in the form of energy of elastic strain of the rock.

The concept of fracture, followed by implied frictional sliding across a fault face, is plausible only for shallow shocks. The coefficient of friction between dry rock faces is of order unity, and at any depth below a few kilometers the overburden pressure exceeds the shear strength of the rock and dry frictional sliding is impossible. If a fluid pore pressure is introduced the effective normal stress across the fault is reduced; this has the effect of lubricating it, allowing the sliding to occur at greater depths. However, because the density of water is only about 35% of the densities of crustal rocks, the equilibrium hydrostatic pressure is no more than this fraction of the lithostatic (overburden) pressure. It follows that the pore pressure effect can only be

---

* This probably remains the most intensively studied earthquake and is widely used in discussions of the earthquake mechanism. Some authors have objected that it is not "typical."
† The velocity of rupture of a fault is now recognized to be less than the velocity of shear waves (Press et al., 1961).

significant if fluid in excess of hydrostatic equilibrium is generated by a meta-morphic process (Raleigh and Paterson, 1965). The difficulty may also be overcome by postulating a process of unstable creep leading to catastrophically rapid shearing of a narrow fault zone (see Section 7.5). With this slight modification the elastic rebound mechanism can reasonably be applied to earthquakes with any depth of focus.

A striking feature of many, but by no means all earthquakes is the sudden relative displacement of the rocks on opposite sides of a fault. This was very apparent in the great San Francisco earthquake and led Reid to suggest that it was the essential feature of all *tectonic* earthquakes and that where no major fault movement was evident the break had simply not reached the surface. The only earthquakes excluded from this category are those of *volcanic* origin, which could have a different mechanism, but are never large. A striking example of a fault displacement apparent at the surface is shown in Fig. 4.5. This is a *transcurrent* fault, typical of those in California, across which the relative movement is horizontal. In California movements on the main NW–SE trending faults are right-lateral or dextral, which means that if one stands on one side of the fault and looks across it, the opposite side is seen

Figure 4.5: Transcurrent displacement of rows of orange trees which occurred during an earthquake in Imperial Valley, California, in September 1950. Photograph by David Scherman, Life Magazine (c) Time Inc.

to move to the right.* The fact that movements on the main faults in California are always in the same sense suggests very strongly that earthquakes are sudden, relatively local jumps in a larger-scale movement between blocks of the crust and upper mantle, as in the statements of Reid quoted earlier. The nature of this movement is considered further in Chapter 7.

The San Francisco earthquake is also the starting point of the dislocation theory of earthquakes, a more detailed mathematical development of the elastic rebound theory, in which the stresses and strains associated with fault displacement are explicitly considered. If a known displacement occurs over a known area of fault face, then the change in stress, strain, or displacement at any point in the medium is calculable. In the general case the mathematics is tedious and resort to numerical methods is necessary, but a simple model suffices to demonstrate the principle. Chinnery (1961) considered the effect of uniform slip occurring over a rectangular fault face, following mathematical developments by J. A. Steketee. However, in several cases, including San Francisco 1906, displacements near to a fault are adequately explained by assuming the fault face to be infinite in one dimension and this greatly simplifies the problem by reducing it to a two-dimensional one. It is worth considering the reasonableness of this approximation in the case of the San Francisco earthquake, which was accompanied by movement over a 300 km length of fault, although significant displacements were confined to a zone only about 10 km wide on either side. Clearly this situation is only explicable if the depth of fault movement was very slight compared with the length and the assumption of an infinite fault is therefore a very good approximation, except near to the ends. The sequence of events in the strain rebound theory, as applied to this situation, is as shown in Fig. 4.6.

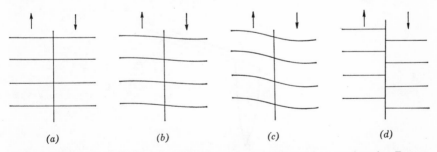

Figure 4.6: Sequence of events in the elastic rebound theory of an earthquake. Due to regional shearing movement in the sense shown, elastic strain is slowly built up from the unstrained state (a) to state (c), at which it is suddenly released across the fault, producing displacements as in (d).

* Fault movements in the opposite sense (left-lateral or sinistral) occur on smaller faults, normal to the trend of the San Andreas system.

Displacements during shallow, transcurrent faulting are those characteristic of screw dislocations, familiar in solid state physics (Cottrell, 1953; see also Kittel, 1966) and represented in Fig. 7.8*b*. In a single, simple dislocation the displacement is uniform across the fault face, as in Fig. 4.7*a*. This model of a fault is mathematically convenient as a starting point in dislocation theory, but is not a physically plausible model because the displacement discontinuity at the bottom of the fault (the dislocation axis) is a singularity, implying

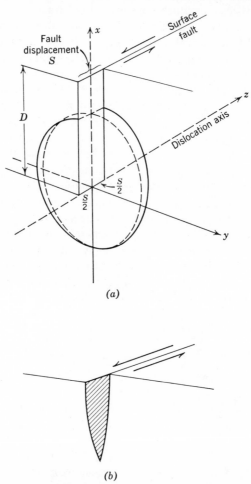

Figure 4.7: Section across a mathematical model of a transcurrent fault. In (*a*) the simple case of uniform relative displacement *S* between opposite faces of the fault is represented. The dashed circle is displaced to the solid, twisted circle. However, the discontinuity in slip at the bottom of the fault is physically unrealistic and demands that the fault movement vary with depth rather as in (*b*).

infinite stresses. The model must be improved by grading the fault displacement to zero at the edges of the fault plane, as in Fig. 4.7b. This is referred to as a compound dislocation because it may be treated as a sum of many small dislocations with progressively displaced axes.

In a simple screw dislocation in an infinite medium, the shear is uniform about a circle centered at the axis of the dislocation as in Fig. 4.7a. If the circle has a radius $r = (x^2 + y^2)^{1/2}$, then the shear strain is $(S/2\pi r)$ for a fault slip S. We are here interested in the resolved component of the shear strain across any plane $y = $ constant, which is $(S/2\pi r) \cdot x/r$. The effect of a free surface at $x = D$ is to modify the stress field so that there are no stresses across the surface and this is represented by the addition of the stress field of hypothetical image dislocation situated at a distance D above the surface (at $x = 2D$). Then the shear strain across any plane ($y = $ constant) parallel to the fault plane ($y = 0$) is, with substitution of $x = D$ to obtain the surface strain,

$$\epsilon_{x=D} = \frac{S}{2\pi} \left[\frac{x}{x^2 + y^2} + \frac{2D - x}{(2D - x)^2 + y^2} \right]_{x=D} = \frac{SD}{\pi(D^2 + y^2)} \quad (4.11)$$

The first term gives strains due to the dislocation itself and the second term accounts for the free surface in terms of the image dislocation. Displacements of surface points are obtained by integrating with respect to y, noting the singularity of the model at $y = 0$:

$$\text{Displacement} = \int_\infty^y \epsilon \, dy = \frac{S}{2} \left(1 - \frac{2}{\pi} \tan^{-1} \frac{y}{D} \right) \quad (4.12)$$

In Fig. 4.8, Eq. 4.12 is plotted on a scale which gives a reasonable fit to the geodetically observed displacements accompanying the San Francisco earthquake of 1906. A particular significance of the fit is that it provides an estimate of the depth of fault movement from the pattern of surface displacement. More elaborate calculations by Sabiha Shamsi and the author, using models of the type represented in Fig. 4.7b, do not affect the displacement curve sufficiently to demonstrate an improvement in matching the observations. In fact, the simple dislocation model gives satisfactory displacement fields at points well removed from the singularities at the boundaries of a fault plane.

In dip-slip faulting, the movement is along the *dip* of the fault plane, as in Fig. 4.9; although vertical movement is an essential feature of dip-slip faulting, the fault face itself is not necessarily vertical and alternative geometries are shown in the figure. If the fault plane extends much farther horizontally than vertically, then this type of fault can be represented by an edge dislocation, as in Fig. 7.8a.* The Alaskan earthquake of 1964 is believed to

---

* In general a dislocation has both edge and screw components and can only be approximated by one of these when the fault plane is so long that effects at its ends are neglected.

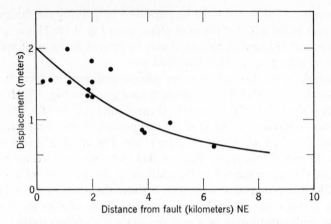

Figure 4.8: Displacement as a function of distance from a transcurrent fault of depth 3.5 km and slip 4 m, according to Eq. 4.12, with data from the comparison of geodetic surveys before and after the 1906 California earthquake (NE side of fault only). The displacements were measured with respect to distant points which were unaffected by the earthquake and the curve represents the deformation of a straight line drawn normal to the fault immediately before the earthquake. Since the earthquake must be presumed to have released, not produced, strain energy, the measured strains are the inverse of the strains released by the shock.

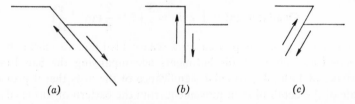

Figure 4.9: (*a*) Normal, (*b*) vertical, and (*c*) reverse (or overthrust) faults of the dip-slip type. In general, fault movements may have both dip-slip and transcurrent components and if neither is dominant, the faulting is termed "oblique."

have been caused by a reverse (thrust) fault movement of the type represented in Fig. 4.9c, but at a very shallow dip angle (9°) and probably not breaking the surface, although if it did so, the break would have been out at sea. Fault plane solutions (discussed below) are ambiguous and for the Alaskan shock, the preferred shallow dipping plane cannot be distinguished from one dipping steeply perpendicular to it. The compelling evidence is obtained from surface displacements. Dislocation models of both types, i.e., with steeply dipping fault planes (Press, 1965, Press and Jackson, 1965) and nearly horizontal fault planes (Savage and Hastie, 1966; Stauder and Bollinger, 1966), have been fitted to the vertical displacement data in Fig. 4.10, but only the latter are

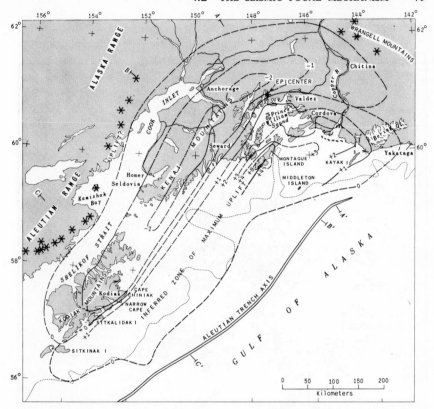

Figure 4.10: Contours of changes in surface elevation which occurred as a result of the Alaskan earthquake of March 1964. Reproduced, by permission from Plafker (1966). The contours are marked in metres and shown by broken lines where they are inferred. The dotted line, approximating the edge of the continental shelf, represents 200 m water depth and volcanoes are indicated by asterisks. (Copyright 1966 by the American Association for the Advancement of Science.) Somewhat larger vertical displacements of the sea floor have been reported by Malloy (1964).

compatible with the horizontal movements in Fig. 4.11. Surface movements of a dislocation model of this earthquake are shown in Fig. 4.12.

Reliable surface indications of the direction of fault movement in a particular earthquake are not generally available, but the direction can often be deduced from studies of the seismic waves which arrive at observatories distributed all around the focus. This (fault plane solution) method is based on a development by P. Byerly and has been reviewed by Stauder (1962). The essential idea is represented in Fig. 4.13, in which the directions and relative magnitudes of *first motions* of the seismograph traces are represented vectorially as a function of azimuth. In this simple picture the fault movement is

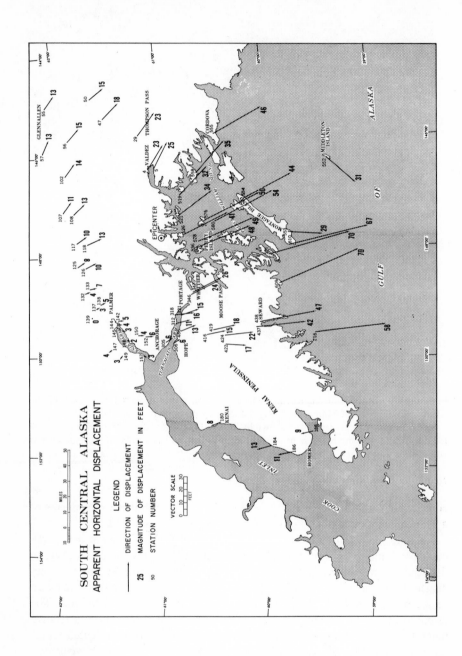

SOUTH CENTRAL ALASKA
APPARENT HORIZONTAL DISPLACEMENT

LEGEND

DIRECTION OF DISPLACEMENT

25   MAGNITUDE OF DISPLACEMENT IN FEET

50   STATION NUMBER

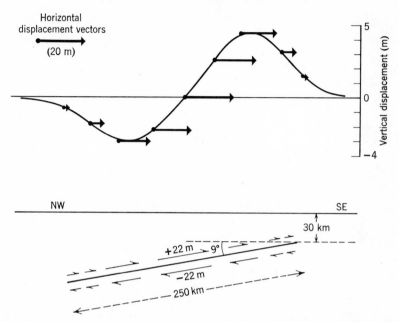

Figure 4.12: Surface displacements due to a dislocation model representing the Alaska earthquake and fitted to the data in Figs. 4.10 and 4.11. The fault plane is here assumed to be so long in the direction normal to the figure that two-dimensional equations of edge dislocations apply.

assumed to be synchronous over the fault face, which is equivalent to assuming a point source, and the first wave to arrive is either a compression or a rarefaction according to the orientation of the observing station with respect to the fault movement; the division of first motions into quadrants gives the *quadrantal pattern*, which is always sought. The quadrants are separated by two nodal planes, the fault plane and an *auxiliary plane* normal to it. The method has an essential ambiguity in that it does not distinguish between the fault plane and the auxiliary plane, but frequently secondary effects, such as azimuthal variations in surface waves, or other arguments (as in the

Figure 4.11: Horizontal displacements accompanying the great Alaskan earthquake of March 1964. Figure reproduced from Small and Parkin (1967), by permission of the Coast and Geodetic Survey, Environmental Science Services Administration, U.S. Department of Commerce. These data are of interest not only because they show the importance of horizontal motion in the Alaska earthquake, but because they are the largest reported geodetic displacements associated with any earthquake. Horizontal movements are difficult to measure accurately but the broad features of the gross movements shown here are not in doubt. It is concluded that the relative movement between fault faces amounted to 40 meters.

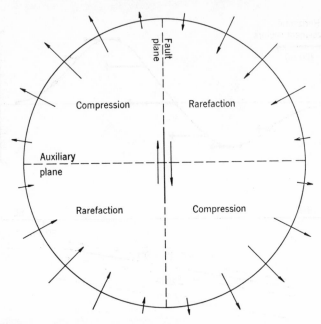

Figure 4.13: Quadrantal pattern of first motions in seismic waves radiated from an idealized (point source) fault movement.

Alaskan case considered above) make only one of the alternatives plausible. In general, of course, the fault plane and the direction of shear may have any orientations and the fault plane problem is a three-dimensional one, with observations made on the spherical surface of the Earth. The detailed deductions are complicated by the refraction of seismic waves (see Sections 4.3 and 4.4) and consequent refraction of the nodal planes, but this is a complication of detail and raises no difficulty of principle.

A detailed consideration of the mechanism of faulting shows that the simple couple represented in Fig. 4.14*a* is unsatisfactory. In the original work of P. Byerly the first motions were considered to arise from the *throw* of the blocks on opposite sides of the fault, but a fundamental difficulty arises from the fact that a single couple has a moment and is incompatible with equilibrium of the stress field before the rupture of a fault. The stress field of a screw dislocation must be represented as a double couple as in Fig. 4.14*b*; more complicated combinations of couples have been suggested to represent particular geological situations. Further, the stress field of an edge dislocation is not one of pure shear, but the requirement that any stress field be in static equilibrium both before and after stress release ensures that it appears as a type II source.

(a)   (b)

Figure 4.14: Single couple with moment and double couple without moment; following H. Honda, these are known as type I and type II sources. Both theoretically and observationally the controversy over mechanism is now resolved in favor of type II.

The validity of fault plane solutions has sometimes been doubted. Following Bridgman (1945), Evison (1963, 1967) has favored volume change accompanying a sudden phase transition as a focal mechanism, although there is no evidence of the existence of a suitable transition. The phase transition hypothesis was inspired partly by the basic objection to the original mechanism of elastic rebound, that is, the inadmissibility of frictional sliding, but this may be overcome by a hypothesis of creep instability (Section 7.5).

The apparent dominance of compressions over rarefactions or vice versa in early first motion studies was also taken to support the volume-change hypothesis. However, a dramatic improvement in first motion data was effected by the availability of records from the long-period instruments of the worldwide network stations installed by the U.S. Coast and Geodetic Survey. These instruments record waves of lengths comparable to the dimensions of active focal regions of earthquakes; they therefore reflect the principal movements of earthquakes, without being seriously affected by local diffraction or structural complexities near to seismic sources. With better data, the fit of conventional fault planes has also improved (Sykes, 1967) and is now very good.

The concept of strength is vital to the theory of earthquake mechanism. In this context strength requires careful definition (see Section 7.3), but if we suppose that it is simply a stress which, when exceeded, causes mechanical failure of a solid, then, in terms of the elastic rebound theory, we expect the stress drop during an earthquake to be independent of the earthquake magnitude. This means that the energy release is proportional to the focal volume from which strain is released. Tsuboi (1956) and Båth and Duda (1964) conclude that this is so, although their reasoning requires estimates of the volumes in which aftershock sequences of major earthquakes are observed and is not completely satisfying. The variation of focal volume with magnitude should be apparent from the variation in frequency spectrum of emitted waves, the dominant wavelengths being related to size of the focal volume. Thus if strength varies with depth in the manner suggested in Section 7.3, the spectra of waves with similar strain amplitudes from earthquakes with different depths will be shifted according to the variation of strength with depth. However, it appears that the dominant frequencies in the spectra of seismic waves are less dependent upon magnitude than would be supposed by the constant strain release hypothesis. This observation is explicable if both strain release and focal volume increase with earthquake magnitude, but the problem is confused by the selective attenuation of the higher-frequency components in seismic waves.

4.3 Elastic Waves and Seismic Rays

Virtually all of our direct information about the interior of the Earth has been derived from observations of the propagation of the elastic waves generated by earthquakes. Certain classes, the surface waves, are guided by the density and velocity layering at and near to the surface and are important in the elucidation of crustal and upper mantle structures,* but of greater general interest are the body waves which penetrate the interior. Since much of the Earth is an elastic solid, two kinds of body wave can be recognized, compressional or P waves and shear or S waves. The designations P and S refer to primary and secondary arrivals; the compressional waves, being faster, arrive first at a recording station. Since both P and S are sound waves (of very low frequencies), their velocities V_p and V_s are given in terms of the ratio of the elastic modulus, appropriate to the particular deformation associated with each wave, to density ρ:

$$V_p = \sqrt{\frac{m}{\rho}} \tag{4.13}$$

$$V_s = \sqrt{\frac{\mu}{\rho}} \tag{4.14}$$

The passage of an S wave involves a pure shear of the medium, so that μ is the ridigity or shear modulus of the medium. The elastic modulus appropriate to P waves is less obvious; we refer to it as the modulus of simple longitudinal extension because the material undergoing successive compressions and dilations in a P wave is subject to lateral constraint. It cannot expand and contract laterally in response to the longitudinal compressions and dilations as can a rod transmitting compressional waves, which are long compared with the rod diameter. m is readily related to the more familiar elastic constants, Young's modulus q, bulk modulus k, rigidity μ, and Poisson's ratio v, by equations for deformation of elastic bodies given, for example, by Joos (1934).

* A monographic treatment of surface waves is by Ewing, Jardetsky, and Press (1957).

We may consider a body to be subjected to three mutually perpendicular, normal stresses σ_1, σ_2, and σ_3. Dealing first with the response to σ_1 only, the material is not laterally constrained and so its longitudinal strain ϵ_1 is simply related to σ_1 by Young's modulus:

$$\epsilon_1 = \frac{\sigma_1}{q} \tag{4.15}$$

and the lateral strain is given by Poisson's ratio:

$$\epsilon_2 = \epsilon_3 = -v\epsilon_1 = -\frac{v\sigma_1}{q} \tag{4.16}$$

Thus by superimposing the strains for all three stresses, we have

$$\epsilon_1 = \frac{1}{q}[\sigma_1 - v(\sigma_2 + \sigma_3)] \tag{4.17}$$

$$\epsilon_2 = \frac{1}{q}[\sigma_2 - v(\sigma_3 + \sigma_1)] \tag{4.18}$$

$$\epsilon_3 = \frac{1}{q}[\sigma_3 - v(\sigma_1 + \sigma_2)] \tag{4.19}$$

Now in the constrained medium σ_2 and σ_3 are self-adjusted to make $\epsilon_2 = 0 = \epsilon_3$ so that

$$\sigma_2 = v(\sigma_1 + \sigma_3) \tag{4.20}$$

and

$$\sigma_3 = v(\sigma_1 + \sigma_2) \tag{4.21}$$

whence

$$(\sigma_2 + \sigma_3) = \frac{2v\sigma_1}{(1 - v)} \tag{4.22}$$

and

$$\epsilon_1 = \frac{\sigma_1}{q}\left[1 - \frac{2v^2}{(1 - v)}\right] \tag{4.23}$$

Thus

$$m = \frac{\sigma_1}{\epsilon_1} = \frac{q(1 - v)}{(1 - 2v)(1 + v)} \tag{4.24}$$

We can now give the more familiar expression for m in terms of k and μ from the relationships between elastic constants:

$$k = \frac{q}{3(1 - 2v)} \tag{4.25}$$

$$\mu = \frac{q}{2(1 + v)} \tag{4.26}$$

whence

$$m = k + \tfrac{4}{3}\mu \tag{4.27}$$

Then, writing Eqs. 4.13 and 4.14 in the usual form,

$$V_p = \sqrt{\frac{k + \tfrac{4}{3}\mu}{\rho}} \tag{4.28}$$

$$V_s = \sqrt{\frac{\mu}{\rho}} \tag{4.29}$$

A rigorous application of elasticity theory to the derivation of Eqs. 4.28 and 4.29 is given by Bullen (1963). The derivation assumes perfect elasticity, which is a reasonable approximation to the truth; anelastic response causes attenuation, as discussed in Chapter 7, but this is sufficiently slight for large earthquakes to be readily observable with seismometers on the opposite side of the Earth. Strictly, adiabatic elastic moduli must be used, but the adiabatic and isothermal rigidities are equal and the two bulk moduli differ by only about 1%; the difference is not yet significant in terms of comparisons of seismological and laboratory measurements of velocity, but could become important with improved understanding of the deep interior. Birch (1938, 1952) has made calculations on the variations of elastic constants with pressure at depth. Other assumptions in the derivation are that the medium is isotropic and homogeneous. Anisotropy results from preferred alignment of noncubic crystals and has been observed to give seismic velocities dependent upon direction in certain parts of the crust (outer few tens of kilometers). It is also indicated by discrepancies in surface wave velocities (Rayleigh versus Love waves) and is almost certainly significant in at least some parts of the mantle, but has not received detailed attention. The problem is reviewed by Anderson (1965). The possibility that regional or local stress causes an anisotropy is considered in Section 4.9.

Since the principal use of seismology is in the elucidation of the Earth's internal structure, the influences on seismic wave motion of departures from homogeneity are of vital interest. Three types of inhomogeneity are recogniz-able:

1. A gradual change of density and elastic constants with depth, due to effects of pressure and temperature on chemically homogeneous material.

2. A sharp boundary between chemically or physically distinct media. " Sharp " here means relative to the wavelengths of seismic waves of interest.

3. A chemical or phase transition which, although not " sharp " in the sense of category (2), gives a much stronger progressive change in properties than (1). All three types cause refraction of seismic waves and sharp boundaries cause also reflections and partial conversion of P waves to S waves or S to P. The laws of reflection and refraction follow from seismic ray theory, in close analogy with geometrical optics. Central to this theory is Fermat's principle of least time, according to which the actual path of light, or of a seismic wave, between two points is shorter in time than any immediately neighboring hypothetical path, i.e., a ray takes the quickest route. This does not imply that there is necessarily only one path; in general there may be several alterna-tive paths, but each of them requires a minimum of transit time relative to small deviations in the path. In practice this means that we can describe wave propagation in terms of rays (optical or seismic), which are everywhere normal to the wavefronts of the waves. In terms of Huygens' construction, illustrated in Fig. 4.15 for the case of a wave refracted at a sharp boundary, each point on a wavefront acts as a source of new wavelets, whose envelope represents the wavefront at a later time. Constructive interference of wavelets occurs only at the envelope and propagation is therefore necessarily normal to the wavefronts.

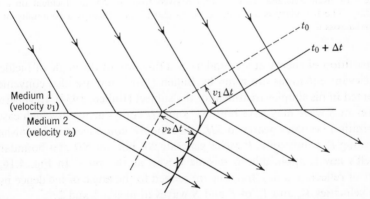

Figure 4.15: Huygens' construction for refraction of a seismic wave at a sharp boundary. Each point in the boundary may be regarded as a source of wavelets whose envelope is the refracted wavefront. The time interval is Δt between the positions of a wavefront indicated by broken and solid lines.

Motion of the medium in compressional or *P* waves is longitudinal so that there is no polarization of a *P* wave. However, *S* waves, being transverse, are polarized and it is necessary to distinguish vertical and horizontal polarizations, termed *SV* and *SH* waves, which behave differently at sharp horizontal boundaries. The horizontal motion of an *SH* wave is transferred across a boundary as a simple refraction with some reflection, but an *SV* wave, which has a component of motion normal to the boundary, generates *P* waves in addition to the refracted and reflected *SV* waves, as illustrated in Fig. 4.16.

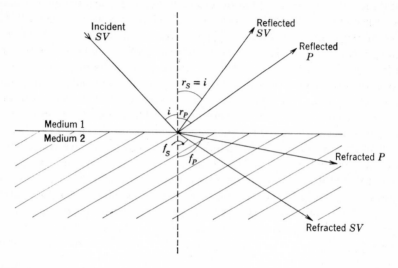

Figure 4.16: Reflected and refracted rays derived from an *SV* ray incident on a plane boundary. The boundary is assumed to be "welded," i.e., there is a continuity of solid material across it.

The partition of energy at a boundary is a function of the angle of incidence; the relevant equations are given by Bullen (1963) and are also conveniently presented in his simpler treatment of the subject (Bullen, 1954). In general, of course, an *S* wave may have both *SV* and *SH* components, in which case the polarization becomes nearer to *SH*, since the *SV* component is diminished in generating *P*. Conversely, *P* waves generate *SV* but not *SH* at a boundary.

Snell's law is applicable to seismic waves, as in optics. In Fig. 4.16, the angles of reflection and refraction are related to the angle of incidence by the wave velocities V_p and V_s of *P* and *S* waves in media 1 and 2:

$$\frac{\sin i}{V_{S_1}} = \frac{\sin r_S}{V_{S_1}} = \frac{\sin r_P}{V_{P_1}} = \frac{\sin f_S}{V_{S_2}} = \frac{\sin f_P}{V_{P_2}} \qquad (4.30)$$

Similar equations can be written by inspection for reflections and refractions of an incident *P* ray or for an *SH* ray (for which only reflected and refracted *SH* occur). In the problem considered in Section 4.4, in which there is a monotonic increase in velocity with depth, Snell's law is applied in differential form. In this context Snell's law can be seen as a consequence of the requirement that, for a wave which crosses a boundary, the component of its velocity parallel to the boundary must be the same on both sides.

A liquid medium, such as the Earth's core, constitutes a special case in which $\mu = 0$ and there is therefore no *S* wave propagation. *S* waves are, however, reflected from the core boundary and *SV* can generate *P*, which propagates through the core.

A limitation of ray theory is that it applies only to wavefronts or obstacles large compared with the wavelength. The effects of small obstacles or sharply curved wavefronts must be described in terms of diffraction. Small-scale inhomogeneities cause scattering. This is important in the outer part of the Earth but appears to be less significant in the deep interior. Waves propagated near to the surface are more affected than those which penetrate deeply.

In addition to the body waves, *P* and *S*, there are surface waves of two kinds, which propagate around the boundaries between layers of different materials or in strong velocity and density gradients near the surface of the Earth. Rayleigh waves may be guided by a single boundary, the free surface, or in a surface wave-guide in which velocity and density increase downward. In Rayleigh wave motion the paths of the particles of the medium are ellipses whose major axes are normally vertical and minor axes are in the direction of propagation of the wave.* The sense of the particle motion in a normally layered Earth is usually retrograde (as a wheel rolling back toward the source). Stoneley waves are a special case of Rayleigh type motion which may propagate along the boundary between two elastic media if their elastic properties are not too different or if one of them is a liquid, as on the ocean floor.

Love waves are *SH* waves (particle motion horizontal and normal to the direction of propagation) which require a wave-guide for propagation. Commonly one boundary of the wave-guide is the Earth's free surface and the other is the normal increase in shear wave velocity, V_s downward in the Earth. However, a low velocity channel internal to the Earth, such as the widespread, probably worldwide uppermantle minimum in shear wave velocity at depths of 100 to 200 km, can also be a guide for Love waves.

Rayleigh waves which propagate at the free surface of a homogeneous medium are nondispersive, with a velocity of about 0.92 V_s in a medium with Poisson's ratio 0.25. With velocity layering Rayleigh waves are dispersive

---

* In some media the minor axis may be vertical and inelastic effects causes the axes to be tilted.

and Love waves are always dispersive because velocity layering or gradients are necessary for their propagation. Surface wave velocity normally increases with wavelength, because V_s increases with depth in the wave-guide and the longer wavelengths penetrate deeper. The shear wave velocities in the adjacent media $V_{s\,max}$, $V_{s\,min}$, set the limits on surface wave velocities. For Love waves $V_{s\,max} > V_L > V_{s\,min}$ and for Rayleigh waves $0.92\,V_{s\,max} > V_R > 0.92\,V_{s\,min}$ approximately. A useful summary of observed surface wave dispersions is by Oliver (1962).

P and S waves from an earthquake both arrive before the surface waves. This is important because surface waves from distant shallow earthquakes normally arrive with much larger amplitudes than the body waves; the energy of a surface wave is spread along an expanding line around the Earth, whereas the wavefront of body waves is an expanding surface and the amplitude thus diminishes more rapidly with distance. Figure 4.17 is reproduced from a seismogram showing arrivals of P, S, and surface waves at Charters Towers, Queensland, from an earthquake off Northern Sumatra ($\Delta = 54.9°$). The recognition of the several different arrivals is a skill acquired by practice. An important recent development is the use of arrays of seismometers, whose outputs can be phased to select arrivals from particular directions and thus to diminish greatly the complications in seismic wave trains. This can make the several different arrivals on a record much more distinct.

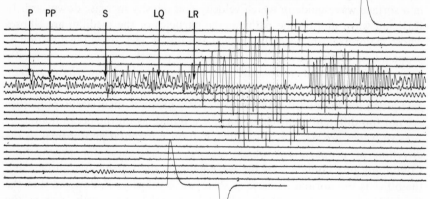

Figure 4.17: Seismogram obtained at Charters Towers, Queensland (Station CTA) showing arrivals of P, PP, S and surface waves LQ, LR, from a magnitude 5.9 earthquake off the west coast of Northern Sumatra (distance 6100 km, $\Delta = 54.9°$) on August 21, 1967. The successive lines are parts of a continuous helical trace on a seismogram which was unwrapped from a recording drum. This recording is from a long period east-west instrument; an upward deflection of the trace represents an eastward movement of the ground, the maximum peak-to-peak ground movement here being about 200μ. Minute marks superimposed upon the trace show that almost 8 mins elapsed between the P and S arrivals. The top and bottom traces show calibration pulses. Figure courtesy of Dr. J. P. Webb and Mr. P. Gaffy.

4.4   Travel Time and Velocity—Depth Curves for Body Waves

If the velocities of elastic waves were uniform within the Earth, then seismic rays would be straight lines, following chords as in Fig. 4.18. The travel time from a surface focus to an observatory at an angular distance Δ from the focus would be

$$T = 2\,\frac{R}{v}\,\sin\frac{\Delta}{2} \qquad (4.31)$$

for waves of velocity v.

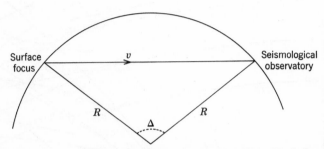

Figure 4.18: Path of a seismic ray through a hypothetical uniform Earth from a surface earthquake to an observatory. In seismology the distance between two surface points is commonly given as the angle Δ between radii from the center to the surface points.

The departures of observed travel times from the simple Eq. 4.31 reflect the nonuniformity of the Earth's interior. In fact, virtually all of the detailed knowledge of internal structure we have has been derived from seismology and appeals ultimately to the observation of seismic travel times.

A vital feature of the travel times is that they increase less strongly with distance than is indicated by Eq. 4.31. Observed $T - \Delta$ curves (Fig. 4.19) are more curved that this equation would suggest. The velocity at depth is thus greater than at the surface and seismic rays are refracted as in Figs. 4.21 to 4.24. Elucidation of the details depends upon accurate travel time data, which have been developed by successive refinements and improvements over many years. The process requires the statistical treatment of data on many earthquakes from numerous seismological observatories; improved control has been possible with large nuclear explosions whose locations, origin times, and depths of focus (zero) are known and do not also have to be deduced from the observed seismic arrivals. A comprehensive tabulation of travel times is given by Jeffreys and Bullen (1940), whose results are represented in Fig. 4.19. The multiplicity of the curves arises from the reflections and partial P to S and S to P conversions at the core-mantle boundary and the surface of the Earth, as shown by the ray paths of Fig. 4.20. In precise travel-time studies, the tables require corrections for the ellipticity of the Earth, which is a decreasing

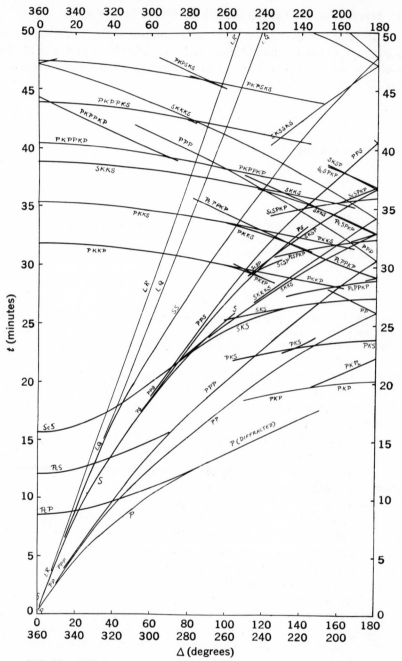

Figure 4.19: The Jeffreys-Bullen travel time curves. Reproduced, by permission, from Jeffreys and Bullen (1940).

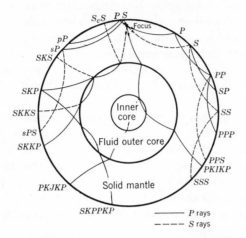

Figure 4.20: Seismic rays whose travel times are represented in Fig. 4.19. Reproduced, by permission, from Bullen (1954).

function of depth, as noted in Section 2.1. The corrections are minimized by using geocentric rather than geographic coordinates. The details are discussed by Jeffreys (1962) and Bullen (1963).

Seismic rays from a near earthquake or an explosion have generally been represented rather simply, as in Fig. 4.21, in terms of a layered structure of the outer part of the Earth or crust. The principal layers of the crust, each of which may be present in varying thickness or absent, are, in order of both depth and increasing seismic velocity, sediments, with very low velocities and high attenuation, a granitic layer, and an "intermediate" layer, possibly of basaltic composition. Underlying the intermediate layer is the mantle, which is distinguished from the crust by a sharp and almost worldwide discontinuity in seismic velocities. This is the Mohorovičić discontinuity, generally abbreviated to "M" layer or "Moho."

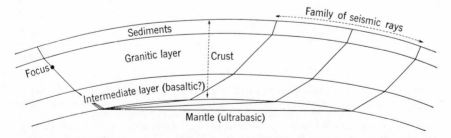

Figure 4.21: Seismic ray from a near earthquake in a layered Earth, showing refraction through characteristic layers of the crust.

The layered model with granitic and intermediate layers is a convenient one for discussions of crustal structures of continental areas. Local details are explored by explosion seismology and broad features by surface wave dispersions. The granitic layer generally amounts to 10 to 20 km of the total continental crustal thickness of about 40 km. Oceanic areas are much simpler, the granite layer being absent and the intermediate (basaltic) layer only about 5 km thick.

Travel times from near earthquakes can be represented in terms of "families" of rays, a "family" being comprised of rays whose points of deepest penetration are all in the same layer. The family represented in Fig. 4.21 has penetrated the Mohorovičić discontinuity to the top of the mantle. The refraction is such that all of the rays in a family have very similar paths in all layers except the lowest, so that the variation of travel time with distance is due to the variable path in the deepest layer penetrated and if the velocity there is v, we have

$$v = \frac{dL}{dT} = a \frac{d\Delta}{dT} \tag{4.32}$$

where $L = a\Delta$ is distance from the epicenter, related to angular distance Δ by the radius of the Earth a. If a sufficient array of seismometers is used, as in seismic exploration, the velocities in a sequence of layers within which velocity increases downward can be determined, although the problem is generally complicated by variable thicknesses of the layers and by heterogeneity, particularly in the upper crust.

Following closely the treatment of Bullen (1963), we may also derive the $T - \Delta$ relationship for remote earthquakes (teleseisms), which give information about the deeper parts of the Earth. In near earthquake studies the Earth may be assumed flat, but in studying teleseisms a spherically layered Earth is considered. The layering need not be discrete (as in an onion), but it is convenient to envisage it that way initially. Figure 4.22 shows the geometry of a ray in a three-layer Earth in which the velocities increase inward from the surface.* Applying Snell's law to each of the boundaries A, B in Fig. 4.22:

$$\frac{\sin i_1}{v_1} = \frac{\sin f_1}{v_2} \tag{4.33}$$

$$\frac{\sin i_2}{v_2} = \frac{\sin f_2}{v_3} \tag{4.34}$$

---

* The following analysis applies to an Earth in which any decrease in velocity v with depth is so slight that $dv/dr < v/r$ everywhere, r being radius. With this condition the travel time curve is continuous and, given complete travel time data, $v(r)$ may be determined without ambiguity. In the real Earth $dv/dr > v/r$ in the upper mantle and at the core-mantle boundary. The effect on seismic rays is illustrated in Figs. 4.26 and 4.27. For many years it obscured the structure of the upper mantle.

but from the two triangles

$$q = r_1 \sin f_1 = r_2 \sin i_2 \qquad (4.35)$$

so that

$$\frac{r_1 \sin i_1}{v_1} = \frac{r_1 \sin f_1}{v_2} = \frac{r_2 \sin i_2}{v_2} = \frac{r_2 \sin f_2}{v_3}$$

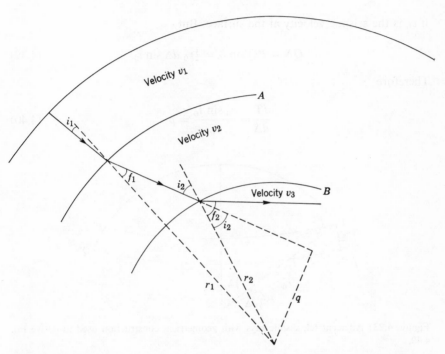

Figure 4.22: Teleseismic ray in a three-layer Earth, with constructions used to show the geometrical significance of the parameter of a seismic ray.

Equation 4.36 could be extended to refractions at any number of boundaries or to gradual refraction in a layer of progressively increasing velocity. Thus for the particular ray considered

$$\frac{r \sin i}{v} = \text{constant} = p \qquad (4.37)$$

where i is now quite generally the angle between the ray and the radius at any point. p is termed the parameter of the ray, a geometrical constant for all points along it. By determining the parameter of a ray we obtain a value of r/v at its deepest point of penetration, where $\sin i = 1$.

A further important relationship for the parameter p follows from a simple consideration of the geometry of two infinitesimally different rays PP' and QQ' in Fig. 4.23. PN is a normal from P to QQ', which means that it is a wavefront, and the difference in travel time between PP' and QQ' is thus

$$dT = 2\frac{QN}{v_0} \tag{4.38}$$

if v_0 is the seismic velocity at the surface. But

$$QN = PQ \sin i_0 = \tfrac{1}{2}r_0 \, d\Delta \sin i_0 \tag{4.39}$$

Therefore

$$\frac{dT}{d\Delta} = \frac{r_0 \sin i_0}{v_0} = p \tag{4.40}$$

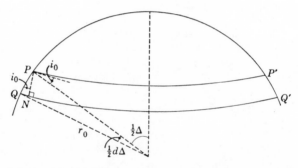

Figure. 4.23: Adjacent teleseismic rays with geometrical construction used to derive Eq. 4.40.

Since T, Δ are tabulated from observations, $dT/d\Delta$ is a measured quantity and thus p is a known function of the angular distance Δ traversed by seismic rays.

The angular distance Δ can be written as an integral. We first note that

$$p = \frac{r \sin i}{v} = \frac{r}{v} \cdot \frac{r d\theta}{ds} \tag{4.41}$$

where θ is the partial angular distance, as in Fig. 4.24, and s is the actual distance measured around the path to the point on the ray specified by θ. Further,

$$(ds)^2 = (dr)^2 + (r \, d\theta)^2 \tag{4.42}$$

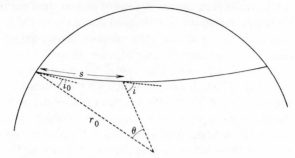

Figure 4.24 Geometry used to obtain the integral expression for Δ.

Eliminating ds from Eqs. 4.41 and 4.42,

$$\left(\frac{r^2 \, d\theta}{vp}\right)^2 = (dr)^2 + (r \, d\theta)^2 \tag{4.43}$$

and introducing, for convenience, a factor $\eta = r/v$, we obtain

$$\frac{d\theta}{dr} = \frac{p}{r(\eta^2 - p^2)^{1/2}} \tag{4.44}$$

Integrating from the deepest point of the ray (r') to the surface (r_0):

$$\tfrac{1}{2}\Delta = \int_r^{ro} \frac{p \, dr}{r(\eta^2 - p^2)^{1/2}} \tag{4.45}$$

Since Δ, p are measured quantities, Eq. 4.45 is an integral equation giving η (and hence v) as a function of r. A method of solution reported by Jeffreys (1962) from a communication by G. Rasch and using simplifications due to E. Wiechert, L. Geiger, and others is given in Appendix B. The result is

$$\int_0^{\Delta_1} \cosh^{-1}\left(\frac{p}{p_1}\right) d\Delta = \pi \ln\left(\frac{r_0}{r_1}\right) \tag{4.46}$$

Equation 4.46 is a convenient form for numerical integration from a tabulation of travel times in equal intervals of Δ, because p is a known function of Δ by Eq. 4.40 and p_1 is the value of p at $\Delta = \Delta_1$. Equation 4.46 thus determines r_1 corresponding to Δ_1 and therefore to $\eta_1 = r_1/v_1$, and so gives $v(r)$ within the range of r down to the maximum penetration of a particular type, or family, of seismic ray, subject to the limitation that the analysis cannot be applied directly to regions in which $dv/dr > v/r$.

Through most of the Earth seismic velocities increase slowly with depth and the preceding analysis is directly applicable. However, important departures from this simple assumption introduce complexities to the ray structure and consequently to the travel time curves. A zone of greater than normal increase in velocity with depth introduces an enhanced curvature to the T-Δ curve; if the velocity increase is strong enough, the situation shown diagrammatically in Fig. 4.25 results. Within a limited range of p, i.e., of sin i_0, Δ decreases with decreasing i_0 instead of increasing. This causes a triplication of arrivals and a T-Δ curve as in Fig. 4.25b. A zone in which velocity decreases with depth causes refraction as in Fig. 4.26a. There is then a range of depths within which no rays have their deepest points and a range of Δ in which arrivals are very weak or not observed. A scale drawing of ray paths and wavefronts within the Earth, in which these effects are apparent, is given in Fig. 4.27; the variation of velocity with depth is given in Fig. 4.28.

The multiplicity of rays results in duplication of data, which allows some double-checking of the velocity results; for example, K velocities must be the same from PKP and SKS travel times. There is now no prospect that the data in Fig. 4.28 will be changed in more than minor ways. However, considerable interest attaches to some of the details, such as the low velocity zone at about

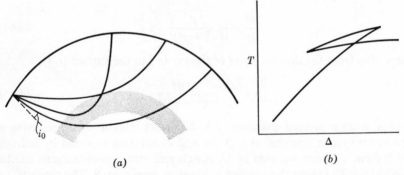

Figure 4.25: (a) Effect on seismic rays of a layer in which velocity increases rapidly with depth in an otherwise "normal" Earth. (b) Corresponding features of the travel-time curves.

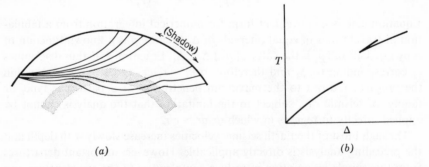

Figure 4.26: The effect of a layer in which velocity decreases with depth.

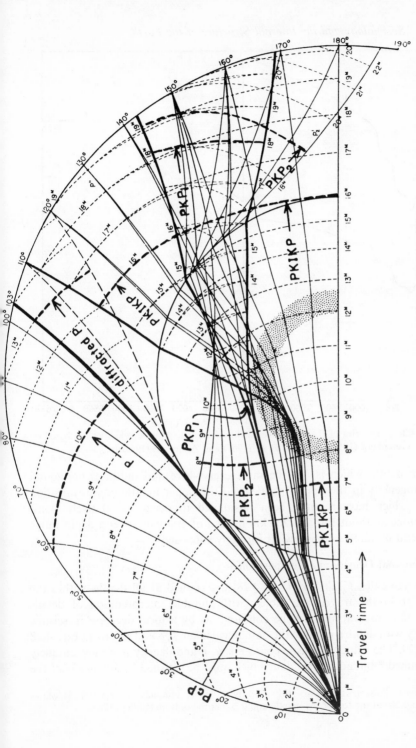

Figure 4.27: Rays and wavefronts for compressional waves in the Earth. Reproduced, by permission, from Gutenberg (1959). (Copyright Academic Press.)

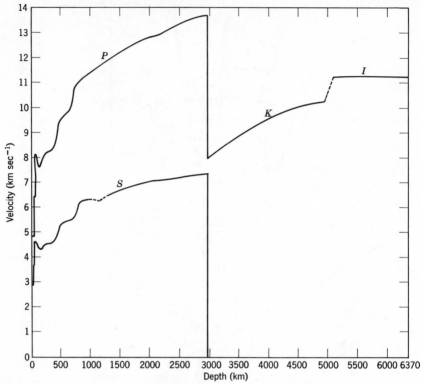

Figure 4.28: Velocities of *P* and *S* waves within the Earth. Mantle velocities are based on data by Gutenberg (1958a), Toksöz et al. (1967) and Anderson (1967b).

150 km depth, which is not the same everywhere and indicates horizontal inhomogeneity in the upper mantle, and the upper mantle phase transition zones, which have recently been postulated to occur as relatively sharp transitions at about 350 and 700 km instead of gradually over a depth range of several hundred kilometers, as formerly supposed.

4.5 Internal Density and Composition

The velocities V_p and V_s (Fig. 4.28) determined at all depths within the Earth from the travel-time tables (with suitable interpretation of details where the observations may be confused, as by a local decrease in seismic velocity with depth), provide the data to calculate k/ρ and μ/ρ from Eqs. 4.28 and 4.29. The values of elastic constants, k, μ and density ρ are not uniquely determined,* but additional data which must be satisfied by any *model* of the

---

* However, Poisson's ratio $v = (3k - 2n)/(6k + 2n)$ is uniquely determined. It varies between 0.269 at the top of the mantle and 0.300 at the bottom (Bullen, 1963).

Earth are quite restrictive. Most recent models are similar to that originally designated A' in K. E. Bullen's extensive work on the densities of the inner zones of the Earth (see Bullen, 1963). The important boundary conditions which the Bullen A, A', and B models satisfied were the total mass of the Earth and its moment of inertia (Section 2.2). The physical requirement of hydrostatic stability was also imposed, i.e., density must increase with depth at a rate not less than that of chemically homogeneous material subjected to increasing compression, with allowance also for a temperature gradient. Where chemical or phase differences occur, the denser material underlies the lighter. More recently, periods of more than 100 identified modes of free oscillation of the Earth (Section 4.6) have been measured. The observations show a satisfactory general agreement with periods calculated for the Bullen model A', but provide data for significant refinements of the model (Dorman, Ewing, and Alsop, 1965; Bullen and Haddon 1967a, 1967b; Anderson 1967b; Press, 1968a, 1968b.)

The internal division of the Earth into distinguishable zones is apparent from the velocity data and is shown in Fig. 4.29, with designations A (crust) to G (innermost core) assigned by Bullen. Also shown are the values of density within the Earth and extrapolated (with certain assumptions concerning compressibility) to zero pressure and atmospheric temperature.

The variation of density due to increasing compression with depth in a chemically homogeneous layer (without phase changes) is calculable by the method of L. H. Adams and E. D. Williamson, to which Birch (1952) has added a correction for the temperature gradient. We consider a spherically symmetrical Earth and write, with neglect of the thermal expansion:

$$\frac{d\rho}{dr} = \frac{d\rho}{dp} \cdot \frac{dp}{dr} = \frac{\rho}{k}(-g\rho) = -\frac{Gm(r)\rho^2}{kr^2} \qquad (4.47)$$

where

$$m(r) = \int_0^r 4\pi r^2 \rho(r)\, dr \qquad (4.48)$$

The negative sign arises because r is radius, measured outward; $m(r)$ is the total mass within radius r, which is responsible for the gravity g at r. The quantity ρ/k is experimentally determined at all depths from the velocity data:

$$\frac{\rho}{k} = (V_p^2 - \tfrac{4}{3}V_s^2)^{-1} = \phi^{-1} \qquad (4.49)$$

so that, by Eq. 4.47, the density gradient is entirely determined by the density distribution itself, i.e., by the particular Earth model considered. In his generalization of the Adams-Williamson equation, Birch (1952) showed that the simple equation (4.47) applied to a medium in which the temperature gradient was equal to the adiabatic gradient (see Chapter 9). When the gradient is not adiabatic an additional term is required to take account of the thermal expansion coefficient, α:

$$\frac{d\rho}{dr} = -\frac{g\rho}{\phi} + \alpha\rho\tau \qquad (4.50)$$

where τ is the difference between the actual and adiabatic gradients, being positive if the actual gradient is greater. Consideration of the second term in Eq. 4.50 is useful in indicating that a high temperature gradient in the top of the mantle is in accord with a relatively low density gradient, but that in the lower mantle the temperature gradient is much smaller.

It is apparent from the velocity-depth curves and Eqs. 4.47 or 4.50 that the whole mantle cannot be homogeneous. Further, in his early work on Earth models, K. E. Bullen (see Bullen, 1963) showed that by applying the Adams-Williamson equation throughout the mantle, with a reasonable density of 3.3 gm cm^{-3} in the outer mantle, and assigning the remainder of the Earth's mass to the core, he needed an unacceptable concentration of mass in the outer core relative to the inner core to satisfy the observed moment of inertia of the Earth. The difficulty is avoided by assigning additional mass to the mantle, with region C (Fig. 4.29), 400 to 900 km below the surface, as a region of inhomogeneity. J. D. Bernal appears to have been first to suggest that the required increase in density through region C might be due to a progressive change of mantle material to high-pressure phase or phases and that the upper and lower mantles might be similar in chemical composition. Subsequent work, particularly by A. E. Ringwood and co-workers (Ringwood and Major, 1966a, 1966b; Ringwood, 1967), has confirmed the essential correctness of this idea. Under pressures exceeding about 150,000 atm (corresponding to depths exceeding 400 km) olivine and similar upper mantle minerals transform to spinel structures in which the relatively large oxygen ions have a close-packed structure and the other ions (Si^{4+}, Mg^{2+}, Fe^{2+}, Fe^{3+}, etc.) are essentially interstitial. In the spinel type structures (including stishovite, a high pressure polymorph of quartz) the silicon ion is in sixfold coordination with oxygen, which at first seems surprising. However, this merely represents an accommodation to strong compression with the large oxygen ions. Some investigators have found evidence from surface wave dispersion and free oscillation periods that the density increases due to phase transitions in the upper mantle are concentrated in relatively narrow depth ranges at 350 to 400 km and 700 km (Anderson, 1967b; Toksöz et al., 1967).

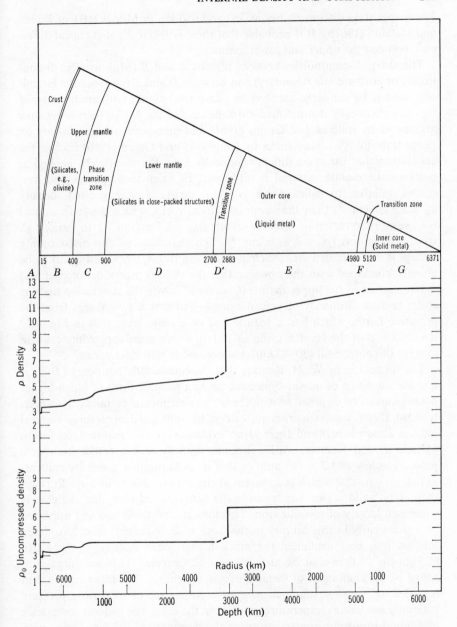

Figure 4.29: Major subdivisions of the Earth's interior with estimated densities. Absolute values of uncompressed density are less certain but indicate clearly the differences in phase and chemical composition.

These appear as discontinuities at 350 and 650 km in Model HB1 of Bullen and Haddon (1967b). It is probable that there is also a slight chemical difference between the upper and lower mantles.

The sharp discontinuities between regions A and B (Mohorovičić discontinuity or crust-mantle boundary) and between D and E (core-mantle boundary) cannot be similarly ascribed to phase transitions; both are boundaries between chemically distinct media. In the case of the Moho thermodynamic arguments by Bullard and Griggs (1961) and high-pressure experiments on phase transitions in basic rocks by Ringwood and Green (1966) lead to the conclusion that the crust differs chemically from the mantle. In geological parlance the mantle material is ultrabasic, i.e., rich in magnesia and iron oxides relative to silica, and is thus denser (uncompressed density $\rho_0 \approx 3.3$ gm cm^{-3}) than the common crustal rocks, which grade from acid (i.e., silica rich) materials, such as granite ($\rho_0 \approx 2.7$ gm cm^{-3}), to basic, e.g., basalt or gabbro ($\rho_0 \approx 2.9$ gm cm^{-3}). The sharpness of the discontinuity (except in some limited areas such as midocean ridges, where it appears to be absent), combined with the evidence that the crust is a product of chemical differentiation in the upper mantle (Chapter 8), invite the conclusion that the entire crust is ultimately of volcanic origin and that it was absent from the primitive Earth, which had a mantle and core only. Note that in Fig. 4.29 the thickness of the crust is given as 15 km, which is an approximate world average of continental (30–40 km) and oceanic (5 km) thicknesses.

The suggestion by W. H. Ramsey that the core-mantle boundary (D-E) is a phase transition of olivine-type minerals to a metallic form of much higher density cannot be disposed of with the same experimental certainty. However, relevant shock wave experiments extend to appropriate pressures (several million atmospheres) and there is no evidence for the required transition (Altschuler and Kormer, 1961). It seems unlikely that it could occur at a pressure as low as 1.3×10^6 atm, or that it could produce a density contrast as dramatic as that which is apparent at the core-mantle boundary. Further, there is no valid reason for rejecting the conventional view that the core is composed largely of metallic iron. This idea is a very old one and appeared prominent early in this century in the work of E. Wiechert, who considered that an iron core explained the much higher mean density of the Earth (5.5 gm cm^{-3}) than that of surface rocks (2.8 gm cm^{-3}); it was largely inspired by the abundance of metallic iron in meteorites. The inner core is then regarded as being of solidified iron—a consequence of increasing pressure with only moderate temperature gradient in the core. The general acceptance of a liquid iron-solid iron transition at the boundary of the inner core, with recent information about the phase diagram of iron at high pressures, provides a very useful "fixed" point for the discussion of the Earth's internal temperatures (Chapter 9).

Identification of the materials in the interior of the Earth depends essentially on the density data and, in particular, on the extrapolation of the observed densities to zero pressure, as at the bottom of Fig. 4.29. This extrapolation is necessarily subject to increasing uncertainty at progressively greater depths and it is of interest to consider some of the laboratory evidence, which indicates that it is nevertheless quite good. Normal elasticity theory assumes that strains are infinitesimal and is therefore inadequate to describe compressions up to 3.5×10^6 atm, which cause first-order changes in density, but Birch (1952) has emphasized the importance of a more general theory of finite strain by Murnaghan (1951) (see also a summary by Takeuchi, 1966, pp. 67–72). This theory is developed in terms of a parameter, ϵ, which Birch calls simply "strain." For a simple hydrostatic compression the density change is

$$\frac{\rho}{\rho_0} = (1 - 2\epsilon)^{3/2} = (1 + 2f)^{3/2} \tag{4.51}$$

where $f = -\epsilon$ is introduced for the convenience of dealing with a positive quantity. In the range of infinitesimal strains, ϵ is indistinguishable from the conventionally defined strain. The particular advantage of this approach is that when Helmholtz free energy Ψ is expressed as a polynomial in f:

$$\Psi = af^2 + bf^3 + \cdots \tag{4.52}$$

all terms beyond the first (quadratic) term are found to be extremely small. The condition that Ψ must be a minimum then yields a universal relationship between pressure p and density ratio ρ/ρ_0:

$$p = \frac{3}{2} k_0 \left[\left(\frac{\rho}{\rho_0} \right)^{7/3} - \left(\frac{\rho}{\rho_0} \right)^{5/3} \right] \left\{ 1 - \xi \left[\left(\frac{\rho}{\rho_0} \right)^{2/3} - 1 \right] + \cdots \right\} \tag{4.53}$$

Here ξ is derived from b in Eq. 4.52 and was found to be negligible from P. W. Bridgmann's experiments on the highly compressible alkali metals, for which ρ/ρ_0 exceeded 2.0, as well as for iron and other less compressible materials (see Birch, 1952). k_0 is the value of bulk modulus at zero pressure; the value under pressure is found by differentiating Eq. 4.53:

$$k = \frac{dp}{d(\rho/\rho_0)} = \frac{3}{2} k_0 \left[\frac{7}{3} \left(\frac{\rho}{\rho_0} \right)^{4/3} - \frac{5}{3} \left(\frac{\rho}{\rho_0} \right)^{2/3} \right] \tag{4.54}$$

Combining (4.53) and (4.54), we obtain

$$\frac{p}{k} = \left(\frac{\rho}{\rho_0} \right) \cdot \frac{(\rho/\rho_0)^{2/3} - 1}{\frac{7}{3}(\rho/\rho_0)^{2/3} - \frac{5}{3}} \tag{4.55}$$

Thus, if the validity of finite strain theory (including higher order terms if necessary) can be accepted into the range of compressions up to millions of atmospheres, ρ_0 can be determined directly from the known variation of p, k, and ρ with depth for any model. The foregoing treatment neglects the complication caused by the temperature variation. This must be accounted for in the manner of Eq. 4.50, noting that the expansion coefficient α is a function of pressure, also given in terms of finite strain theory by Birch (1952). It should be mentioned that finite strain theory is less successful in dealing with the variation of rigidity with compression. In this case the higher order terms in equations corresponding to (4.52) and (4.53) are quite large.

We can now return to the significance of the uncompressed densities plotted in Fig. 4.29, and in particular to the values for the core. Shock wave experiments on iron-nickel alloys (McQueen and Marsh, 1966) confirm the conclusion of finite strain theory that the density of pure iron at core pressures is about 8% greater than the density of the core. Metallic nickel is almost certainly an important additive, but this does not have the effect of reducing the density. Opinion has been divided as to which of the common light elements occurs in the core to the extent of 10–20%. As mentioned in Chapter 1, Ringwood (1966a) favors elemental silicon, but Alder (1966) preferred MgO, and sulfur must also be regarded as a strong candidate.

The density of the inner core is subject to wide uncertainty. It constitutes a very small fraction of the Earth's mass and an even smaller fraction of the moment of inertia, so that there is only a poor control on the density in model calculations. Similarly, free oscillation periods are relatively insensitive to its density and elasticity. However, a direct approach, using the amplitudes of phases such as *PKiKP*, which are reflected from the inner core, is now possible with the refined techniques of seismometer arrays. A slight increase in density at the boundary is favored and indeed must occur if it represents a pressure-induced transition of core material to the solid state. However, there is no requirement that the inner core be chemically identical to the outer core; solubility of the minor constituents may differ markedly between solid and liquid; during the presumed progressive solidification of the inner core, a process akin to zone refining, whereby minor constituents are concentrated either in the solid or in the liquid, may be presumed to have occurred.

There is a further interesting conclusion of finite strain theory. Compressibilities of materials are reduced under high compression and the compressibilities of different materials become more nearly equal. The higher the pressure, the more true this becomes. At the pressures of the core and lower mantle the differences in compressibility between materials such as silicate and liquid iron are slight compared with the differences observed in laboratory measurements. This accords with the conclusions of Bullen (1963), whose Earth model *A* had so nearly equal compressibilities for the bottom of the mantle and top of the core that the possibility of exact equality had to be allowed. Bullen developed a second model (B) in which a central assumption was continuity of compressibility across the core-mantle boundary. Although we cannot expect the compressibilities to be exactly equal, the near equality is a useful restraint upon Earth models.

In the present discussion, and in seismology generally, spherical (or ellipsoidal) symmetry is assumed. While really gross asymmetry can be discounted, horizontal inhomogeneity in the crust is obvious and it appears to extend well down into the upper mantle at least. The problem is discussed by Ringwood (1962), Cook (1962), and Clark and Ringwood (1964). Interesting seismological evidence has been obtained by Carder et al. (1964), Otsuka (1966), Bolt and Nuttli (1966), Cleary and Hales (1966) and Hales and Doyle (1967). Bolt and Nuttli found that wavefronts of seismic waves traveling across an array of seismographs in California departed in a regular way from normals to the great circles through the epicenters from which they originated. The deviation was sensitive to the direction of the epicenter, the maximum value being 11°. Cleary and Hales (1966) found that delays in seismic arrivals at observatories in central United States were similarly sensitive to the directions of the epicenters. In terms of total travel time the effect is very small, but in terms of inhomogeneities in the mantle it is probably very significant. In particular, Hales and Doyle (1967) found that early arrivals in shield areas and late arrivals in orogenic areas were more pronounced for S than for P waves, implying a significant reduction in rigidity under orogenic areas where enhanced heat flow indicates higher temperatures. Much more data from many areas will be needed to study the inhomogeneities, including evidence of correlation with other parameters such as heat flow and gravity (Toksöz and Arkani-Hamed, 1967).

Lateral inhomogeneity in the upper mantle is clearly related to the presence of a low velocity layer. That such a layer is expected in a chemically homogeneous region of the upper mantle can be seen directly from the data on the pressure and temperature dependences of the wave velocities V_p and V_s, as summarized by Anderson et al. (1968). Taking forsterite (Mg_2SiO_4) as more representative of the mantle than the other minerals they consider, the relevant values are as follows:

For compressional waves:

$$\left(\frac{\partial V_p}{\partial p}\right)_T = 10.3 \times 10^{-3} \text{ km sec}^{-1} \text{ kb}^{-1} \tag{4.56}$$

$$\left(\frac{\partial V_p}{\partial T}\right)_p = -4.1 \times 10^{-4} \text{ km sec}^{-1} \text{ }^{\circ}\text{C}^{-1} \tag{4.57}$$

For shear waves:

$$\left(\frac{\partial V_s}{\partial p}\right)_T = 2.45 \times 10^{-3} \text{ km sec}^{-1} \text{ kb}^{-1} \tag{4.58}$$

$$\left(\frac{\partial V_s}{\partial T}\right)_p = -2.9 \times 10^{-4} \text{ km sec}^{-1} \, {}^\circ\text{C}^{-1} \tag{4.59}$$

A low velocity layer occurs if (dV/dz) is negative, where z is depth measured positive downward, and since

$$\frac{dV}{dz} = \left(\frac{\partial V}{\partial p}\right)_T \frac{dp}{dz} + \left(\frac{\partial V}{\partial T}\right)_p \frac{dT}{dz} \tag{4.60}$$

and $dp/dz = 0.33$ kb km^{-1} in the upper mantle, the temperature gradients which must be exceeded to form a low velocity layer are 8.3°C km^{-1} for compressional waves and 2.8°C km^{-1} for shear waves. These values refer only to a particular mineral but the conclusion is a more general one. The temperature gradient in the upper mantle (see Fig. 9.4) may be high enough to give a low velocity layer for P waves and is virtually certain to do so for S waves. A definitive profile for both V_p and V_s could give an improved temperature profile; in particular we can suggest immediately that the temperature gradient is lower under such areas as Pre-Cambrian shields, where the low velocity layer is weak for V_s and probably absent for V_p.

4.6   Free Oscillations

The excitation of free oscillations in the Earth by large earthquakes has been compared to the ringing of a bell. Recognition that the Earth as a whole could, if suitably excited, undergo free vibrations in an indefinite number of modes extends back 100 years. It was a product of the development of the theory of elasticity by Lord Kelvin, H. Lamb, A. E. H. Love, Lord Rayleigh, and others. The early work, which is mainly concerned with the oscillations of a hypothetical uniform Earth, has been reviewed by Stoneley (1961). Calculations in terms of more realistic Earth models were hardly possible without electronic computers and were in any case of limited interest until it became apparent that free oscillations were observable.

Interest in the subject was renewed by H. Benioff's development of instruments to observe very long-period seismic waves. From an examination of his records following the Kamchatka earthquake of 1952, Benioff tentatively identified a 57-min period as a fundamental mode of free oscillation. In the following few years instrumental improvements were directed specifically to the observation of free oscillations and Alterman et al. (1959) calculated the periods of oscillation for realistic Earth models. When the next really large earthquake, capable of exciting oscillations of observable amplitude, occurred in Chile in May 1960, several geophysical laboratories were able to record the oscillations. Bullen (1963, p. 260) has described the scientific excitement at the Helsinki meeting of the International Union of Geodesy and Geophysics, later in 1960, when representatives of several groups met and compared their preliminary results. This meeting effectively established free oscillations as a new branch of seismology. Further observations following an earthquake in Kamchatka in 1963, and more particularly the great Alaskan earthquake of March 1964, have confirmed and amplified the 1960 results, with some minor but apparently significant discrepancies (Slichter, et al., 1966) which remain to be clarified. Since the oscillations die away in a few days (even less for the shorter periods), it is remarkable how much has been accomplished in the few brief periods of recording.*

The deformations of the Earth in free oscillation are indicated in Fig. 4.30 for the fundamental modes of the three different types. The simplest to envisage are the *radial* oscillations, i.e., those in which particle motion is purely radial (Fig. 4.30*a*). The fundamental is an alternating compression and rarefaction of the whole Earth and there is an infinite series of overtones for which spherical nodal surfaces occur within the Earth. The radial oscillations are a special case of spheroidal oscillations, in which, in general, particle motion has both radial and tangential components and the Earth's surface is deformed in the manner of the spherical harmonic functions $[P_l^m(\cos \theta) \cos m\lambda]$ (see Appendix A). The tesseral harmonics, in which $m > 0$, have not been distinguished observationally from the zonal harmonics $P_l(\cos \theta)$, with which their periods are degenerate, and the superscript m is dropped. Then $l = 0$ [i.e., $P_0(\cos \theta) = 1$] gives the case of radial oscillations; $l = 1$ [i.e., $P_1(\cos \theta) = \cos \theta$] is precluded by the fact that it represents a net translation of the surface and therefore of the center of gravity; $l = 2$ represents the "football" mode, in which the Earth deformation is alternately prolate and oblate (Fig. 4.30*b*), and all higher l occur. The form of the motion in the general case (for $m = 0$) can be represented by the r and θ components of displacement, u and v, where

$$u = U(r)P_l(\cos \theta) \sin \omega t$$

$$v = V(r) \frac{\partial P_l(\cos \theta)}{\partial \theta} \sin \omega t \tag{4.61}$$

where U and V are related radial distributions of the motion within the Earth and $\sin(\omega t)$ indicates an oscillation whose frequency $(\omega/2\pi)$ is related also to U, V and determined by the internal elasticity and density structure. The "colatitude" θ refers not to geographic coordinates but to a pole determined

---

* Observational data have been summarized by Slichter (1967) and S. W. Smith (1967).

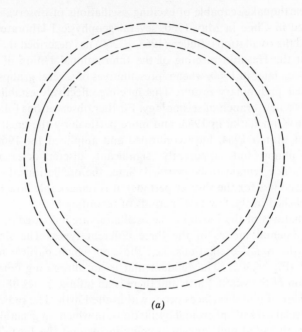

(a)

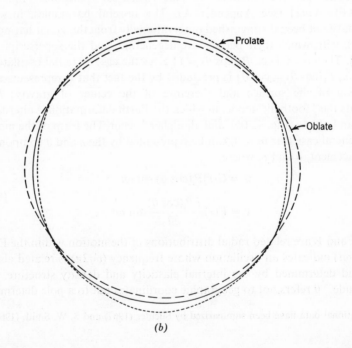

(b)

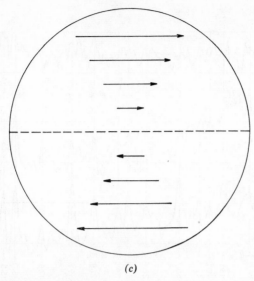

(c)

Figure 4.30: The simplest representative modes of free oscillation. (a) Radial oscillation $_0S_0$. (b) Spheroidal "football" mode $_0S_2$. (c) Instantaneous motion in toroidal mode $_0T_2$.

by the geometry of the exciting earthquake. The radial functions U and V can take an indefinite number of distinct forms, each representing a fundamental oscillation of order l or one of a series of overtones in which spherical surfaces within the Earth are nodes of the motion. The number of nodal surfaces is represented by a prefix n, so that the general representation of a spheroidal oscillation is $_nS_l{}^m$, or, since interest is restricted to zonal harmonics with $m = 0$, simply as $_nS_l$. The fundamental radial oscillation is $_0S_0$ and the fundamental "football" mode is $_0S_2$; spheroidal modes $_0S_3$, $_0S_4$, ..., are motions with increasingly subdivided zonal distributions and each of these modes has overtones with internal nodal surfaces.

Oscillations of an essentially different type, in which the motion is entirely circumferential, with no radial component, are known as torsional or toroidal oscillations. In this case there are no displacements in the r, θ directions but only in the λ direction:

$$w = \frac{W(r)}{\sin \theta} \frac{\partial P_l(\cos \theta)}{\partial \theta} \sin \omega t \qquad (4.62)$$

where, as before, θ is measured from a "pole" of the motion which may be anywhere on the Earth. The simplest of these modes, $_0T_2$, is a twist between two hemispheres which oscillate in a rotational sense with a nodal plane between them (Fig. 4.30c).* Higher modes are due to a finer subdivision of the

* $_0T_1$ does not vanish by Eq. 4.62, but implies a variation in the rate of rotation of the whole Earth and is incompatible with conservation of angular momentum.

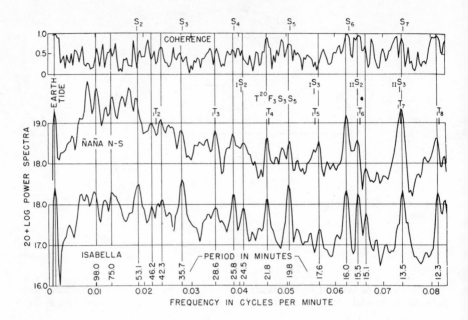

Earth into 3, 4, etc., zones with opposite motions and, as in the spheroidal case, there are overtones with internal nodal surfaces whose number is represented by the prefix n. The general toroidal oscillation is $_nT_l^m$, with $m = 0$ in most cases of interest.

There is a direct relationship between the higher-order free oscillations and long-period surface waves. The oscillations are simply standing waves (surface waves in the more familiar cases) and may be regarded as a superposition of oppositely traveling surface waves, Rayleigh waves (vertical polarization) in the case of the spheroidal modes and Love waves (horizontal polarization) for toroidal modes. This connection emphasizes an important feature of the oscillations: the lower modes, with periods up to 53 min, stress the whole Earth, but successively higher-frequency oscillations are increasingly concentrated in the outer part of the Earth, so that those with periods of a few minutes effectively stress only the upper mantle. The periods of the free oscillations are thus determined by the densities and elasticities of the internal layers with different weightings for the layers according to the mode of oscillation. It is this feature which makes free oscillations very useful in clarifying the details of internal structure.

Calculation of free oscillation periods for any particular Earth model is mathematically heavy work. Except for the special case of radial oscillations, the toroidal periods are easier to compute than the spheroidal ones, because

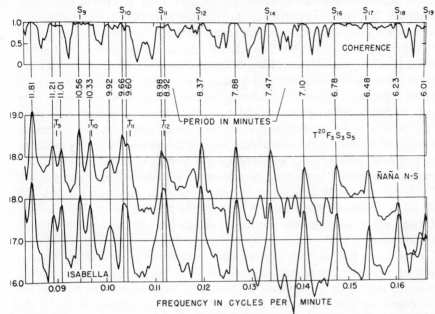

Figure 4.31: Power spectra of records obtained by Benioff et al. (1961) at Isabella, California and Nana, Peru during a 5-day period immediately after the Chilean earthquake in May, 1960. The top curve is the coherence between the two records, a quantitative measure of the reliability or significance of the values in the spectra. The general agreement is obviously excellent. Figure reproduced by permission of the authors and the American Geophysical Union.

toroidal modes involve a pure shear with no dilatation and therefore no change of density. Gravity does not contribute to the restoring forces on displaced particles as it does in the case of the spheroidal modes, in which there are radial displacements and density changes. Calculation of the free periods may be reduced to the solution of a second-order ordinary differential equation in the cases of radial or toroidal oscillations, to one of the fourth order for spheroidal oscillations if the effect of gravity is neglected, and one of sixth order if gravity must be allowed for. However, even the simplest case requires a lengthy numerical integration of the equations of motion through the assumed density and elasticity structure and the use of a digital computer is necessary if the computation is to proceed very far. For the development of the equations in suitable form, and a discussion of the method, reference should be made to Alterman et al. (1959).

Comparison of the periods computed for a particular Earth model with those of identified modes in the harmonic analysis of a free oscillation record, such as that in Fig. 4.31, allows the model to be refined by a succession of minor adjustments which bring its periods into closer agreement with those observed. Bolt (1964) has reviewed the situation after the Chilean earthquake records had been digested and more recent work has been reported by Bullen and Haddon (1967a, 1967b). The internal structure is sufficiently well established to allow unambiguous identification of the well separated frequencies of the lower modes. Identification is aided, particularly for the higher modes, by the experimental distinction between spheroidal and toroidal modes: strain and pendulum seismometers record both, but recording gravity meters see only the spheroidal oscillations because the toroidal modes involve no density changes or vertical accelerations. Also, certain modes may be poorly recorded by particular observatories which happen to be near to nodal lines for those modes. The term "terrestrial spectroscopy" was coined by Pekeris et al. (1961) for the study of free oscillation spectra; as they pointed out, the similarity to atomic spectroscopy includes the mechanical analogue of the Zeeman effect, an observed splitting of certain lines by the Earth's rotation and ellipticity.

Damping of free oscillations gives useful evidence of anelasticity in the deep interior. Assuming frequency-independence of the intrinsic Q of mantle material, the higher Qs observed for the lower modes reflect the high Qs for the deeper parts of the mantle (see Section 7.4). Of particular interest is the extraordinarily high Q for the $_0S_0$ mode, reported by Slichter et al. (1966) to exceed 25,000 following the Alaskan earthquake.* Since this mode is purely compressive, it may be inferred that damping of stress waves is primarily due to the shear component of stress.

4.7 Microseisms

In addition to the discrete seismic events associated with earthquakes, seismic records show a more or less continuous background of movement, having a dominant period usually of 5 to 10 sec and variable amplitude of ground motion with a maximum exceeding 10 μ (Brune and Oliver, 1959). There is a variety of sources for these microseisms, some of which are demonstrably local in origin (wind action on trees, etc.), but the more important are due to the action of storm waves at sea (Deacon, 1947; Darbyshire, 1962). There are two microseismic effects of ocean waves: the beating of waves on shorelines, which can be appreciably coherent by virtue of the refraction of waves so that they tend to be parallel to coastlines, and the interference of sea waves over deep water. The two may be distinguished by the frequencies of the microseisms they generate (Haubrich et al., 1963; see also Hasselmann, 1963, for a theoretical discussion). The effect of waves on

---

* A Q as high as this cannot be measured reliably, because of the possibility of excitation by many small events over a period of several weeks.

a coastline is to generate microseisms with the same period as the waves and is readily understood, but the interference effect gives microseisms of half the ocean-wave period. Interfering water waves cause a pressure fluctuation at the bottom, even in deep water; this is not apparent in the first-order theory of wave motion but was found as a second-order effect by M. Miche, whose calculations were developed into a complete theory of microseism generation by Longuet-Higgins (1950). The simple treatment by Longuet-Higgins and Ursell (1948) is followed here.

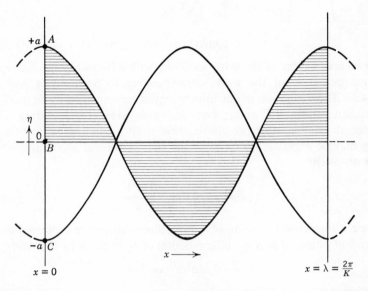

Figure 4.32: One wavelength of a standing ocean wave with successive positions $ABCBA\ldots$ Shaded areas represent the displacement of water from which the oscillation in potential energy is calculated.

Consider the case of a sinusoidal standing wave in which the vertical displacement η of the water surface* is given by

$$\eta = a \cos kx \cos \omega t \qquad (4.63)$$

Profiles of the water surface at $t = (2\pi/\omega)n$, $(2\pi/\omega)(n + \frac{1}{4})$, $(2\pi/\omega)(n + \frac{1}{2})$, where n is any integer, are shown in Fig. 4.32. The total potential energy of

* Wavelength $(2\pi/k)$ and frequency $(\omega/2\pi)$ are of course related, but water waves are dispersive and the relationship is not obvious. For the general case, in water of depth h,

$$\omega^2 = gk \tanh (kh)$$

This is of interest in Section 4.8.

the column of water bounded by the planes $x = 0$, $2\pi/k$, per unit length of wave measured along the wave crests, i.e., normal to the plane of the diagram, oscillates between the undisturbed value in position B to a higher value in positions A, C; the difference V is the work done in raising water within the area marked by horizontal shading to the areas marked by vertical shading. If the water surface in a particular length dx stands at a height η, the average height of the water above the reference plane is $\eta/2$ and its mass is $\rho\eta\,dx$ (which is of course negative where η is negative). Thus

$$V = \int_0^{\lambda = 2\pi/k} \frac{\eta}{2} g\rho\eta\,dx = \tfrac{1}{2}g\rho \int_0^{\lambda} \eta^2\,dx$$

$$= \tfrac{1}{4}\rho g\lambda a^2 \cos^2 \omega t = \tfrac{1}{8}\rho g\lambda a^2(1 + \cos 2\omega t) \quad (4.64)$$

The potential energy of the water oscillates with a frequency 2ω, having two maxima per cycle of the wave, corresponding to positions A and C in Fig. 4.32. The oscillation in potential energy is accompanied by an oscillation in pressure within the water; considering an area corresponding to one wavelength of unit width, as above, there is an oscillating vertical force F, which may be expressed in terms of the sum of vertical accelerations of all water drops dm:

$$F = \int \frac{d^2z}{dt^2}\,dm = \frac{1}{g}\frac{d^2V}{dt^2} \quad (4.65)$$

where z is the vertical position of a drop. Since this can be expressed directly in terms of potential energy, differentiation of V, as given by Eq. 4.64, yields

$$F = -\tfrac{1}{2}\rho\lambda a^2\omega^2 \cos 2\omega t \quad (4.66)$$

or, referred to unit area, the oscillatory pressure is

$$p = -\tfrac{1}{2}\rho a^2\omega^2 \cos 2\omega t \quad (4.67)$$

This is the average pressure acting over the whole area and is therefore propagated downward to indefinitely great depths and is apparent as a pressure fluctuation on the ocean floor, generating microseisms. Its frequency is twice that of the surface wave and the strength of the pressure oscillation depends upon the square of the wave amplitude, i.e., it is a second-order effect.

We may note that there is also a first-order pressure fluctuation having the same frequency as the wave but that, unlike the second-order effect, it is attenuated with depth as rapidly as the particle motion, which hardly extends to depths greater than one wavelength. The second-order oscillation does not invoke any particle motion at depth, except that arising from compression of the water, as considered by Longuet-Higgins (1950), and yielding of the bottom.

The ideal situation of a standing wave extending to infinity is of course quite unlike the actual ocean surface. However, the Longuet-Higgins mechanism will operate whenever two wave trains with similar frequency components, traveling in approximately opposite directions, interfere over an area whose dimensions are comparable to or larger than the depth of water; in other words, the extent of the coherence of the interfering waves must be comparable to the depth of the water. The fact that ocean waves are trochoidal, not sinusoidal, does not materially affect the argument.

The association between occurrences of large-amplitude microseisms and cyclonic disturbances at sea has been noted by numerous authors and reviewed by Deacon (1947). Whereas large microseisms always appear to be associated with storms (which may be well removed from coastlines), the reverse is not always true because the necessary condition of oppositely traveling wave trains may not occur. In particular, cyclonic disturbances generate microseisms but monsoon winds, which maintain an approximately constant direction, do not. However, the essential feature of the standing wave theory, which is confirmed by the observations, is the frequency doubling. The dominant period of ocean swell is generally 10–20 sec and for microseisms 5–10 sec; moreover, variations in both have been observed to correspond.

Some years ago meteorologists took interest in microseisms as possible indicators of hurricanes at sea. Tripartite stations were established to explore techniques of storm location, but the results were generally discouraging. Now that meteorological satellites are in operation, interest in microseisms is virtually restricted to the signal-to-noise problem of reducing their effects on seismic records. Since they are principally Rayleigh waves (vertical polarization is expected from the mode of generation), some advantage is obtained by using seismometers in deep bore-holes, where surface noise is less, but holes can hardly be deep enough to give much relief from noise with periods of order 5 sec. Another approach is to use arrays of seismometers coupled either electrically or in the analysis of records.

4.8 Tsunamis

Sea waves of long wavelength are commonly generated by submarine earthquakes and are simply due to sudden subsidence or uplift of large areas of sea floor, or possibly a combination of both subsidence and uplift in adjacent areas. The waves sweep across open ocean at high speed and have caused severe damage to coastal areas thousands of miles from the earthquakes which generated them. The Pacific ocean is particularly affected because so much of its perimeter is seismically active. These waves are commonly referred to as "tidal waves" but are quite unrelated to tides and the Japanese word tsunami is in universal scientific usage. Van Dorn (1965) has given a comprehensive review of the phenomenon. The Chilean earthquake of May 1960 was particularly effective in tsunami generation, producing waves with amplitudes of many feet all around the Pacific (e.g. Fig. 4.33). However, as in other similar cases, the amplitude was variable and by no means directly related to distance from the shock.

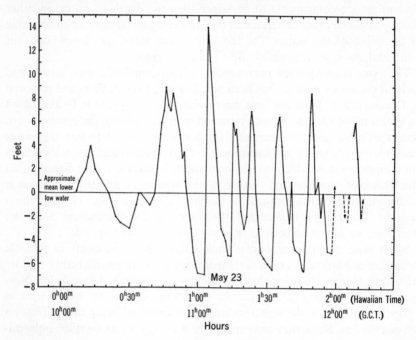

Figure 4.33: Tsunami at Hawaii from the Chilean earthquake of May 1960, from a sequence of observations at Wailuku River bridge by U.S. Geological Survey. Reproduced, by permission, from Eaton et al. (1961).

To understand the apparently fickle manner of tsunami propagation and the selection of certain sections of coastline for waves of destructive amplitudes it is necessary to recognize the depth-dependence of wave velocity. This is a feature of shallow water waves, i.e., those with wavelengths much greater than the depth of water. The velocity of this class of wave may be derived by assuming equipartition of the potential and kinetic energies of the wave motion (see also Proudman, 1953, pp. 247–251). The potential energy of a single wavelength, for a "thickness" b measured along the crests, is, as in Section 4.7 (Eq. 4.64):

$$V = \tfrac{1}{4}\rho g \lambda a^2 b \qquad (4.68)$$

the cosine term being now equated to unity because we are considering a traveling wave of constant shape. The condition that wavelength be very much greater than depth ensures that vertical motion of the water is slight compared with the horizontal motion. The kinetic energy is therefore calculable in terms of the horizontal velocities of the water particles, which must be such that in half of one wave period τ, the horizontal motion replaces troughs by crests and vice versa. The volume of water transferred between adjacent quarter-wavelengths of the path in time $\tau/2$ is $(a\lambda b/\pi)$, and, since the cross-sectional area of moving water is hb for depth h of water, the required mean water velocity during the half-cycle of the wave is

$$\bar{v} = \frac{2}{\pi}\frac{a\lambda}{h\tau} \tag{4.69}$$

Since the velocity is sinusoidal the peak value is

$$v_0 = \frac{\pi}{2}\bar{v} = \frac{a\lambda}{h\tau} \tag{4.70}$$

and the mean square velocity is therefore

$$\overline{v^2} = \tfrac{1}{2}v_0{}^2 = \frac{1}{2}\left(\frac{a\lambda}{h\tau}\right)^2 \tag{4.71}$$

Since the total mass of water in one wavelength is

$$m = \rho\lambda hb \tag{4.72}$$

its kinetic energy is

$$\tfrac{1}{2}m\overline{v^2} = \frac{1}{4}\frac{\rho\lambda^3 ba^2}{h\tau^2} \tag{4.73}$$

Equating kinetic and potential energies (Eqs. 4.68 and 4.73), we obtain the wave velocity

$$c = \frac{\lambda}{\tau} = (gh)^{1/2} \tag{4.74}$$

Equation 4.74 is a special case of that given earlier (footnote on p. 115) since $c = \omega/k$ and $kh \ll 1$ is assumed. With this assumption, the velocity depends upon depth only, being 0.22 km sec^{-1} (790 km hour^{-1}) in ocean 5 km deep. The periods of tsunamis are 10–40 min; thus the wavelengths in deep water are 100 km or more, and are quite unnoticeable at sea, but as they approach a coastline the depth and consequently the velocity decrease, so that the waves increase in height as the wavelength contracts. Further, the waves are refracted by the submarine topography so that contours of the ocean floor focus the waves on to particular sections of the shore. Also, oscillations (seiches) can be excited in harbors and estuaries which have suitable dimensions. Eaton et al. (1961) considered that the higher-frequency fluctuations which followed the onset of the tsunami after about an hour in the record shown in Fig. 4.33 were due to the excitation of local oscillations but a very similar record of the Chilean tsunami was obtained by the author from a wave recorder on Norfolk Island, which is a sufficiently small perturbation of the ocean floor that the recording was virtually the (smaller amplitude) open ocean wave. Apparently the propagation is dispersive, longer wavelengths arriving first, and the simple analysis which leads to Eq. 4.74 is incomplete; the important neglected factor is the slope of the ocean floor, which is presumed to be responsible for the dispersion.

4.9   The Earthquake Prediction Problem

It is paradoxical that decades of seismological study have given us a detailed picture of the inaccessible interior parts of the Earth but very little idea of what happens at an earthquake focus during and before a shock, or why shocks occur when and where they do. Specific prediction of time, place, and magnitude is impossible with our present knowledge* and the deficiency is essentially in the understanding of the earthquake mechanism. The mechanics of the problem are discussed in Section 4.2 and the origin of the stresses which are responsible in Section 7.5, but the details are too vague or uncertain to indicate a system of earthquake prediction, although the underlying pattern of mantle motions which are ultimately responsible appears now to be reasonably clear (Section 7.1). It has even been considered that the problem is unsolvable in principle, but in the past few years enough encouraging evidence has accumulated to stimulate renewed interest. There are detailed plans for major research programs in Japan (Tsuboi et al., 1962, Rikitake, 1966a) and the United States and a feeling of hopefulness is apparent in the proceedings of a joint U.S.-Japan conference in 1964 (Hagiwara and Oliver, 1964). Several semipopular discussions cover the subject (Stacey, 1964a; Oliver, 1966; Press and Brace, 1966).

If the elastic rebound theory is valid, then substantial elastic strains occur in the focal region before an earthquake. The search for a method of prediction is essentially a search for a satisfactory method of detecting and recognizing these strains. The peak values of elastic strain are likely to be of order 10^{-4}, corresponding to a stress of about 100 kg cm^{-2}, but this may be apparent only very locally and immediately before an earthquake. Generally, measurements of strains of about 10^{-6} are required. Since linear strain meters to the design of Benioff (1959) and liquid level tilt-meters, as used particularly in Japan, are available to measure accurately tidal strains of

---

* Statistical statements to the effect that major earthquakes will continue to occur on the recognized active belts of Fig. 4.1, but hardly at all elsewhere, are the basis for general precuations but do not constitute prediction.

peak-to-peak amplitude about 5×10^{-8}, the observations would appear superficially to present no difficulty, but in practice very elaborate precautions are necessary to avoid instrumental drift, including strain ageing of walls of tunnels used for the instruments. In an installation in a seismically inactive area Major et al. (1964) found a secular strain of 0.2×10^{-6} per year and in active areas secular strain may be substantially greater without having an obviously direct, causal connection with earthquakes. The same difficulty besets large-scale observations of strain from geodetic surveying and comparison of tide gauge records. Also, meteorological disturbances such as heavy rainfall can cause temporary strains of order 10^{-6}. The problem therefore is in recognizing strains which are genuine precursors of earthquakes. Sassa and Nishimura (1956) (see Figs. 4.34 to 4.36)

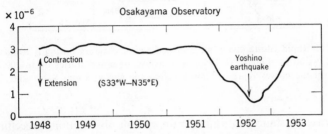

Figure 4.34: Variation in ground strain at an observatory 90 km from the epicenter of the Yoshino earthquake in 1952. From Sassa and Nishimura (1956).

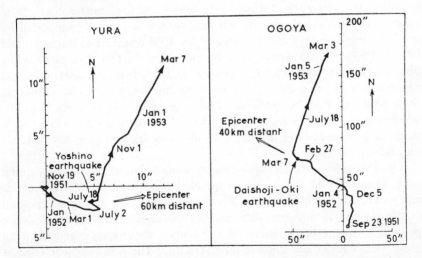

Figure 4.35: Vector plots of progressive changes in ground tilt at two Japanese observatories. The distance of each experimental point from the origin of the figure represents the magnitude of the tilt accumulated since the starting date and the direction is the direction of downward tilt. A tangent to such a vector curve represents the instantaneous direction of tilting movement. From Sassa and Nishimura (1965).

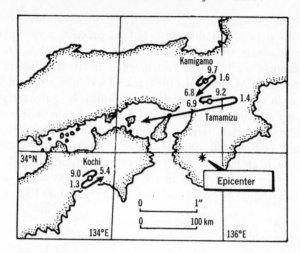

Figure 4.36: Rapid tilting giving an S-shaped vector diagram during a 10-hour interval before the Nanki earthquake in 1950, as measured at observatories up to 200 km away. Numbers on the curves represent hours before the shock. From Sassa and Nishimura (1956).

have reported instances in which strain and tilt observations preceding earthquakes appear in principle to have been usable for prediction, the essential additional requirement being a sufficiently dense network of observing stations to outline the geographical extent of the strains and to eliminate strictly local effects and instrumental defects. This conclusion is generally supported by other Japanese observations reported at the 1964 conference (Hagiwara and Oliver, 1964), but allowance must be made for the fact that encouraging observations are more readily reported than discouraging ones. Geodetic observations of "anomalous" strains preceding earthquakes in California have been reported by Hoffman (1968). Geodetic observations have the advantage of greater independence of local disturbances than smaller scale strain measurements, but they are discontinuous and can generally be made only infrequently.

A basic problem is to distinguish elastic strain (from which rock may spring back when released during an earthquake) from inelastic deformation or creep, as considered in Chapter 7. What is really needed is a direct observation of stress, which is very difficult to make. However, indirect observations are possible by virtue of the effects of stress upon the magnetic and electrical properties of rocks and upon seismic velocities. The piezomagnetic effect is the best understood of these effects; stress modifies the magnetocrystalline anisotropy of magnetic minerals, so that an anisotropy of susceptibility appears when stress is applied to a rock (Stacey, 1963a) and remanence is correspondingly deflected. The change amounts to about 1% in magnetite-bearing igneous rocks for a stress of 100 kg cm^{-2}, and, since the magnetization is commonly about 10^{-3} emu, we expect to have local variations of about 10^{-5} emu associated with seismic activity. An order of magnitude estimate of the consequent time-dependent magnetic anomaly at the surface of the Earth due to this effect is obtained by supposing that measurements of the field anomaly ΔF are made immediately above a spherical volume of rock, in which the increment in magnetization ΔI is 10^{-5} emu. Then

$$\Delta F = \tfrac{4}{3}\pi\Delta I f \approx 5 \text{ to } 10 \times 10^{-5} \text{ oe} = 5 \text{ to } 10 \text{ gammas} \qquad (4.75)$$

where $1 \leqslant f \leqslant 2$, depending upon the orientation of the magnetization This is, of course, a gross simplification but calculations based on simple, plausible stress patterns (Stacey, 1964b) suggest that seismomagnetic anomalies with peak intensities of about 10 gammas (10^{-4} oe) would be expected if the stressed rocks had magnetizations of 10^{-3} emu and proportionately more for stronger magnetizations. This is measurable with suitable instruments, such as proton precession magnetometers, differentially connected to cancel geomagnetic disturbances of magnetospheric origin. A comparison by Kato and Utashiro (1949) of the daily values of magnetic declination at two observatories, 80 km (Katsuura) and 460 km (Kakioka) from the Nankaido earthquake of 1946, is shown in Fig. 4.37. Since the horizontal field intensity in the area is about 0.3 oe, a change in declination of 3′ of arc represents a field increment of about 25 gammas. More sophisticated instruments, such as those now operated in California by Breiner (1964), can be expected to give improved observations of the same type and Breiner and Kovach (1967) have reported

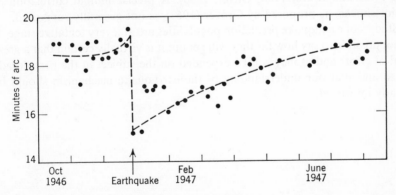

Figure 4.37: Differences between values of magnetic declination at two Japanese observatories 80 km and 460 km from an earthquake. Redrawn from Kato and Utashiro (1949). Scatter of the observations does not allow any significance to be attached to the change in field, suggested by the broken line, before the earthquake, but this effect is now sought with much improved instruments in closer networks.

local magnetic changes, evidently due to deep-seated stresses, which precede creep increments on the San Andreas fault in California.

Electrical properties of rocks are also affected by stress but the direct observation of stress effects in the ground is so hampered by the disturbing effects of ground water that it is not clear that useful data are obtainable. More hopeful is the observation of the stress dependence of seismic velocity. This is well demonstrated in the laboratory (Tocher, 1957; Volarovich et al., 1963) and is to be sought in a Japanese experiment (Tsuboi et al., 1962) in which explosions repeated at monthly intervals will be timed at seismic stations 200 to 300 km distant. Possibly much shorter-range observations would be more useful as seismic rays traveling several hundred kilometers penetrate well into the mantle and the laboratory observations suggest that under substantial hydrostatic pressures the effect on seismic velocities of superimposed deviatoric stresses is greatly reduced.

A completely independent approach is based on possible time-space correlations between earthquakes. Knopoff (1964c) has studied the statistics of earthquake occurrences in Southern California and has shown that they are not strictly independent events. This merely suggests that the chance of an earthquake is higher immediately after a shock than after a long dormant period, but occurrences of aftershocks which gradually diminish in a few weeks or months following major earthquakes are well known and, at least in some cases, major shocks are preceded by foreshocks. The possibility that statistically characteristic sequences of microforeshocks are detectable with sensitive seismometers is suggested by laboratory observations on rock fracture and suitable instrumental networks are proposed in the United States (see Press and Brace, 1966; Oliver, 1966). A precise mutual correlation of major shocks, as proposed by Grover (1967), is not generally supported.

Studies of earthquake prediction possibilities are at a very tentative stage. It is not possible to say how far they will get but it is virtually certain that a great deal of effort and money will be expended on the subject in the next decade or so and that our understanding of the earthquake mechanism should be greatly improved.

Chapter 5

THE GEOMAGNETIC FIELD

"There are too many ways in which the core can be made to convect to permit an unambiguous interpretation."

ELSASSER, 1963, p. 29.

5.1 The Main Field

In this section the geomagnetic field is considered as a static entity, whose features must be explicable in terms of any satisfactory theory of its origin. The variation of these features with time is discussed in Section 5.2.

To a useful first approximation the field may be represented by that of a magnetic dipole situated at the center of the Earth. The dipole moment is about 8×10^{25} emu and the axis is inclined at about $11°$ to the Earth's geographic axis. The equations which describe a dipole field are readily obtained by writing down the magnetic potential in the field of a dipole. Consider the potential at a point P, as in Fig. 5.1, situated at (r, θ) in a spherical polar

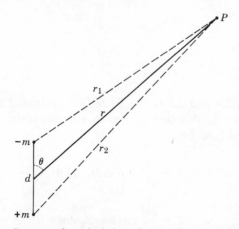

Figure 5.1: Geometry for calculation of magnetic potential of a dipole.

coordinate system (r, θ, λ) with origin at the center of a magnetic dipole of moment M. The moment is supposed to be due to poles of strength $\pm m$, separated by a distance d along the axis of the coordinate system, so that

$$M = md \tag{5.1}$$

The potential at P is thus

$$V = \frac{(-m)}{r_1} + \frac{(+m)}{r_2} = -\frac{m}{[r^2 + (d/2)^2 - rd\cos\theta]^{1/2}}$$

$$+ \frac{m}{[r^2 + (d/2)^2 + rd\cos\theta]^{1/2}} \tag{5.2}$$

Then allowing d to become infinitesimal relative to r, while maintaining a finite moment (md), V reduces to

$$V = -\frac{m}{r}\left[\left(1 - \frac{d}{r}\cos\theta\right)^{-1/2} - \left(1 + \frac{d}{r}\cos\theta\right)^{-1/2}\right]$$

$$= -\frac{md}{r^2}\cos\theta = -\frac{M}{r^2}\cos\theta \tag{5.3}$$

The field at P is obtained by taking the gradient of V. We can consider P to be at the surface of a spherical Earth with the dipole at the center and then we are interested in the horizontal and vertical components of the field, H and Z, relative to the surface at P:

$$H = \frac{1}{r}\frac{\partial V}{\partial\theta} = \frac{M}{r^3}\sin\theta \tag{5.4}$$

$$Z = \frac{\partial V}{\partial r} = \frac{2M}{r^3}\cos\theta \tag{5.5}$$

Applying Eqs. 5.4 and 5.5 to the Earth, the horizontal and vertical field components at the Earth's surface, due to a dipole which gives the closest fit to the observed field, are

$$H = \frac{M}{a^3}\sin\theta = H_0\sin\theta \tag{5.6}$$

$$Z = \frac{2M}{a^3}\cos\theta = Z_0\cos\theta \tag{5.7}$$

where

$$H_0 = \frac{Z_0}{2} = 0.312 \, \text{oe} \qquad (5.8)$$

From Eqs. 5.4 and 5.5, the total field intensity F is obtained:

$$F = (H^2 + Z^2)^{1/2} = \frac{M}{r^3}(1 + 3\cos^2\theta)^{1/2} \qquad (5.9)$$

and the angle of inclination I of the total field to the horizontal:

$$\tan I = \frac{Z}{H} = 2\cot\theta = 2\tan\phi \qquad (5.10)$$

Here θ is magnetic colatitude and ϕ or $(90° - \theta)$ is magnetic latitude. Equation 5.10 is basic to the calculation of paleomagnetic pole positions (Chapter 6), since if the direction of magnetization of a rock records the ancient field direction, the ancient magnetic latitude is obtained from this equation (assuming, of course, that the field is dipolar.) The inclination of the field, given by Eq. 5.10, is the differential form of the equation for a magnetic line of force:

$$\tan I = \frac{dr}{r\,d\theta} = 2\cot\theta \qquad (5.11)$$

which integrates to

$$\frac{r}{a} = \frac{\sin^2\theta}{\sin^2\theta_a} \qquad (5.12)$$

where θ_a is the colatitude at which the line of force crosses radius a.

The equations for a dipole field apply equally to the field outside a uniformly magnetized sphere. That the Earth's field resembled that of a magnetized sphere was recognized by William Gilbert, physician to Queen Elizabeth I, whose contemporaries supposed the magnetic alignment of a lodestone, or natural compass, to be an extraterrestrial influence. Gilbert made the first quantitative study of geomagnetism* and his name has been given to one of the paleomagnetic polarity epochs, shown in Fig. 6.11. However, we cannot conclude that the Earth is in fact a uniformly magnetized sphere; further evidence of the origin of the field is obtained by considering departures from the dipole character.

---

* Chapman and Bartels (1940) have given translations (from Latin) of relevant portions of Gilbert's treatise *De Magnete*.

The potential of the geomagnetic field can be represented as an infinite series of spherical harmonic functions, of which Eq. 5.3 is the first term. A comprehensive discussion of this method of analyzing geomagnetic data is given by Chapman and Bartels (1940). The general expression for potential is obtained in Appendix A*:

$$V = \frac{1}{a} \sum_{l=1}^{\infty} \sum_{m=0}^{l} \left\{ \begin{array}{l} \left[C_l^m \left(\frac{a}{r}\right)^{l+1} + C_l'^m \left(\frac{r}{a}\right)^l \right] \cos m\lambda \\[2mm] + \left[S_l^m \left(\frac{a}{r}\right)^{l+1} + S_l'^m \left(\frac{r}{a}\right)^l \right] \sin m\lambda \end{array} \right\} P_l^m(\cos \theta) \quad (5.13)$$

where θ, λ are the coordinates of magnetic colatitude and longitude and a is the Earth radius. V itself is not directly observable but components of the field northward (X), eastward (Y), and downward (Z) are measured over the Earth's surface ($r = a$). Then relationships

$$X = \left(\frac{1}{r} \frac{\partial V}{\partial \theta} \right)_{r=a}$$

$$Y = \left(\frac{1}{r \sin \theta} \frac{\partial V}{\partial \lambda} \right)_{r=a} \qquad (5.14)$$

$$Z = \left(\frac{\partial V}{\partial r} \right)_{r=a}$$

allow the coefficients in Eq. 5.13 to be determined up to an order limited by the detail of the observations. Note that to determine separately the primed and unprimed coefficients in (5.13), we need to obtain both V and $(\partial V/\partial r)$. The vertical field Z gives $(\partial V/\partial_r)_{r=a}$ directly but V is only indirectly obtainable from X and Y, its horizontal derivatives. C. F. Gauss first applied this method to the analysis of the geomagnetic field and showed that the coefficients C_l^m and S_l^m described a field of internal origin, and that $C_l'^m$ and $S_l'^m$ represented a field of external origin, which he found to be nonexistent. We now know that the external field is not totally absent; a field amounting to perhaps 3×10^{-4} oe (30 gammas) at the surface at geomagnetically quiet times, and several times as much during magnetic storms, is due to a "ring current" or drift of charged particles spiraling about geomagnetic field lines at several earth radii. However, Gauss's conclusion that the main field is of internal origin is valid and in the present context interest is restricted to the coefficients C_l^m and S_l^m.

---

* Note that there is no $l = 0$ term; this would correspond to a magnetic monopole within the Earth.

In geomagnetism it is convenient to use the Gauss coefficients $g_l{}^m$ and $h_l{}^m$, which are

$$g_l{}^m = \frac{C_l{}^m}{a^2}$$
$$h_l{}^m = \frac{S_l{}^m}{a^2}$$

(5.15)

The Gauss coefficients have the dimensions of magnetic field, whereas $C_l{}^m$ and $S_l{}^m$ have dimensions of magnetic pole strength (or mass in the case of gravitational potential). Equation 5.13 is then rewritten:

$$V = a \sum_{l=1}^{\infty} \left(\frac{a}{r}\right)^{l+1} \sum_{m=0}^{l} (g_l{}^m \cos m\lambda + h_l{}^m \sin m\lambda) P_l{}^m(\cos \theta) \qquad (5.16)$$

and Eqs. 5.14 apply as before.

Extensive tabulations of the geomagnetic field and its change during the period 1905–1945, with harmonic analyses, are given by Vestine and his colleagues (Vestine et al., 1947) and more recent harmonic analyses are by Finch and Leaton (1957) and Vestine et al. (1963). The most interesting and perhaps the most significant aspect of these analyses is the separation of the best fitting dipole field from the remaining, nondipole part of the field. In particular, the nature of the nondipole field is more apparent when the stronger dipole field is removed. This is emphasized by the work of Bullard et al. (1950), who used the Vestine data and produced contour maps of the nondipole field, one of which is reproduced as Fig. 5.2. This is the field of the Earth, as represented by all spherical harmonic terms except $g_1{}^0$, $g_1{}^1$, $h_1{}^1$, which are due to components of the central dipole moment along the Earth's geographic axis ($g_1{}^0 a^3 = -0.3057a^3$), in the equatorial plane through the Greenwich meridian ($g_1{}^1 a^3 = -0.0211a^3$), and normal to both ($h_1{}^1 a^3 = 0.0581a^3$).* It should also be noted that the best-fitting dipole field is centered about 300 km off the Earth's center. This is known as the eccentric dipole field.

Before discussing the significance of the features apparent in Fig. 5.2, it is worth considering the extent to which such an analysis fully represents the geomagnetic field. Inevitably the limitation of a spherical harmonic analysis to a finite number of terms smoothes out the finer details. However, Alldredge et al. (1963) (see also Alldredge 1965) carried out a Fourier analysis of a single profile of magnetic field around the world, extending the analysis to terms corresponding to wavelengths of 10 miles. They found that strong harmonic terms appeared only down to wavelengths of about 3000

---

* Note that $[(g_1{}^0)^2 + (g_1{}^1)^2 + (h_1{}^1)^2]^{1/2} = H_0$ as given by Eq. 5.8.

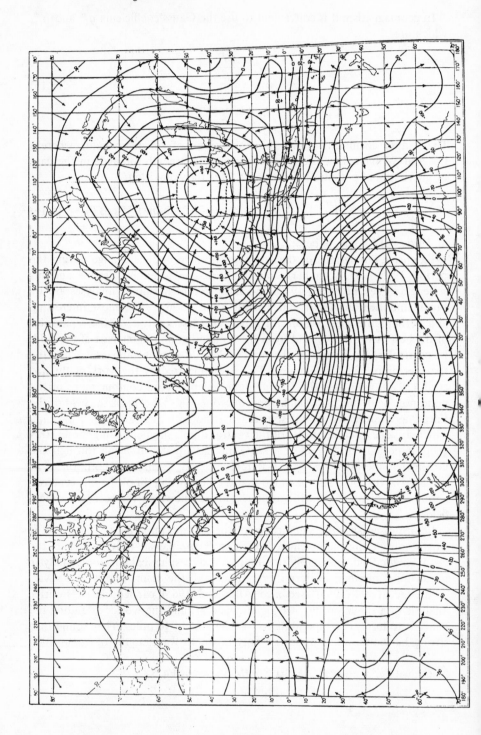

miles. Between about 2000 and 200 miles the spectrum was very weak, but stronger variations with wavelengths of 10 to 100 miles were apparent (Fig. 5.3). The fine detail in the field structure, which gave the higher harmonics in this analysis, must have been due to magnetic sources near to the Earth's surface because their scale is too small to correspond to deep sources.

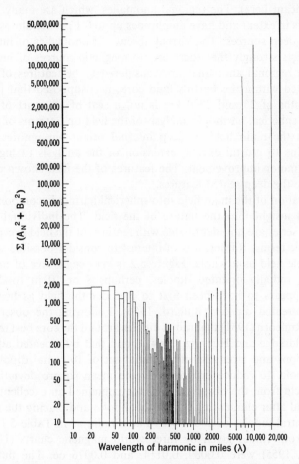

Figure 5.3: Spectrum of variations in geomagnetic field on a profile encircling the Earth. The values plotted are sums of the squares of amplitudes of Fourier components in wavelength intervals of 10 nautical miles. Note that the scale is logarithmic, as a very wide range is covered, and that the minimum indicates a virtual absence of components in the wavelength range 200 to 2000 miles. Reproduced, by permission, from Alldredge et al. (1963).

Figure 5.2: The nondipole field for 1945. Contours give the vertical component in intervals of 0.02 oe and arrows represent the horizontal component. Reproduced, by permission, from Bullard et al. (1950).

They represent local magnetic anomalies up to about 100 km in extent and rarely exceeding 10^{-2} oe. These terms are reasonably attributed to the magnetization of rocks in the outer 25 km of the Earth; below this depth temperatures are higher than the Curie point of magnetite (580°C) and rocks are therefore nonmagnetic (strictly they are paramagnetic, but paramagnetism is not significant here). The regional anomalies, which are many thousands of kilometers in extent and have amplitudes up to 0.1 oe, are almost certainly due to very deep sources. The virtual absence of anomalies of intermediate extent suggests strongly that there are no magnetic sources of intermediate depths. The regional anomalies are considered to be features of the main field generated within the Earth's fluid core; it is supposed that the mantle between depths of 25 and 2900 km is not a seat of any part of the main field.* Thus spherical harmonic analysis of the field up to terms of order 5 or 6 represents the main field of deep internal origin but ignores the local anomalies due to crustal effects; extension of the analysis to higher terms effects only minor improvement. The features of the field shown in Fig. 5.2 are therefore the deep-seated features.

The separation of the main field into spherical harmonic components gives only limited insight into the nature of the field. The individual harmonic terms in the series are not identifiable with features of the field except perhaps for the dipole terms. Rather, it is of interest to consider possible sources for the nondipole field as a whole. Figure 5.2 is very suggestive of an array of deep-seated, radially oriented dipoles, perhaps 8 or 10 in total number. E. Halley appears to have been first to consider the field as being due to several deep-seated dipoles (Bullard, 1956). Following the observation by Lowes and Runcorn (1951) that radial dipoles around the core best represented the field, Alldredge and Hurwitz (1964) found that by repeated adjustments of the positions and strengths of an initial set of 10 radial dipoles, plus a centered dipole, to reduce progressively the mean square deviation of the computed field from the observed field, they obtained an excellent fit to the observed field after eliminating two of the dipoles and placing the remaining eight noncentral dipoles at about 0.25 Earth radii, as in Table 5.1. The rms deviations from values given by three world magnetic charts (U.S., 1945, 1955; U.K., 1955) were 0.0089, 0.0114, and 0.0076 oe. The fit is clearly excellent, bearing in mind uncertainties in the charts, and confirms that a set of radial dipoles does reasonably represent the field.

Two further points are worth noting. The strength of the centered dipole in the best-fitting radial dipole model is about twice as great as the dipole obtained from spherical harmonic analysis. Thus if physical significance is

---

* By virtue of its electrical conductivity the mantle may modify the more rapidly varying features of the field (see Section 5.3).

TABLE 5.1: MAGNETIC MOMENTS M AND POSITIONS OF RADIAL DIPOLES AT 0.25 EARTH RADII WHICH, WITH THE CENTRAL DIPOLE, GIVE THE BEST FIT TO THE FIELD, AS REPRESENTED BY THE ADMIRALTY CHART FOR 1955. (Moments are expressed in units of Ma^{-3} emu cm^{-3}, where a is the Earth's radius.)

	Ma^{-3}	Colatitude (deg)	East Longitude (deg)
Central dipole M_0	−0.69711	23.6	208.3
Radial dipoles at	0.10250	13.7	341.9
0.25a	0.11440	46.0	179.9
	−0.02724	54.9	40.1
	0.07704	77.4	241.7
	0.02879	91.3	120.8
	−0.09469	139.8	319.3
	−0.11795	141.1	43.0
	0.04103	102.9	180.1

From Alldredge and Hurwitz (1964).

attributed to the radial dipoles, they cannot be treated independently of the centered dipole and the dipole from harmonic analysis has no particular significance. Further, the best-fitting radial dipoles are deeper than intuition would have suggested, about halfway between the core-mantle boundary and the center of the Earth. However, this is presumably a result of supposing that the sources are ideal dipoles and not extended volumes of the core. The fitting of a dipole to the field of an extended magnetized body normally places the dipole deeper than the body itself. We are therefore justified in supposing that the nondipole field (and perhaps also the dipole field) originate in the outer core.

A comparison of the terrestrial magnetic field with the fields of other planets is possible. Evidence of planetary magnetic fields has been reviewed by Kern and Vestine (1963), Hide (1966), and E. J. Smith (1967). Space probes have failed to find any magnetic field associated with our three nearest neighbors in space. Upper limits for the magnetic moments M are

$M_{\text{Moon}} < 10^{-5} M_{\text{Earth}}$ (Ness et al., 1967)

$M_{\text{Mars}} < 3 \times 10^{-4} M_{\text{Earth}}$ (E. J. Smith, 1967)

$M_{\text{Venus}} < 3 \times 10^{-3} M_{\text{Earth}}$ (TASS, 1967; Van Allen et al., 1967)

It is noted in Section 1.5 that the moment of inertia of Mars indicated by the motions of its satellites shows that the concentration of mass at the center is much less than in the case of the Earth. Its core is therefore small and the absence of a field is compatible with the assumption that the Earth's core is responsible for the terrestrial field. We have no information on the core of Venus, but apparently Venus is rotating very slowly. Thus if planetary rotation is another necessary condition for the establishment of a magnetic field, we do not expect Venus to have a field. Of the giant planets, Jupiter and Saturn emit radio waves suggestive of extensive magnetic fields with trapped charged particles (see review by Hide, 1966a, and papers in Hindmarsh et al., 1967). The compositions of these planets are quite different from that of the Earth (see Chapter 1) but they are presumed to have substantial fluid parts. Similarly, the Sun has a magnetic field, apparently associated with fluid motions, which cause dramatic local intensifications of the field in sunspots.

It is of interest to enquire what is the magnetic moment of the smallest terrestrial-type dynamo. Possibly the smallest of the radial dipoles listed in Table 5.1 represents an approximate limit below which the dynamo action is not self-sustaining in a body with fluid velocities similar to those of the Earth's core. If so, then this limit is about 0.08 M_{Earth}, and the moments of the Moon, Mars, and Venus are certainly smaller than this. Thus the Earth is probably unique among the terrestrial planets in having a magnetic field. It is nevertheless possible that the other planets had fields at earlier stages in their histories. The observation that numerous chondritic meteorites have magnetic moments of apparently extraterrestrial origin (Stacey, 1967e) is suggestive of a parent body with a terrestrial-type field, but the origin of the meteorites is still obscure and the necessary field may have been associated with the solar nebula during the early stages of formation of the solar system, as required by the theory of Alfvén (1954) (see Section 1.1).

5.2   Secular Variation and the Westward Drift

It has been known for over 400 years that the main geomagnetic field is not steady but undergoes a secular variation, which is coherent over large areas of the Earth. The original discovery is attributed to H. Gellibrand, who in 1635, recognized a steady progressive change in magnetic declination, or angle between magnetic north and geographic north, at London (Chapman and Bartels, 1940, p. 910). Vector plots showing the variations of inclination and declination over long periods of time indicate apparent cyclic variations at many observatories. This type of curve was first plotted for London by L. A. Bauer and has subsequently been used by a number of other authors. Curves for London and Paris are shown in Fig. 5.4. Detail is poor for the early years but the general parallelism of the curves is clear; the scale of the secular variation is therefore large compared with the distance between London and Paris. It is important to note that we are considering variations in field which are much stronger than the transient changes due to disturbances in the magnetosphere or ionosphere. Such disturbance fields are discussed here in only one connection—they give information about the electrical conductivity of the upper mantle (Section 5.3).

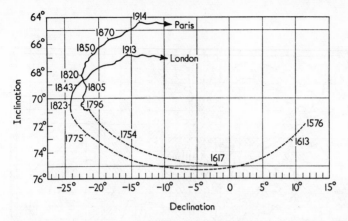

Figure 5.4: Vector plots of the inclinations and declinations of the geomagnetic field at London and Paris over several centuries. This plot is due originally to C. Gaibar-Puertas and is reproduced from Runcorn (1962). Another version appears as Fig. 6.7.

The worldwide pattern of secular variation has been plotted in various ways by Vestine et al. (1947a), one of whose figures is reproduced as Fig. 5.5. The general form is very similar to the nondipole field (Fig. 5.2) and can also be represented by a pattern of deep-seated sources. Comparison of Figs. 5.2 and 5.5 suggests that strong centers of secular change tend to be situated west of the strong centers of the nondipole field, implying that the nondipole field is shifting westward. As early as 1692 E. Halley published his conclusion that the secular variation was due to a steady westward drift of the geomagnetic field (Bullard, 1956). Halley supposed that part of the field was due to magnetic poles situated in an inner core and he thus came very close to present-day ideas; he pointed out that if the core rotated more slowly than the outer shell of the Earth, its field would drift westward. A detailed analysis of the westward drift was made by Bullard et al. (1950), using Vestine's tabulation of the field over a 40-year period (1905–1945). They concluded that the nondipole field was drifting westward at an average rate of $0.18°$ year^{-1} at all latitudes, and for all harmonic components, but with a wide scatter in calculated rates; they found that the dipole field, i.e., the equatorial component, was drifting very little if at all, but Vestine (1953) reported a westward drift of $0.30°$ year^{-1} of the eccentric (off center) dipole. The substantial scatter in the drift at different latitudes, or between different harmonic components, appears as uncertainty in the analysis but was due to a genuine variability in the features of the field rather than to inadequacy of the data. The features, as shown in Fig. 5.2, form, deform, and disappear rather like eddies in a stream of water.

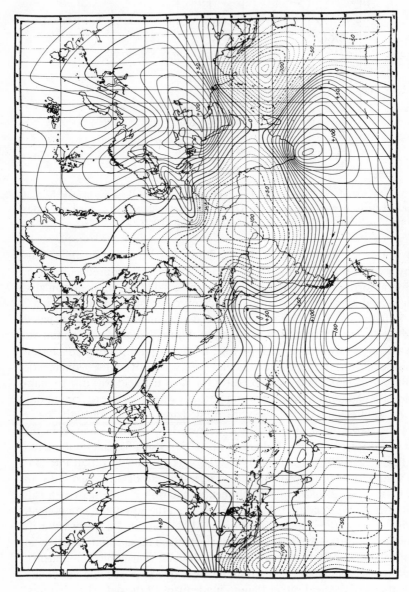

Figure 5.5: Contours of rate of secular change in intensity of the vertical component of the geomagnetic field at epoch 1942.5. Values are in gammas per year. Figure reproduced, by permission, from Vestine et al. (1947a).

As noted in Section 5.1, it is probably unwise to attach too strong a physical significance to the separation of the dipole and nondipole fields. The dipole field itself is not constant but contributes to the secular variation. From a study of the field as a whole, Nagata (1965) concluded that there were five distinguishable features of the present secular variation:

1. A decrease in the moment of the dipole field by 0.05% year^{-1}.
2. A westward precessional rotation of the dipole at 0.05° of longitude per year.
3. A rotation of the dipole toward the geographic axis at 0.02° of latitude per year.
4. The westward drift of the nondipole field at 0.2° of longitude per year.
5. Growth and decay of features of the nondipole field, giving changes which average about 10 gammas (10^{-4} oe) year^{-1}.

There are, of course, alternative ways of presenting these conclusions. We may refer to the decreasing strength of the axial dipole, i.e., the component of the dipole field parallel to the geographic axis and to a decrease in strength, at a slightly greater rate, of the westward-drifting equatorial component of the dipole field. All analyses agree in finding a westward drift of the non-dipole field, but the rate appears to depend upon how the data are analyzed. Thus Alldredge and Hurwitz (1964) examined the motion of the dipoles in their radial dipole model of the field and found that in the interval 1945–1955 the centered dipole grew by 22% and drifted west at more than 1° year^{-1}. Excluding the central dipole and also a very small one which apparently moved rapidly, they obtained a mean westward drift of 0.25° year^{-1} but this mean probably has no significance because the variability was very great— four of the nine dipoles drifted eastward. This may indicate that the model is not a good one, but a more likely alternative is that the westward drift itself is much less significant than has been supposed.

Another interesting way of presenting the secular variation is due to As (1967), who used the concept of a dipole vector G to represent the central dipole required to give the observed field at any point on the surface. The magnitude of the vector is

$$G = \left[H^2 + \left(\frac{Z}{2} \right)^2 \right]^{1/2} \qquad (5.17)$$

H and Z being the horizontal and vertical components of the surface field, which are given by Eqs. 5.6 and 5.7 for an ideal dipole field. For every point in a dipole field, G is constant in magnitude and direction and for the Earth's field the vectors are distributed in a cone whose angle represents the deviation of the field from that of an ideal dipole. As finds that the angle of the cone has increased by a factor of two in the past 200 years; since the dipole field has been decreasing relatively slowly, this means that the nondipole field has increased dramatically.

As (1967) obtained some confirmation of a suggestion, originally due to Vestine (1953), that the secular variation was correlated with the fluctuations in the rate of rotation of the Earth (Section 2.5). Vestine considered the westward drift of the eccentric dipole field in the periods 1905–1925 and 1925–1945 and concluded that acceleration of the rotation (of the mantle) was accompanied by accelerated westward drift. This conclusion is not entirely dependent upon a physical significance of the average westward drift, but only upon the fluctuations in its rate. It is virtually impossible to be sure of correlations such as this. Munk and MacDonald (1960b) concluded that it was real. If so, it is the most direct evidence that fluctuations in rotation are caused by an angular momentum exchange by electromagnetic coupling of the core and mantle.

The motion of particular features of the geomagnetic field—especially the westward drift—is widely interpreted as evidence of motion in the Earth's fluid core and is basic to the discussions of Elsasser (1950) and Bullard et al. (1950). The Bullard model of the core is a convenient starting point for a quantitative assessment of the significance of the drift. The core is considered to be undergoing convective, i.e., radial motion, but conservation of angular momentum in the rising and sinking columns of material ensures that there is also a differential rotation. Thus if we consider convection to start in a fluid core which is initially rotating coherently, the rising material carries with it relatively little angular momentum and its angular velocity is reduced, while the sinking material, also conserving its angular momentum, increases in angular velocity. The outer part of the core then rotates more slowly than the inner part. Bullard et al. (1950) identified the features of the nondipole field (Fig. 5.2) with large-scale eddies in the outer part of the core, which rotates more slowly than the inner part. The dipole field was considered to be deeper-seated and was thus associated with the inner part of the core. The motion of the mantle is then determined by a dynamic balance in the coupling to the dipole and nondipole fields and is thus intermediate between the slower rotation of the outer core and faster rotation of the inner core.

Although this model is important as an introduction to several new concepts, in this form it is inadequate to explain the secular variation as summarized by Nagata (1965). The dipole field appears to drift westward rather than eastward, which would be required by the balance between inner and outer core rotations. Further, the remarkable growth of the nondipole field relative to the dipole field in the past 200 years would cause progressively stronger coupling to the outer core and thus progressively diminish the rate of westward drift, by a total of perhaps 0.1° per year over 200 years. Conservation of angular momentum requires that about 10% of this change be due to slowing of mantle rotation; this means the mantle rotation would have been slowed down at a rate

$$\frac{d\omega}{dt} \approx -0.01° \text{ year}^{-1} \text{ per 200 years} \approx -1.0 \times 10^{-21} \text{ rad sec}^{-2}$$

Since this is twice the rate of slowing by tidal friction (Section 2.4) it must be regarded as quite improbable. Nevertheless, the basic idea of a dynamic equilibrium in the coupling of the core and mantle is important. There must be a balance between the westward- and eastward-drifting components of the field (or a continuous adjustment toward balance, with a time constant of a few years). These components may be substantially independent, as the radial dipoles of Alldredge and Hurwitz (1964) appear to be. It may simply happen by chance that the westward-drifting components of the field penetrate to the Earth's surface more effectively at the present time and that an eastward drift would have been observable at various times in the past; the archeomagnetic data of Aitken and Weaver (1965) (see Fig. 6.7) suggest that this is so. The cyclic secular variation of Fig. 5.4, which would be produced by westward drift of a nondipole field, is apparently characteristic of the past few centuries only.

A completely independent approach to the problem of the westward drift has been made by Hide (1966b) and Rikitake (1966b), who envisage the drift as due to propagating hydromagnetic waves rather than bodily motion of the core. However, the necessity for these arguments depends upon the significance of the drift as a permanent feature of the field, which the foregoing arguments leave very much in doubt.

The absence of significant features of the nondipole field over the Pacific region has attracted some attention (Cox, 1962). Paleomagnetic work on Hawaiian lavas by Doell and Cox (1965) suggests that the secular variation has been consistently low in the area for hundreds of thousands of years and the usual implication is that the lower mantle, on the Pacific side, contains a layer of highly conducting material, presumed to be metallic iron, and that this acts as an electromagnetic screen (see, for example, Creer, 1963). However, Tarling's (1965) measurements on Hawaiian lavas, although less extensive than those of Doell and Cox, are sufficient to raise doubt about the screening of the Pacific; moreover, the supposition that the secular variation is currently small in the Pacific is not borne out by the observations at Honolulu magnetic observatory. It is only the direction of the field that has been approximate constant. During the period 1902–1938, the magnitude of the field decreased by 800 gammas in the horizontal component and 1600 gammas in the vertical component; this coincides with the world-average secular variation of about 50 gammas year^{-1}. We can therefore regard the secular variation as a general phenomenon, common to the whole Earth.

In anticipation of the discussion in Section 5.4 it is useful to refer to some of the conclusions of Chapter 6. Paleomagnetism extends the record of the geomagnetic field over an enormously greater time scale, although with much less precision. Over the past few thousand years the intensity of the field has changed by a factor of two or more. The extreme fluctuations of the field appear to be reversals of the dipole field, the most recent of which may have been only a few thousand years ago. Allowing for the ambiguity in polarity, the field appears to be that of an axial dipole when averaged over tens of thousands of years, i.e., on this time scale the magnetic and geographic axes coincide.

5.3 Electrical Conduction in the Core and Mantle

The temperatures and pressures of the Earth's deep interior are not reproducible in the laboratory, except in transient, shock-wave experiments, and the chemical composition is uncertain, particularly in the case of the core. Thus estimates of electrical resistivity in the core and lower mantle, based on extrapolations of laboratory data, are subject to considerable uncertainties. Observations of secular variation, and also the electromagnetic core-mantle coupling, as evidenced by fluctuations in rotation (Section 2.5), allow estimates to be made of lower mantle conductivity. As Tozer (1959) has shown, these are in satisfactory accord with the onset of semiconduction in silicates at elevated temperatures. Evidence of the core conductivity is less direct, but the core is presumed to be metallic and the known range of conductivities of metals is limited. The best estimate is thus obtained by extrapolation from laboratory observations; however, this still leaves a significant margin of uncertainty.

Consider first the resistivity of the core. Most estimates have been simple extrapolations of the resistivity of pure iron to a high temperature and pressure, but the effect of alloying elements is likely to be much greater than the effects of temperature and pressure. Stacey (1967d) used Powell's (1953) value of the resistivity of liquid iron at its melting point (139 $\mu\Omega$ cm or 1.39×10^5 emu) and estimated that this was increased by a factor 1.6 at the core temperature of about 3600°C and a further factor of about 1.3 by compression to 2×10^6 atm. The introduction of about 10% nickel increases the resistivity of iron by a factor 3, but more important is to make a reasonable guess for the effect of the light constituent which is required by the estimates of core density (see Sections 1.5 and 4.5). If we accept Ringwood's (1959, 1966) suggestion that the core contains as much as 15% silicon, then data on transformer core material indicate that the resistivity is increased by a factor of about 10. An alternative suggestion by Alder (1966), that the light constituent is MgO, is likely to have at least as great an effect. Then, assuming that the total effect of alloying is to increase resistivity by a factor 10, the preferred value of the resistivity of the core is 3×10^6 emu. The total range of admissible values is 10^6 to 2×10^7 emu. The lower limit is substantially higher than most earlier estimates and is quite unfavorable to convective theories of the geomagnetic dynamo, as noted in Section 5.4. This problem requires experimental work on the resistivities of iron alloys in the liquid state.

There are three independent observations from which the electrical conductivity at various depths in the mantle can be deduced; geomagnetic variations of magnetospheric origin, secular variation and core-mantle coupling. All depend upon the integrated effects of currents flowing in a range of depths, so that there can be no unique interpretation. However, when all of the data are considered together, a reasonably unambiguous conductivity profile of the mantle results. Current estimates of conductivity at depth give uncertainties of a factor of about 10 and this must be regarded as satisfactory. There seems to be no possibility of estimating conductivity with the same precision as that with which the mechanical properties of the Earth are determined from seismology. Electrical re⁻istivities of solids, from metallic conductors to good insulators, encompass an enormous range, more than $10^{24}:1$ at ordinary temperatures, so that to specify the resistivity of a material within a factor 10 does in fact imply a reasonably exact knowledge of its electrical state.

The skin effect is a well-established phenomenon in electromagnetism. A fluctuating electromagnetic field induces currents in a conductor, which oppose changes in the field (Lenz's law of electromagnetic induction). It follows that penetration of the conductor by the field is opposed by the electrical conduction and therefore that the fluctuations do not penetrate a conductor as effectively as they would penetrate an insulator, in which they are not attenuated. For derivation of the basic equations of this effect reference should be made to a text on electromagnetism* and for a comprehensive review of applications to geophysics see Rikitake (1966c). The principle is conveniently described in terms of the penetration of a plane electromagnetic wave of angular frequency ω, into a semi-infinite conductor of conductivity σ, bounded by the plane $z = 0$, and extending indefinitely in the positive z direction. Then if the magnetic field strength at the surface is $H_0 \sin \omega t$, the strength at depth z is

$$H = H_0 e^{-\alpha z} \sin(\omega t - \alpha z) \qquad (5.18)$$

where

$$\alpha = \frac{1}{z_0} = (2\pi\omega\sigma)^{1/2}$$

* E.g., Harnwell (1949), but note that Harnwell uses the MKS system and not the cgs system, which is followed here.

and z_0 is known as the skin depth, or depth at which the amplitude is reduced to $1/e$ of its surface value. In this problem the permeability is taken to be unity and neglected; field strength H (oersteds) is numerically equal to magnetic intensity or flux density B (gauss). Only at depths less than about 25 km can any known minerals be ferromagnetic (or ferrimagnetic); at greater depths the temperatures are well above their Curie points. It is to be noted that not only is the field at depth attenuated but that it has a phase lag relative to that at the surface. It follows that the induced currents modify the field observed outside the conductor, which can be represented as a sum of two fields, one of external origin and the other internally generated by induction, and that there is a phase difference between these two components. The phase difference is sensitive to any variation of conductivity with depth.

The plane wave approximation is not directly applicable to the Earth as a whole; here we consider a spherical conductor, in a simple case with uniform conductivity, but in a better approximation nonuniform but spherically symmetrical. Induction in a conducting sphere within the Earth by geomagnetic variations was considered in detail by S. Chapman and co-workers, the calculation of greatest interest being that of a sphere with radially varying conductivity (Lahiri and Price, 1939). The spherical symmetry of the problem allows the components of a disturbance field which are of external and internal origins to be separated by spherical harmonic analysis. Referring to Eq. 5.13 (or Eq. A.11 in Appendix A), the coefficients C_l^m and S_l^m describe the spatial variations of a field whose origin is internal to the Earth; $C_1'^m$ and $S_1'^m$ represent a field of external origin. In the analysis of the main geomagnetic field the primed coefficients are found to be zero, but in representing disturbance fields, i.e., diurnal variations, magnetic storms, and similar effects, the primed coefficients are larger because the disturbance fields are generated outside the solid Earth, in the magnetosphere or magnetic envelope to which the geomagnetic field is confined by its interaction with the interplanetary plasma. However, the induced currents within the Earth cause fields which, being of internal origin, are represented by the unprimed coefficients in Eq. 5.13. Thus both primed and unprimed coefficients are required to represent disturbance fields and the ratios $C_l^m/C_1'^m$, $S_l^m/S_1'^m$ give the ratio of internal (induced) field to external (inducing) field for any harmonic component. These ratios are determined by spherical harmonic analyses of disturbance fields, as are also the phase differences between the components of the field represented by primed coefficients and the corresponding unprimed coefficients.

The most reliable data and physically the simplest situation to analyze is the diurnal variation, which has a 24-hour period so that the phase difference between the induced and inducing fields appears as a phase difference in longitude as the diurnal magnetic wave travels around the Earth. The diurnal variation was given greatest weight in the analysis of Lahiri and Price (1939), who investigated the consequences of supposing that conductivity below a certain depth within the Earth could be represented by an inverse power of radius r (relative to the Earth radius a):

For

$$< qa, \qquad \sigma = k\left(\frac{qa}{r}\right)^m$$

$$r > qa, \qquad \sigma = 0$$

$$\text{(5.19)}$$

For a number of chosen values of m, they calculated values of q and k which gave agreement with the observed amplitude ratio and phase lag of external and internal diurnal fields. Then sets of values which did not also give agreement with magnetic storm data were discounted. The range of admissible values is represented by the extreme distributions d and e in Fig. 5.6. It is convenient to regard the intersection of these curves as an approximate fixed point on the scale of mantle conductivity, 1.5×10^{-12} emu at a depth of 600 km. The important conclusion is that at this depth conductivity increases rapidly with depth.

The difficulty caused by the presence of the oceans, whose conductivity is about 4×10^{-11} emu, becomes apparent when we consider the relatively low conductivity of the mantle. Although conduction in a spherical ocean of uniform thickness is calculable, the actual, irregular oceans present a very difficult problem at short periods, for which the ocean effect is important. From a study of geomagnetic variations at continental margins Parkinson (1964) concluded that the oceans appeared to cause greater deflections of geomagnetic disturbance fields than could be accounted for by the conductivity of the sea water alone; he suggested that the upper mantle was a better conductor under the oceans than under continents. This conclusion is of considerable interest in the discussion of the temperatures within the Earth (Chapter 9). Evidence of lateral inhomogeneity of electrical conductivity is also being studied over quite limited continental areas, where the ocean effect cannot be responsible (Kertz, 1964; Schmucker, 1964).

Both the ocean effect and upper mantle inhomogeneities become insignificant in the study of longer period geomagnetic variations, which penetrate deeper into the mantle. It appears reasonable to suppose that the seismological observation that the lower mantle appears more homogeneous than the upper mantle can be extended to conductivity observations. On this basis Banks and Bullard (1966) used the small annual variation in geomagnetic field to estimate the conductivity to be about 2×10^{-11} emu at a depth of about $0.2a = 1275$ km. Yukutake (1965) has extended the method of harmonic analysis to what is presumably the longest possible period, the 11-year cycle associated with solar activity, and found a conductivity of about 6×10^{-10} emu at a depth of 1600 km. These two values are shown as fixed points in the conductivity profile in Fig. 5.6, but it should be noted that they are both very approximate. Their estimation depended upon the separation of very small cyclic changes in field from much larger (and irregular) secular variations.

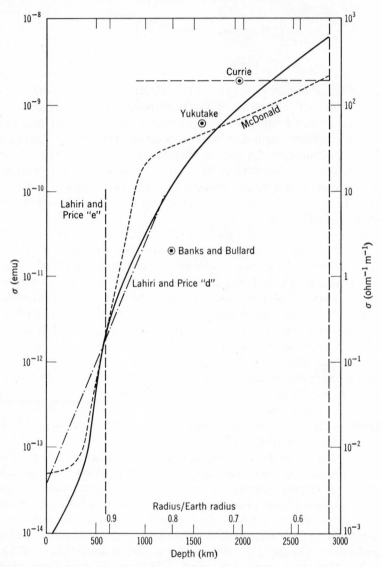

Figure 5.6: Electrical conductivity of the mantle. The author's preferred curve is shown as a solid line. Also shown are distributions by Lahiri and Price (1939) and McDonald (1957) and data by Yukutake (1965), Banks and Bullard (1966), and Currie (1968). Uncertainties are of order a factor of 10.

The penetration of the mantle by the geomagnetic secular variation is also an electromagnetic skin-effect problem, but in this case the source field and conductivity distribution are both unknown and we have only the observed secular variation. Estimates of mantle conductivity from the secular variation spectrum depend upon assumed forms for the secular variation at the surface of the core. As Runcorn (1955) showed, if the secular variation at the bottom of the mantle had a white spectrum and the observed diminution of higher frequencies at the surface of the Earth was due solely to attenuation by a layer of thickness L and uniform conductivity σ, then by Eq. 5.18 the amplitude spectrum at the surface has the form

$$H = H_0 \exp[-(2\pi\omega\sigma)^{1/2}L] \qquad (5.20)$$

By fitting the constants of this equation to a plot of the observed secular variation spectrum Runcorn (1955) obtained

$$\sigma L^2 = 1.1 \times 10^9 \text{ emu} \qquad (5.21)$$

so that if the layer is considered to have a thickness of 1000 km, its conductivity is 1.1×10^{-7} emu.

There are, however, several respects in which this analysis is too simple. First, the sources of secular variation are localized and are attenuated with distance purely geometrically as well as electromagnetically. Although we do not know the actual secular variation spectrum at the core, we can be sure that the higher-frequency components are relatively more localized and therefore that, even without electromagnetic screening the higher frequencies will appear more attenuated at the surface. The preceding estimate of conductivity is therefore certainly too high. These difficulties were overcome in a more sophisticated calculation by McDonald (1957), who assumed a spatially random distribution of secular variation sources at the core surface and allowed for geometrical spreading as well as electromagnetic screening. McDonald assumed a power law variation of conductivity with radius in the lower mantle similar to that assumed by Lahiri and Price (1939) for the upper mantle. In his final conductivity profile, reproduced in Fig. 5.6, McDonald joined his lower mantle conductivity curve to an upper mantle curve intermediate between the Lahiri and Price d and e conductivity distributions. It seems probable that McDonald's curve is valid within a factor of about 10. Currie (1968) reconsidered the secular variation spectrum and favored a rather higher mean value for the conductivity of the lower mantle (2×10^{-9} emu) than did McDonald.

The final piece of evidence of conduction in the mantle is the electromagnetic core-mantle coupling. This was considered in detail by Rochester (1960) on the basis of E. C. Bullard's model of the geomagnetic field. To obtain a coupling time constant as short as 10 years, Rochester found it necessary to assume a mean conductivity in the lower mantle appreciably greater than McDonald's preferred value but still well within the range of uncertainties. The coupling problem is discussed in Section 2.5, where it is suggested that the coupling time constant is appreciably shorter even than 10 years. For this reason the author's preferred curve (solid line in Fig. 5.6) is steeper in the lower mantle than the McDonald curve. The greater conductivity at the bottom of the mantle gives enhanced coupling, but as far as attenuation of secular variation is concerned, some compensation is allowed in the middle of the mantle.

The range of conductivities represented in Fig. 5.6 is clearly indicative of semiconduction processes, whose relevance to the mantle has been reviewed by Tozer (1959). The basic physics of semiconduction is discussed in texts on solid state physics (e.g., Kittel, 1966), and it will suffice here to quote the significant results.

Semiconductors are characterized by the existence of an energy gap between the highest filled (valence) electron states and the next available states (the conduction band). An electron in a completely full band cannot give electrical conduction because to conduct it must respond to an applied electric field, i.e., it must change from one state to another with greater momentum in the direction of the field; if there are no vacant states accessible to it, it cannot make this change and a material in which this is so for all electrons is an insulator. However, in a semiconductor the energy gap E_g between the filled valence states and empty conduction states is sufficiently small, relative to the available thermal energy kT at temperature T that a few electrons are excited into the conduction band. The conductivity is proportional to the number of electrons thus excited and is a strong function of temperature. A material in which this occurs is an intrinsic semiconductor. Normally, however, impurities in a simple semiconductor give rise to additional electron energy levels within the gap. These impurity levels may be occupied by electrons at low temperatures, in which case they are known as donor levels, because they give electrons into the conduction band more readily than does the valence band; alternatively, the impurity levels may be vacant at low temperatures and are then known as acceptors, because electrons may be excited into them from the valence band. Conduction may occur either by virtue of electrons in the conduction band of "holes" in the valence band, or both. The situation is represented diagrammatically in Fig. 5.7.

Both the intrinsic and extrinsic mechanisms are processes of electronic conduction. Also possible at high temperatures is ionic conduction, in which ions move bodily under the influence of an electric field. All three processes are described by very similar equations. Conductivity is proportional to the number of charge carriers, which increases with temperature according the Boltzmann distribution. Thus the number of intrinsic charge carriers is

$$n_i = A \exp\left(-\frac{E_g}{2kT}\right) \qquad (5.22)$$

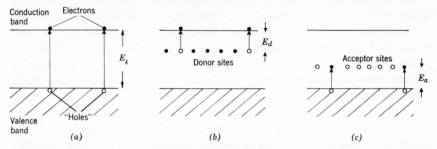

Figure 5.7: Illustration of energy levels in intrinsic and extrinsic semi-conductors. Since the energy gaps, E_d and E_a for the excitation of electrons to or from impurity levels, are smaller than the energy gap between the valence and conduction bands, extrinsic (impurity) conduction occurs more readily than intrinsic conduction at low or moderate temperatures. However, at high temperatures impurity conduction saturates by virtue of the finite number of available states and intrinsic conduction takes over. (a) Intrinsic semiconductor. (b) n-type extrinsic (or impurity) semiconductor. (c) p-type extrinsic (or impurity) semiconductor.

where A is a factor which depends upon temperature (as $T^{3/2}$ for simple band forms) and is of order 10^{20} cm^{-3}. Allowing for the saturation of extrinsic conduction due to the finite number N of impurity sites, the number n_e of extrinsic charge carriers is given by

$$n_e = (N - n_e)^{1/2} A' \exp\left(-\frac{E}{2kT}\right) \qquad (5.23)$$

where E is E_d or E_a from Fig. 5.7 and A' is a constant equal to $A^{1/2}$ in the absence of any differences in effective masses of the charge carriers. We may assume mobilities of electrons or holes to be determined primarily by phonon scattering, as in metallic conduction, in which case they vary approximately as T^{-1}; the temperature dependence of conductivity in both intrinsic and extrinsic cases is thus dominated by the exponential terms. The temperature dependence of ionic conductivity σ_3 is similarly represented by a Boltzmann term with an activation energy, in this case the energy of ionic diffusion Q. The general form of conductivity as a function of temperature is therefore a sum of three exponential terms, representing intrinsic, extrinsic, and ionic conduction, respectively:

$$\sigma = \sigma_i + \sigma_e + \sigma_3 = \sigma_{i_0} \exp\left(-\frac{E_g}{2kT}\right) + \sigma_{e_0} \exp\left(-\frac{E}{kT}\right) + \sigma_{3_0} \exp\left(-\frac{Q}{kT}\right)$$

$$(5.24)$$

Normally in any limited temperature range one of the terms in Eq. 5.24 is dominant, but over the extreme range of temperatures in the mantle all three processes must be taken into account. The problem is further complicated by the fact that the activation energies are functions of pressure. In the outer 40 km of the earth (i.e., below 1000°C), extrinsic conductivity must be dominant and virtually all laboratory measurements of conductivities of rocks and minerals at elevated temperatures and pressures have been concerned mainly with extrinsic conduction. Below this level intrinsic and ionic conduction become significant; the latter is probably dominant initially but its activation energy Q increases with pressure (the ions are held more firmly in place in the crystal lattices at high pressures), so that in the deeper parts of the mantle ionic conduction is inhibited.* The relatively high conductivity of the lower mantle is probably due mainly to intrinsic electronic conduction, although we can be by no means certain of this because "impurity" concentrations may be very high indeed in the mantle by comparison with normal semi-conductor experience—so much so that the distinction between intrinsic and extrinsic conductivity may be meaningless.

The work of Lahiri and Price (1939) indicated that conductivity rose steeply in the mantle at about 600 km and this has led to the suggestion that the rise is due largely to the transitions to denser phases which occur between about 350 and 800 km. The McDonald conductivity distribution (Fig. 5.6) emphasizes this, indicating that a conductivity rise by a factor of order 1000 is associated with the transitions. Akimoto and Fujisawa (1965) measured the electrical conductivities of olivine (low-pressure) and spinel (high-pressure) forms of Fe_2SiO_4 and found the denser, spinel form to be more highly conducting by a factor of order 100. The change was attributed to a reduction in band energy gap by the greater compaction, but the conduction which they observed was associated with quite a low activation energy and was evidently extrinsic in character. Although it is reasonable to suppose that the instrinsic energy gap E_g decreases with both steady compression and phase transformation of mantle material to high density form, there is no direct evidence to support this supposition. But we can appeal to the estimated conductivity at the bottom of the mantle, which is presumed to be predominantly intrinsic:

$$\sigma = \sigma_{i_0} \exp\left(-\frac{E_g}{2kT}\right) \tag{5.25}$$

where σ_{i_0} is a slight function of E_g; by extrapolation from observations on germanium and silicon the value $\sigma_{i_0} = 7 \times 10^{-6}$ emu (7000 ohm^{-1} cm^{-1} or 7×10^5 ohm^{-1} m^{-1}) is obtained. Then if the conductivity is taken to be 6×10^{-9} emu (600 ohm^{-1} m^{-1}) and the temperature 3700°K at the bottom

---

* Tozer (1959) discusses this point in detail but he assumes a value of σ_{i_0} which is much lower than that obtained by extrapolation from observations on germanium and silicon; this makes ionic conductivity appear to be favored over a greater range of depths than in the present discussion.

of the mantle (see Chapter 9), the activation energy is found to have the surprisingly large value 4.4 ev.* Lower assumed conductivity gives an even higher activation energy. Although there is considerable uncertainty in this estimate, it gives no support to the contention that the band gap is narrowed by the phase transition to close-packed (spinel) minerals. The matter deserves further investigation, but the inflections in McDonald's conductivity profile of the mantle have been eliminated from the smooth "preferred" curve in Fig. 5.6.

5.4   Generation of the Main Field

Study of the secular variation reveals that the features of the main field move both with respect to the surface of the Earth and with respect to one another. If the average westward drift is taken to be representative of the rate of motion, then the velocities observed at the surface are of order 20 km year^{-1}. This is a million times faster than prolonged large-scale motion evident from geology and virtually precludes the participation of the solid part of the Earth. Since the large-scale features of the field indicate a deep internal origin and seismological evidence (Chapter 4) favors a fluid core, we can reasonably suppose the secular variation to be a consequence of core motions. Further, geochemical and density considerations (Sections 1.5 and 4.5) are consistent with a core composition largely of iron, a good electrical conductor. A moving fluid conductor and a magnetic field exercise a mutual control on one another. The study of this problem is known as magnetohydrodynamics or hydromagnetics, and is important not only in connection with the magnetic fields of the Earth and other astronomical bodies, but also in laboratory plasma physics, in which the conductor is a tenuous, highly ionized gas. The concept is the same, although the problem of the Earth's core is in one respect simpler—it is virtually incompressible.

Magnetohydrodynamic theories of the Earth's field are developments of an idea, due originally to J. Larmor, who pointed out that suitable internal motion of a large fluid conductor could cause it to act as a self-exciting dynamo. There are still many unknowns and no completely adequate theory, but the principles appear to be understood and there is no promising alternative to magnetohydrodynamic dynamo action as the origin of the Earth's main field. Elsasser (1950, 1956), Inglis (1955), Hide and Roberts (1961), and Rikitake (1966c) have reviewed the subject.

The necessity for a mechanism of regeneration of the field is seen by making an order-of-magnitude calculation of the time-constant for free decay of electric currents in the core. We can consider the simple case of a core-sized

---

* The energy gap determined from the ultraviolet absorption edge in olivine (the low-pressure type mineral for the mantle) is nearer to 3 ev.

conductor in the form of an anchor ring, to which standard formulas for inductance L and loop resistance R may be applied (see, for example, Harnwell, 1949 p. 330). Assuming that the mean radius of the loop is r_1 and the radius of its cross section is r_2, then, approximately,

$$L = 4\pi r_1 \left[\ln\left(8\,\frac{r_1}{r_2} \right) - \frac{7}{4} \right] \text{ emu} \tag{5.26}$$

$$R = 2\rho\,\frac{r_1}{r_2^{\,2}} \text{ emu} \tag{5.27}$$

where resistivity $\rho \approx 3 \times 10^6$ emu* in the core. The time constant for decay of current in the loop is then

$$\tau = \frac{L}{R} = \frac{2\pi}{\rho}\,r_2^{\,2} \left[\ln\left(8\,\frac{r_1}{r_2} \right) - \frac{7}{4} \right] \text{ sec} \tag{5.28}$$

If the loop just fits within the boundaries of the outer, fluid core, whose inner and outer radii are 1300 km and 3500 km, we have $r_1 = 2.4 \times 10^8$ cm, $r_2 = 1.1 \times 10^8$ cm, so that $\tau \approx 3 \times 10^{10}$ sec $= 10^3$ year. This is a rough estimate, based on a simplified model of core currents. However, it is compatible with the present-day observations of secular variation, which show that the dipole moment of the Earth is changing by 0.05 % year^{-1}, a rate which could be represented by a time constant of 2000 years. It shows that the geomagnetic field cannot be a vestige of the Earth's early history, several thousand million years ago, but it must be continuously regenerated. By Eq. 5.28 the time constant for decay of the current in a loop of a particular shape [i.e., fixed dimension ratio (r_1/r_2)] is proportional to the square of the size. Thus we expect the features of the nondipole field to be subject to more rapid decay.

The rate of decay of the field has been calculated for ohmic dissipation by core currents. This is almost certainly the dominant dissipation mechanism because viscous drag must be negligible for such a large, slow-moving body, assuming any reasonable value of viscosity. However, we cannot calculate a realistic rate of ohmic dissipation without a better picture of the current patterns. Maintenance of the observed poloidal field requires the presence of additional fields of toroidal form, which would be produced, for example, by currents having the geometrical form of wire wound on to an anchor ring. Toroidal magnetic fields are confined to the current-carrying conductors in

---

* See Appendix B for conversion of units. The value of resistivity used here is the one preferred by the author (Stacey, 1967d; see also Section 5.3) and is 10 times higher than the value usually assumed. For this reason the above calculation leads to a time constant 10 times shorter than that usually quoted.

which they are generated, so their existence is deduced and cannot be observed at the Earth's surface, but they are very important to the problem of energy dissipation: the geomagnetic dynamo theories of Elsasser (1950, 1956) and Bullard (1949; see also Bullard and Gellman, 1954) require toroidal fields 10 to 100 times as strong as the poloidal fields and consequent dissipations of order 10^2 to 10^4 times as great. With this in mind we can calculate the ohmic dissipation of a simple toroidal current (producing a poloidal field) in the form of a ring with the dimensions of the core. The mean area of the current loop is

$$A = \pi r_1^2 \tag{5.29}$$

and the magnetic moment of the Earth is

$$M = iA = 8 \times 10^{25} \text{ emu} \tag{5.30}$$

from which the current i is

$$i = \frac{M}{\pi r_1^2} = 4.4 \times 10^8 \text{ emu} = 4.4 \times 10^9 \text{ amp} \tag{5.31}$$

Then, using the resistance of the loop, as given by Eq. 5.27, the rate of energy dissipation is

$$\frac{dE}{dt} = i^2 R = \frac{2M^2 \rho}{\pi^2 r_1^3 r_2^2} \tag{5.32}$$

Applying Eq. 5.32 to the single loop required to give the dipole field, we obtain $dE/dt = 2.3 \times 10^{16}$ ergs sec^{-1}. It is not possible to separate the ohmic dissipations of the currents producing the dipole and nondipole fields, but as an approximation, corresponding expressions for the largest possible, i.e., least dissipative, current loops which could fit into the outer core to produce the radial dipoles of Alldredge and Hurwitz (Table 5.1) give a total dissipation equal to that of the main loop. The minimum possible value of the total ohmic losses by currents producing the observed field is therefore 5×10^{16} ergs sec^{-1}. Since an energy dissipation of 10^{19} ergs sec^{-1}, or perhaps even more, would be entirely compatible with other geophysical observations,[*] there is evidently no difficulty in accommodating ohmic dissipation associated with toroidal magnetic fields about 10 times as strong as the observed poloidal field. However, toroidal fields 100 times as strong are not acceptable. It should be noted that this conclusion is a consequence of the value chosen for core resistivity.

---

[*] We require the core dissipation to be substantially less than the geothermal flux at the surface, 3×10^{20} ergs sec^{-1}. The energy balance of the core is considered in Section 9.4.

A more detailed discussion of the geomagnetic dynamo requires an explicit assumption concerning the ultimate driving mechanism. The theories of W. M. Elsasser, E. C. Bullard, and others assume that convective motion of the core is responsible. This demands a substantial heat flux from the core into the mantle, which is itself a problem to explain. As noted in Section 1.5 it is most unlikely that the core contains appreciable radioactivity and Verhoogen (1961) postulated that the necessary heat was derived from the latent heat of progressive solidification of the inner core (considered further in Section 9.4). No problem of power source arises if the Earth's rotation provides the driving mechanism; the possible availability of 10^{19} ergs sec^{-1} of rotational energy is pointed out in Section 2.4. Bullard (1949) noted that, since the core is ellipsoidal, if it does not follow the mantle precession exactly, then internal motions must occur and that these could be more than adequate to drive the geomagnetic dynamo. However, he assumed that the core does follow the mantle precession and thus favored convection as the power source for the dynamo. Malkus (1963, 1968) reconsidered the problem of the different precessional torques acting on the core and mantle and concluded that they provided a satisfactory dynamo mechanism, whereas the thermodynamic efficiency of the convective process was only very doubtfully adequate. This conclusion is strongly reinforced by the revised estimate of resistivity used here (Stacey, 1967d). Another alternative, which must be allowed, is that the dynamo currents are of thermoelectric origin, although the adequacy of this mechanism has not been demonstrated. At the present time no final choice can be made but several general considerations can be noted.

The dominance of the dipole field and its axial character (at least when averaged over 10^4 years or more) require that the Earth's rotation exercise a controlling influence. In the convection theories this is accomplished by the action of the Coriolis force on the core motions. We can assume either that the dipole field results from a Coriolis-controlled statistical bias of smaller-scale core motions (convective cells) or that large-scale motion responsible for the dipole field (such as that generated by the precession effect) is unstable and breaks up into eddies (the nondipole field).

The general principles of magnetohydrodynamic dynamo action are used also to explain the magnetic fields of stars, including the Sun, and the giant planets Jupiter and Saturn, although the basic driving mechanism may be different in these cases. The magnetic field of the Sun has been studied quite intensively, using the Zeeman splitting of spectral lines. It is locally intensified in sunspots and is generally very variable (Babcock and Babcock, 1955). The abundance of thermal energy makes the convective type of dynamo action most probable in the Sun.

Formal theories of magnetohydrodynamics (see, for example, Elsasser, 1956) express in a more general way the conclusion represented by Eq. 5.28:

a magnetic field can diffuse out of an electrical conductor with a time constant proportional to the square of its dimensions l and to its conductivity (or reciprocal of resistivity ρ). The characteristic time constant is

$$\tau = \frac{l^2}{\eta} \tag{5.33}$$

where η is termed the magnetic diffusivity and is given by

$$\eta = \frac{\rho}{4\pi\mu} \tag{5.34}$$

the permeability μ being very close to unity, except in ferromagnetics, with which we are not concerned here. Taking $\rho = 3 \times 10^6$ emu, we have $\eta = 2.4 \times 10^5$ cm^2 sec^{-1} and, for the whole core, $\tau \approx 10^3$ yr,* as already noted. The time constant represented by Eq. 5.33 can be compared with the characteristic time τ' for the internal motion of a fluid conductor of dimension l with velocity differences v:

$$\tau' = \frac{l}{v} \tag{5.35}$$

Then if $\tau' \ll \tau$, i.e., if

$$lv \gg \eta \tag{5.36}$$

the field does not have time to diffuse out of the fluid and is carried along (and deformed) by the fluid motion. The field is then said to be "frozen in" to the conductor. This effect is basic to all magnetohydrodynamic theories of the geomagnetic dynamo. That it can lead to an intensification of a magnetic field is shown in Fig. 5.8. The applicability of the inequality (5.36) to the Earth's core is demonstrated by considering the features of the nondipole field to originate in core volumes of order 1000 km in radius with internal velocities of about 10 km yr^{-1}(3×10^{-2} cm sec^{-1}), corresponding to the extrapolation of the westward drift down to the core, so that

$$lv \approx 3 \times 10^6 \text{ cm}^2 \text{ sec}^{-1} \approx 12.5 \, \eta \tag{5.37}$$

Roberts and Scott (1965) have emphasized that, for this reason, the secular variation must be due primarily to core motion and not diffusion of field lines.

---

* Taking l equal to the half-thickness of the outer core.

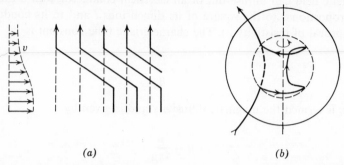

(a)   (b)

Figure 5.8: (a) The deformation of a magnetic field by a velocity shear intensifies the field (the strength of the field being represented by closeness of the lines of force). (b) Differential rotation of the inner and outer parts of the core, which is a consequence of convection, draws out the field lines of an initial poloidal field to produce a toroidal field. Both figures after Elsasser (1950).

The inequality represented by Eq. 5.36 is frequently represented in terms of a "magnetic Reynolds' number" R_m:

$$R_m = \frac{lv}{\eta} \qquad (5.38)$$

which must be substantially greater than unity for regenerative dynamo action to occur. $R_m = 10$ is commonly regarded as the lower limit for dynamo action so that the scale of the features of the nondipole field has a natural explanation. Equation 5.37 gives $R_m \approx 12.5$ for features of the nondipole field, assuming that the fluid velocity v is equal to the observed velocity of the eddies or features of the nondipole field. The fluid velocity could, of course, be greater, in which case $R_m > 12.5$ and the inequality (5.36) is even more strongly satisfied. However, there is no evidence for features of the nondipole field smaller in scale (at core level) than 1000 km or so and it appears possible that any smaller features would not be self-sustaining by virtue of an inadequate magnetic Reynolds' number, in which case the fluid velocity is unlikely to exceed greatly the eddy velocity.

Theoretical models have been devised to show that self-exciting dynamo action is possible in spherical conducting fluids with suitable internal motions; of particular interest is a laboratory model by Lowes and Wilkinson (1963, 1967, 1968), which demonstrates the principle, using metal cylinders rotating inside a metal casting, in the manner of Fig. 5.9a. Electrical connection of the cylinders to the casting was accomplished by filling the spaces with mercury. To scale their dynamo to laboratory size with attainable speeds of rotation, Lowes and Wilkinson used a ferromagnetic metal of high permeability, but this is merely a necessary experimental technique and was not responsible for the sudden onset of a self-excited field at a critical speed of rotation, as shown in Fig. 5.9*b*.

The essential feature of the Lowes-Wilkinson model is the feedback between the two magnetically connected cylinders. An initial axial field in one causes it to act as a homopolar disk, generating an emf between its axis and periphery. The consequent circulating current produces a magnetic field of toroidal form which links with cylinder 2, from which a further current flow is initiated. This produces a second toroidal field linking with cylinder 1 and reinforcing the initial field if the rotations are in the correct sense, as shown in Fig. 5.9*a* or both reversed. For both of the correct combinations of cylinder rotations, a field of either sign may be self-maintained,

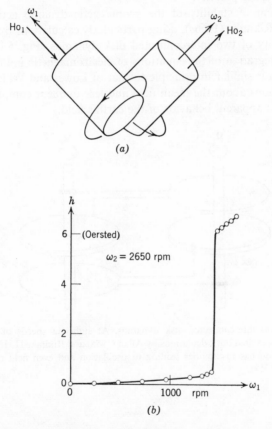

(a)

(b)

Figure 5.9: (*a*) Rotating cylinders in the Lowes-Wilkinson laboratory model of geomagnetic dynamo action. (*b*) Externally observed field as a function of rotational speed of cylinder 1, with cylinder 2 maintained at a fixed speed. Reproduced, by permission, from Lowes and Wilkinson (1963).

the sign being determined by an initial, stray excitation. This is relevant to apparent alternations of polarity of the geomagnetic field—a field of either sign may be generated with equal probabilities. This principle of regeneration operates similarly in the Bullard and Gellman (1954) theory of the geomagnetic dynamo, in which the two motions are (1) radial, convective movements in the core and (2) consequent differential rotation of the inner and outer parts of the core due to conservation of angular momentum (as in the Bullard theory of the westward drift, considered in Section 5.2). Inglis (1955) devised a visual analogue of the Bullard process, which, being based upon plausible core motions, is somewhat more complicated than the Lowes-Wilkinson model, but is similar in principle and, like all self-excited dynamos, it can be excited with either polarity.

The problem of stability of the geomagnetic dynamo action has been reviewed by Rikitake (1966c), using particularly calculations by Allan (1958) on the stability of two interconnected disk dynamos (Fig. 5.10). Repeated numerical integration of the equations of electromagnetic induction for this model, which is similar in principle to that of Lowes and Wilkinson (1963), show oscillations about the mean field and less frequent complete reversals, not unlike the apparent behavior of the earth's field.

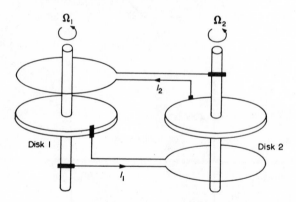

Figure 5.10: Two interconnected disk dynamos. At sufficient speeds of rotation these dynamos are self-excited but calculations by Allan (1958) and Rikitake (1958) show that the current generation has instabilities leading to oscillation and even field reversal. Figure after Rikitake (1958).

Chapter 6

PALEOMAGNETISM—PREHISTORY OF THE EARTH'S MAGNETIC FIELD

"If the facts are correctly observed there must be some means of explaining and coordinating them ...."

BULLARD, 1965, p. 323.

6.1 Introduction

The direct record of the secular variation of the geomagnetic field extends over about four centuries to the present time, with comprehensive cover of the Earth's surface only very recently achieved. While the present observations allow us to infer a great deal about processes occurring within the Earth, we add another dimension to the subject by expanding the time scale of observation back by a factor of 10^6 or so to the remote geological past. This is paleomagnetism or "fossil magnetism," the study of the ancient magnetic field by measurement of the magnetizations of rocks.

As must be expected, we obtain results of geological interest when we consider the field on this extended time scale. Paleomagnetism has given quantitative evidence on problems previously subject to qualitative discussion only and, in the case of continental drift, to violent disagreement. It has caused a revolution in geophysical thinking which should not be underestimated. Convection within the solid mantle of the Earth is now a subject of intense interest and much effort is being expended in comprehending surface features, crustal and upper mantle structures, satellite geoid data, and heat flow in terms of deep-seated convective cells responsible for continental drift. It is doubtful whether much of this would be taken seriously if paleomagnetism had not already established the respectability of the drift as a scientific hypothesis.

A useful review and analysis of paleomagnetism is by Cox and Doell (1960) and a comprehensive treatment with a compilation of data obtained

157

up to 1963 has been given by Irving (1964). Reviews incorporating more recent data are by Creer (1965), McElhinny et al. (1968) and McElhinny (in press). The basic physical concepts of rock magnetism are discussed in Nagata's *Rock Magnetism* (1953, 1961); and theoretical aspects have been reviewed by Néel (1955) and Stacey (1963a).

6.2 Magnetic Minerals in Rocks

Most rocks contain a small percentage (0.1 to 10%) of iron oxide and sulfide minerals which have ferromagnetic or, more correctly, ferrimagnetic properties. These minerals occur as small grains dispersed through the magnetically inert, i.e., paramagnetic or diamagnetic, matrix provided by the more common silicate minerals which make up the bulk of the rocks. The physical study of rock magnetism is thus concerned with the properties of individual grains of strongly magnetic material which are sufficiently well dispersed to be magnetically independent of one another. They are normally so diluted by the nonmagnetic minerals that the magnetism of rocks is slight, but it is readily measurable in most cases. Since the magnetic properties of rocks are due to ferromagnetic components, rocks have all of the normal properties of ferromagnetics—coercivity, remanence, magnetostriction, etc. It is the property of remanence which is of interest in paleomagnetism.

The mineralogy of rock magnetism has been reviewed by Nicholls (1955). It is complicated by a multiplicity of phases and solid solutions of iron oxides, particularly with titanium dioxide; most magnetic minerals are within the ternary system $FeO—Fe_2O_3—TiO_2$. For many purposes it is sufficient to distinguish two types of mineral: the strongly magnetic cubic oxides magnetite (Fe_3O_4), maghemite (γ-Fe_2O_3), and the solid solutions of magnetite with ulvospinel (Fe_2TiO_4), which are known as titanomagnetites; and the more weakly magnetic, rhombohedral minerals based on hematite (α-Fe_2O_3) and its solid solutions with ilmenite ($FeTiO_3$). Pyrrhotite (FeS_x, $1 < x < 2.14$) is magnetically similar to hematite.

Magnetite is representative of the cubic minerals with spontaneous magnetizations comparable to the familiar ferromagnetic metals (Fe, Co, Ni). The Fe^{2+} and Fe^{3+} ions are arranged interstitially in a face-centered cubic oxygen lattice on two types of lattice site, A in fourfold coordination with the oxygen ions and B in sixfold coordination, such that antiferromagnetic coupling between the A and B ions causes antiparallel alignment of their magnetic moments. But the B ions are twice as numerous as the A ions, so that the lattice has a strong spontaneous magnetization (480 emu cm^{-3}), as in commercially developed ferrites. Hematite is representative of the more weakly magnetic, uniaxial minerals, in which the oppositely magnetized sublattices of interacting Fe^{3+} ions are equally balanced, i.e., antiferromagnetic, but canted at a small angle to give a slight spontaneous magnetization perpendicular to the ion moments, as in Fig. 6.1. Its spontaneous magnetization (2.2 emu cm^{-3}) is confined to the basal plane except at temperatures below $-25°C$.

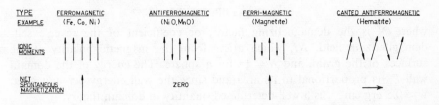

Figure 6.1: Four basic patterns of alignment of atomic magnetic moments by mutual interaction. In paramagnetics the interactions are too weak to cause mutual alignment, and in diamagnetics there are no atomic moments in the absence of a magnetic field.

The strong spontaneous magnetization of magnetite gives rise to magnetostatic forces which cause the magnetic structure to be divided into domains, locally magnetized to saturation but, except in the case of extremely fine grains, arranged to minimize the magnetic moment of the whole body of magnetite. The principles are well understood in terms of ferromagnetic domain theory, as reviewed by Kittel (1949) and applied to rocks by Stacey (1963a), although the grain size is normally so small that ferromagnetic domains in rocks have rarely been subject to direct experimental observation. The multiplicity of domains depends upon grain size in the manner of Fig. 6.2; the properties of the single domain and multidomain structures being essentially different, it is of interest to calculate the critical size for the transition from the single domain to two domain structures.

Following Kittel (1949), the magnetostatic energies E_1, E_2 of a grain, of diameter d [volume $(\pi/6)d^3$] and saturation magnetization I_s, in the single domain and two domain states (Fig. 6.2a and b), are (assuming that the

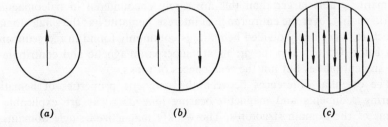

(a) $\qquad\qquad\qquad$ (b) $\qquad\qquad\qquad$ (c)

Figure 6.2: Domain structures of magnetic grains. (a) Single domain. (b) Two-domain grain. (c) Multidomain. Large multidomains may have a more complex subdivision than is indicated by (c). Generally magnetite behaves as (c) and hematite as (a).

domain wall represents a sharp transition between the two domains):

$$E_1 = \tfrac{1}{2}NI_s^2\left(\frac{\pi}{6}d^3\right)$$

$$E_2 = \tfrac{1}{2}E_1 \text{ approximately}$$

(6.1)

where N is the demagnetizing factor, or coefficient of the reverse self-demagnetizing field, NI, which arises from the magnetic polarity at the surfaces of the grain, and $N = \tfrac{4}{3}\pi$ for a sphere. The energy of the domain wall E_w is proportional to its area and since the wall energy per unit area, $w \approx 0.5$ erg cm^{-2}, is a well-determined quantity in domain theory,

$$E_w = \frac{\pi}{4}d^2w$$

(6.2)

At the critical grain diameter d_0 for transition between the two states of a spherical grain, their energies are equal:

$$E_1 = E_2 + E_w$$

(6.3)

from which

$$d_0 = \frac{9}{2\pi}\frac{w}{I_s^2}$$

(6.4)

Substituting numerical values for magnetite we obtain $d_0 = 3 \times 10^{-6}$ cm (0.03 μ). Grains of this size are of little interest because they are so small that they are superparamagnetic, i.e., they are unable to retain remanence because their magnetic moments are continuously reoriented thermally and they behave as classical paramagnetics. The magnetite grains normally observed in rocks are larger than this and are therefore multidomains. However, the very much smaller spontaneous magnetization of hematite gives a much larger value of d_0 by Eq. 6.4. In this case the properties are less well understood and we do not have as reliable a value of w, but it is unlikely to be grossly different from 0.5 erg cm^{-2}, in which case we obtain $d_0 = 0.15$ cm. Hematite grains larger than this are rarely encountered in paleomagnetic work, so that, over the entire range of interest, hematite has the single domain structure. Very finely divided hematite is commonly found in red sediments, but below 0.5 to 1.0 μ the grains are superparamagnetic and contribute to the susceptibilities but not the remanences of rocks.

The essential differences between the magnetic properties of hematite-bearing sediments and magnetite-bearing igneous rocks are explicable in terms of the domain structures. The weakly magnetized single domains of hematite are more highly coercive than the strongly magnetized multi-domains of magnetite. However, the vital property of thermoremanence, the acquisition of remanence by cooling in a small field, is similar in the two cases.

6.3   Remanent Magnetism in Rocks

The remanence acquired, in the time scale of laboratory experiments, by a rock exposed to a magnetic field of order 0.5 oe at room temperature is quite negligible. Remanence is produced by changes in domain structure, rotation of the magnetic moments in single domains, or domain wall movements in multidomains. These changes are opposed by potential barriers which can be overcome only by the application of fields comparable to the coercive force, 20 to over 100 oe for magnetite (Parry, 1965) and usually several hundred oersteds for fine-grained hematite. However, as the temperature is raised, the barrier energies decrease and the energy of thermal agitation becomes sufficient to cause spontaneous changes in domain structure and the magnetization approaches a state of dynamic equilibrium in the external field. A magnetization can thus be induced reversibly at high temperature but becomes a remanence when the rock cools and the potential barriers become effective. This is known as thermoremanent magnetization, commonly abbreviated to TRM. The mechanisms by which TRM is acquired by ideal single domains and ideal multidomains are quite different. Both are explicable in terms of simple theories, but ideal cases are rare and in most rocks thermoremanence has some features of both basic types.

In a multidomain grain the potential barriers opposing changes in remanence arise from crystal imperfections, which cause local stresses and variations in spontaneous magnetization so that the energy of a domain wall is a function of its position, as in Fig. 6.3. The wall moves in discrete (Barkhausen) jumps from one potential minimum to the next, each jump giving rise to a change in the remanent magnetization of the grain. As the temperature is raised the spontaneous magnetization I_s decreases and the potential barriers, which are generally proportional to I_s^2, decrease more strongly, both becoming zero at the Curie point (580°C for pure magnetite). At a temperature below the Curie point, known as the *blocking temperature*, thermal energy is just sufficient to move the domain walls across the potential barriers; when a grain is cooled below its blocking temperature, the magnetization at that temperature becomes a TRM.

In rock magnetism we are interested mainly in very small grains, which have high coercivities and acquire stable remanences. In such grains there are only a few stable positions for each domain wall, as in Fig. 6.3, and in general none of these positions corresponds to zero magnetic moment for the grain. Thus a small multidomain grain cannot be demagnetized, but must always have a certain minimum magnetic moment determined by the Barkhausen discreteness in the positions of its domain walls. The moment may be reversed, for example, by moving the wall from B to C in Fig. 6.3, but it cannot be destroyed because there is no stable position for the wall between B and C. Thus the grain behaves very similarly to single domain grains, whose moments

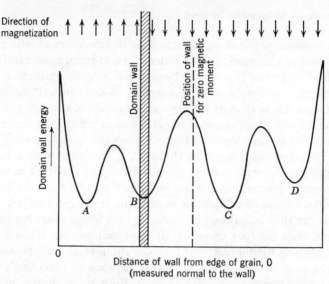

Figure 6.3: Energy of a domain wall as a function of its position in a small magnetic grain. The stable positions of the wall are at the potential energy minima *A, B, C, D*.

may be reversed but not destroyed.* Such grains are termed pseudo single-domain grains. Their thermoremanence appears to dominate the TRM of rocks which are of interest in paleomagnetism (Stacey, 1967a); the mechanism of acquisition for TRM can be described in terms of the theory for single domains.

The theory of thermoremanence in single domains is due to Néel (1955). We consider the simple case of an assembly of identical grains whose magnetic moments may be oriented either one way or in the opposite direction but not between. For true single domains this means that we assume a grain anisotropy parallel for all grains; in the case of small multidomains we assume their domain structures are all oriented the same way. At the blocking temperature T_B, which is assumed to be the same for all grains, thermal agitation causes spontaneous reversal of the moments μ_B and the net alignment of moments in a field H (parallel to the magnetic axis) is given by the Boltzmann probabilities P_+ and P_- for finding a particular grain parallel or antiparallel to H:

$$P_+ = \frac{\exp(\mu_B H/kT_B)}{\exp(\mu_B H/kT_B) + \exp(-\mu_B H/kT_B)} \tag{6.5}$$

$$P_- = \frac{\exp(-\mu_B H/kT_B)}{\exp(\mu_B H/kT_B) + \exp(-\mu_B H/kT_B)} \tag{6.6}$$

* The moments do, of course, disappear on heating above the Curie point, but they re-appear spontaneously on recooling.

If there are N such grains per unit volume of rock, then its magnetization at the blocking temperature is

$$I_B = N\mu_B(P_+ - P_-) = I_{SB} \tanh\left(\frac{\mu_B H}{kT_B}\right) \tag{6.7}$$

where $I_{SB} = N\mu_B$ is the saturation magnetization at the blocking temperature. On cooling to room temperature all of the magnetic moments remain oriented as at the blocking temperature and the magnetization increases by the factor I_{SO}/I_{SB}, where I_{SO} is the room temperature saturation magnetization, so that the thermoremanence is

$$I_{TRM} = I_{SO} \tanh\left(\frac{\mu_B H}{kT_B}\right) \tag{6.8}$$

The general case of a randomly oriented assembly of grains leads to an integration solvable only in terms of an infinite series but the general behavior of I_{TRM} is the same.

Equation 6.8 is the basic equation in the theory of TRM. Generally, but not invariably, $(\mu_B H)$ is sufficiently small in the field range of interest (0–1 oe) for the linear approximation to be valid:

$$I_{TRM} = I_{SO} \frac{\mu_B H}{kT_B} \tag{6.9}$$

The parameters μ_B and T_B can be expressed in terms of the temperature dependence of spontaneous magnetization and the energies of the barriers opposing reversal of the grain moments, but here the theories diverge into alternative models appropriate to different grain sizes. In particular, we are less interested in true single domains than in the pseudo single-domain moments of small multidomains. However, the important common feature is that below the blocking temperature the relaxation time for spontaneous decay of TRM increases very rapidly with decreasing temperature. If we represent the energy barrier by E, the probability of reversal of a grain moment in time dt is

$$dP = Ce^{-E/kT} dt \tag{6.10}$$

where C is a frequency factor of order 10^{10} sec^{-1}. The relaxation time is then

$$\tau = \frac{1}{C} e^{E/kT} \tag{6.11}$$

which is comparable to the time scale of cooling, say 1000 sec, if $E/kT \approx 30$. This is the situation at the blocking temperature. Then if we consider grains with blocking temperatures at about 800°K, a few tens of degrees below the Curie point of magnetite, cooling to atmospheric temperature, about 300°K, the relaxation time increases to

$$\tau_0 = 1000 \exp\left[30\left(\frac{800}{300} - 1\right)\right] \sec \approx 10^{17} \text{ years} \qquad (6.12)$$

even without allowing for the intrinsic increase in barrier energy at the lower temperature. Thus the appeal of paleomagnetism is to the occurrence of magnetite grains with high blocking temperatures, which are magnetically stable even on the geological time scale of 10^9 years, or to hematite grains large enough to be well removed from the superparamagnetic range.

The magnetization of sediments requires another explanation because sediments have not normally been heated sufficiently to acquire TRM. Those sediments which are accumulations of debris from the weathering of igneous rocks frequently contain grains of magnetite whose magnetic moments are partially aligned by the Earth's field during or immediately after deposition and thus impart a detrital remanent magnetization (DRM) to the deposit, as is observed in varved clays. However, the processes of compaction and solidification of a sediment are frequently accompanied by chemical changes in which at least some magnetite is oxidized to hematite. The hematite is formed chemically at low temperature in the Earth's field and if the grain size becomes large enough it may acquire chemical remanent magnetization (CRM) in the process. As a single-domain grain of hematite grows, the energy barrier opposing spontaneous (thermal) reversal of its magnetic moment increases, so that it becomes magnetically stable at a critical size, at which the barrier energy is of the order 25 kT. It has a "blocking size" at constant temperature, just as it has a blocking temperature for a particular size and the CRM acquired by an assembly of developing grains is determined by the Boltzmann distribution of grain moments at the blocking size. The theory of CRM is thus essentially the same as the theory of TRM. Chemical remanence has been demonstrated in magnetite by laboratory reduction of hematite and its stability is similar to that of thermoremanence. Chemical remanence in the hematite of sedimentary rocks is presumed to occur similarly during the oxidation of magnetite.

The natural remanent magnetizations (NRM) of rocks have primary components of thermal or chemical origin which are normally sufficiently stable for palaeomagnetic work. However, secondary components of lesser stability are frequently superimposed and have to be removed before the primary components can be measured. Secondary magnetization occurs because not all of the domains or domain walls in a rock are equally stable. The less stable ones may respond to a changing field, moderate heating during burial, or possibly to tectonic stresses and thus modify the total remanence by adding a secondary component. In unmetamorphosed rocks the primary magnetization is, however, not usually destroyed because the stable domains or domain walls are unaffected by the secondary magnetization.

Several geological tests have been applied to determine the stability of total NRM (Graham, 1949), but laboratory tests are more readily applied in most cases and are now in universal use. The commonest method is partial demagnetization in a slowly decreasing alternating magnetic field of initial magnitude chosen to remove the secondary magnetization but leave most of the primary component. During this process a specimen is simultaneously rotated about two or three mutually perpendicular axes at different rates, much slower than the field alternation, so that it is presented to the demagnetizing field in all orientations. The induction of a spurious remanence is minimized by placing the demagnetizing coil in field-free space, usually produced by a large set of Helmholtz coils which cancel the Earth's field. A good example of the effect of partial demagnetization in reducing the scatter of directions of NRM is shown in Fig. 6.4. Laboratory tests are also applied to determine whether rocks are intrinsically isotropic, as they must be if the directions of remanence are to be significant. A collection of papers on techniques applied in the laboratory tests has been assembled by Collinson et al. (1967).

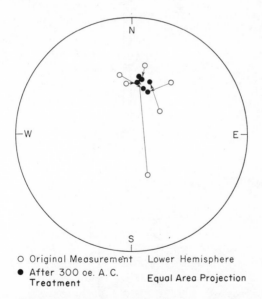

O Original Measurement   Lower Hemisphere
● After 300 oe. A. C.
  Treatment   Equal Area Projection

Figure 6.4: Effects of partial demagnetization on directions of natural remanence in six specimens from a single lava flow. Open circles represent directions before treatment and solid circles after partial demagnetization in an alternating field which was decreased slowly from 300 oe. Figure from Cox and Doell (1960).

The intensities of natural magnetizations of rocks cover a very wide range with no definite limits, but most are between 10^{-2} and 10^{-6} emu. Igneous rocks are generally in the upper part of the range and sediments in the lower part. The measurement of 10^{-2} emu presents no difficulty but to measure the direction and magnitude of a remanence of 10^{-6} emu requires a quite sophisticated technique. The most widely used instrument is the astatic magnetometer illustrated in Fig. 6.5a; also used is the spinner or rock-generator magnetometer (Fig. 6.5b), which probably has a higher ultimate sensitivity, although this is rarely needed.

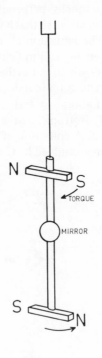

(a)

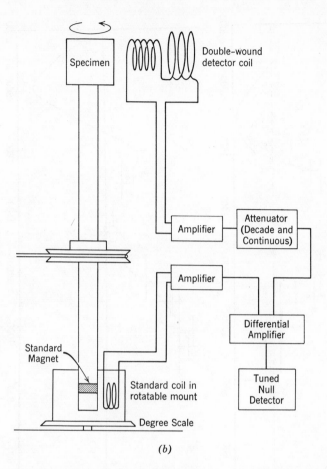

(b)

Figure 6.5: The measurement of magnetic remanence in rocks. (a) The astatic magneto-meter. (b) A spinner-magnetometer. A specimen must be placed in various orientations in either instrument to determine its moment vectorially.

6.4   Archeomagnetism and the Secular Variation

Proceeding backwards in time past the beginning of the direct record of secular variation (1540), we come to the period from which dated archeological remains can be studied. Pottery and, more usefully, bricks from pottery kilns whose last dates of firing can be estimated from the carbon-14 contents of ashes have thermoremanent magnetizations which can be measured by the usual techniques of paleomagnetism, subject to the conditions that the samples may have awkward shapes and in some cases only nondestructive tests are allowable. Measurements on such materials constitute a special branch of

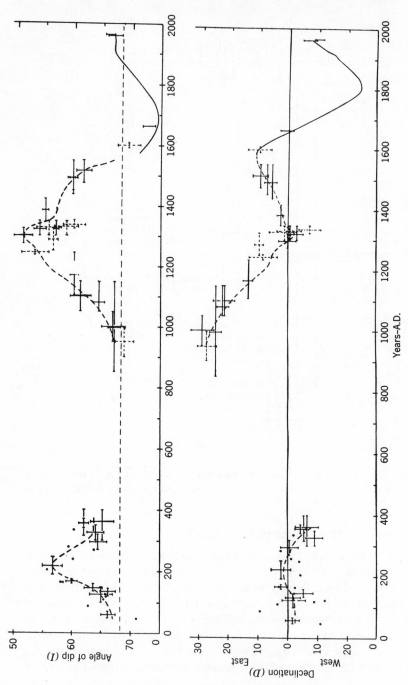

Figure 6.6: Secular variations in magnetic inclination and declination at London, deduced from the remanent magnetism of archeological specimens and compared with the direct observations since 1540. Uncertainties in measured values and in dates are indicated. Broken crosses and dots represent less reliable values. Figure from Aitken and Weaver (1965), by permission of the Society of Terrestrial Magnetism and Electricity, Japan, and the authors.

paleomagnetism known as *archeomagnetism*, pioneered by G. Folgerhaiter in France and now a subject of detailed study also in England, the United States, the Soviet Union, Czechoslovakia, and Japan.

The measurements are concerned semi-independently with two aspects of the past geomagnetic field, its direction and magnitude. The directions in England, as measured by M. J. Aitken and co-workers, are plotted in Figs. 6.6 and 6.7. In Fig. 6.6 the data on inclination and declination obtained from measurements on bricks from reliably dated kilns are plotted independently to show the magnitude of the uncertainty in the measurements. In Fig. 6.7, the data are combined to give an extension of Bauer's representation of the secular variation at London (see Fig. 5.4). Although the time span represented is not continuous, the data are sufficient to dismiss the idea of a simple cyclic secular variation. Further, measurements of the intensity of the geomagnetic field in the past show that the secular variation has involved first-order changes in the dipole field as well as the nondipole field. With certain necessary precautions, developed by J. G. Koenigsberger and studied in detail by Thellier and Thellier (1959), the paleointensities can be obtained by comparing the NRM of baked clays (pots, kiln bricks, etc.), with laboratory induced TRM in the same samples. Due to the superposition of dipole and nondipole fields, intensity measurements from a single area do not uniquely represent the strength of the dipole field, but intensity data from widely separated areas show a mutually consistent trend over the last 2000 years

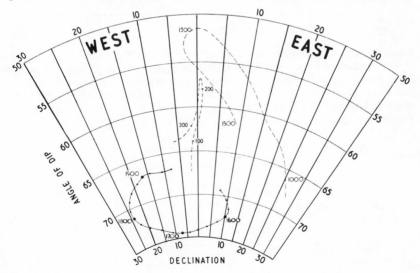

Figure 6.7: A combination of the data of Fig. 6.6 showing that the cyclic secular variation of the past four centuries is not representative of the historical past. Reproduced by permission from Aitken and Weaver (1965).

(Fig. 6.8). The intensity variation in Czechoslovakia over a more extended time scale is shown in Fig. 6.9. Since the total change represented exceeds a factor 2, whereas the present nondipole field is only 20% of the total, it is almost certain that the changes of Fig. 6.9 represent the dipole field.

The archeomagnetic evidence leaves us in doubt about the significance of the separation of the geomagnetic field into dipole and nondipole components. The secular variation involves both together and is not due simply to the non-dipole field, although it is reasonable to suppose that fluctuations of the dipole field may be slower (see Section 5.2). This does not necessarily mean that the westward drift of the nondipole field is not real or even that it is not a significant indication of the nature of the core motions, but merely that very precise and detailed data are needed before it can be distinguished from other contributions to the secular variation.

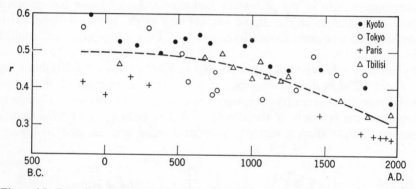

Figure 6.8: Intensity variation of the geomagnetic field over 2000 years from measurement of four archeomagnetic groups. All values have been adjusted to give field intensities at the equator, assuming that the field was a simple dipole, a procedure which is expected to introduce up to 20% errors in individual observations but cannot affect the trend apparent in all of the data taken together. Figure from Sasajima (1965) by permission of the Society of Terrestrial Magnetism and Electricity, Japan, and the author.

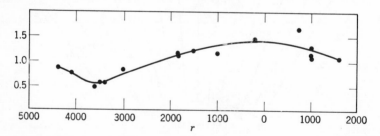

Figure 6.9: Secular variation in the intensity of the geomagnetic field in Czechoslovakia over a period of 45 centuries. Figure from Bucha (1965) by permission of the Society of Terrestrial Magnetism and Electricity, Japan, and the author.

The time constant for major changes in the dipole field appears from Fig. 6.9 to be a few thousand years. This is consistent with other evidence, in particular the reversals of the field, which are discussed in Section 6.6, and may be regarded as the extreme fluctuations of the dipole field. It is rather longer than, although of the same order as, the time constant for decay of an unmaintained electric current in the core, which, by the approximate calculation in Section 5.4, appears to be about 1000 years.

6.5 Paleomagnetic Poles and the Axial Dipole Hypothesis

Paleomagnetic results are normally expressed in terms of pole positions. The pole position deduced from measurements on a particular rock means the coordinates of the axis of a hypothetical dipole field which would produce in the rock a local field parallel to the measured direction of its natural remanence. The pole must lie on a great circle defined by the ancient magnetic declination and is situated at an angular distance θ_m, which is the ancient geomagnetic colatitude and is related to the measured paleomagnetic angle of dip I by a result from Section 5.1:

$$\cot \theta_m = \tfrac{1}{2} \tan I \qquad (6.9)$$

Measurements on a single rock sample give a poor estimate of the orientation of the ancient dipole field because the local field has also a nondipole component for which there is no means of applying a correction. However, by taking an average pole position from a number of measurements on a geological formation, which is sufficiently extensive for its history of cooling or deposition to cover a time span of several thousand years, we expect to average out the effect of the variable nondipole field. The procedures of rock collection and statistical treatment of data necessary to ensure that a proper average is obtained are based on an analysis by R. A. Fisher and are discussed by Irving (1964, Chapter 4).

A satisfactory mean pole position for the period represented by the British archaeomagnetic collections is obtained by averaging the data of Figs. 6.6 and 6.7. Irving (1964) gives its position as 82°N, 172°E, which is well removed from the present geomagnetic pole (78.5°N, 69°W) and is nearer to the geographic pole. A more significant average is obtained by using the archeomagnetic pole as one of several, covering a longer period, each pole being, of course, an average of numerous measurements. This has been done by Cox and Doell (1960) and Irving (1964) with variously selected data, always with the same general conclusion that the paleomagnetic poles cluster about the geographic pole, with the present geomagnetic pole to one side of the distribution, as in Fig. 6.10. This conclusion, first reached by Torreson et al. (1949) on the basis of measurements of magnetization in sedimentary rocks, led to a hypothesis upon which the conclusions derived from

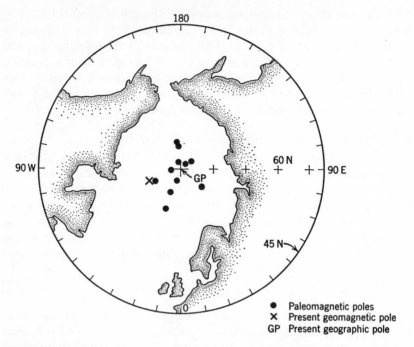

Figure 6.10: Paleomagnetic poles for the past 7000 years, plotted on a polar equatorial projection north of 45°N. Figure reproduced, by permission, from Irving (1964).

paleomagnetism now largely depend. This is the Geocentric Axial Dipole Hypothesis, according to which the geomagnetic field has always been predominantly dipolar and, when averaged over a sufficient time (of order 10,000 years), the axis of the dipole coincided with the geographic or rotational axis of the Earth, and the dipole field was centered at the center of the Earth. A number of tests have been applied to data from remote as well as recent geological periods and there is now no reason to doubt the validity of this hypothesis. The most important consequence is that paleomagnetic poles are ancient geographic or rotational poles and not merely geomagnetic poles.

Apart from its significance in paleomagnetism, the observation that the Earth's field averages to an axial dipole has an important bearing on the theory of the origin of the field. On the geological time scale the inclination of the dipole to the geographic axis appears merely as a transient excursion from the state of symmetry about the rotational axis. This means that we have no reason to seek a fundamental asymmetry in the Earth or in the geomagnetic dynamo mechanism as the cause of the present angle of the dipole field. Although the principles of symmetry are normally applied intuitively to physical cause and effect situations, it is worth considering them explicitly in the present connection. The principles were enunciated by Pierre Curie, whose statements have been summarized by Paterson and Weiss (1961) in the following terms:

"The symmetry of any physical system must include those symmetry elements that are common to all the independent factors (physical fields and physical properties of the medium) that contribute to the system, and it may include additional symmetry elements; however any symmetry elements absent in the system must be absent in at least one of the contributing factors."*

As far as we understand the relevant contributing causes of the geomagnetic field they are simply the rotation of the Earth, which has purely axial symmetry, and various motions, temperature gradients, etc., in the core, which, apart from the effect of rotation, has spherical symmetry. If the geomagnetic field were to deviate in a consistent way from symmetry with respect to the rotational axis, then we would have a field with lower symmetry than the combination of the causes we know about. The principles of symmetry would then compel us to seek an additional contributing cause of lower symmetry. That there is no consistent inclination of the dipole field means that there is no necessity for any such additional basic cause. The principles of symmetry are relevant also to the consideration, in the following section, of the polarity of the geomagnetic field.

6.6   Reversals of the Geomagnetic Field

Sequences of specimens from single geological formations, i.e., a series of lava flows or sedimentary layers, through which there are progressive increases in geological age, are frequently found to have repeated alternations of magnetic polarity. The obvious explanation that the geomagnetic field underwent repeated reversals was recognized very early in the history of paleomagnetism (see Irving, 1964, p. 8), but doubt was raised by the discovery at Mt. Haruna, Japan, of a dacitic lava which acquired laboratory TRM in a sense opposite to the field in which it cooled. It became important to know how common such self-reversals might be; Néel (1955) considered theoretically several mechanisms by which rocks could undergo self-reversal of remanence, not all of which were subject to direct laboratory tests, but known

---

* Doubt about the validity of Curie's principle, as here stated, has been expressed by Elsasser (1966). However, this is a misinterpretation. Elsasser's concern is with unstable deformation of a body, which *locally* appears to have lower symmetry than the combination of its causes, although if viewed as an average of many individual events it does not. The problem is analogous to that of the geomagnetic field, which only complies with Curie's principle if viewed on a long time scale.

self-reversing processes are responsible for very few of the reversed TRMs of rocks and evidently reversals of the field have occured many times. A comprehensive but elementary discussion of the reversals problem is given by Cox et al. (1967).

The only self-reversing mechanism known to occur naturally is due to the re-ordering of Fe^{3+} and Ti^{4+} ions in solid solutions of hematite and ilmenite For compositions near to 50 mole $\%$ of the two minerals the disordered structure has a weak canted (hematite-type) ferromagnetism and the ordered structure is ferrimagnetic (as magnetite). Details of the transition have been worked out by Ishikawa and Syono (1963), who found that reverse TRM was a feature of an intermediate metastable state and that it did not appear in specimens which were either completely ordered or completely disordered. The reversal arises from a negative (antiferromagnetic) coupling between the ions and it is convenient to think in terms of a simple model of two interacting (*A* and *B*) atomic sublattices, which preserve the orientations of their spontaneous magnetizations during the ordering process, but the one which is initially weaker becomes more strongly magnetic during the process of ionic redistribution. This reverses the spontaneous magnetization of the lattice and consequently also the remanence. A similar process of self-reversal has been found by Carmichael (1961) in natural hematite with 15 to 25% ilmenite, which had been subjected to laboratory heating.

Although known self-reversals are very few, there may exist others, too sluggish to be observed in the laboratory but important on the geological time scale. A magnetic reversal may be due to the chemical replacement of ions, although this is likely to be important only in metamorphic or weathered rocks, which are avoided in paleomagnetism. More important is the correlation, claimed by several authors, especially Wilson and Watkins (1967), between natural magnetic polarity and the oxidation of sequences of lava flows in which there are alternating polarities. The reversely magnetized rocks are more highly oxidized, although otherwise similar. The significance of these correlations is not yet clear.

The evidence for reality of the reversals of the geomagnetic field, which is now very strong, is of three kinds:

1. There is a correlation between reversals observed on different continents and in different rock types, both igneous and sedimentary.

2. Sediments which were baked by contact with later igneous rocks have acquired TRM with the same polarity as the igneous rocks in almost all cases, regardless of the polarity of remanence in the unbaked sediments, making self-reversal appear rare (see Table 6.1).

3. The actual process of reversal has been traced, both in rapid sequences of lavas and in deep sea cores.

Since self-reversals are apparently rare, the polarity of the geomagnetic field can be determined without significant ambiguity where several independent measurements agree. The polarity of the field for the past 4 million years was established in the first instance by measurements on igneous rocks dated by the potassium-argon method (Section 8.2); measurements on cores from ocean sediments have confirmed the main polarity epochs and extended the data, as well as clarifying the details of shorter-polarity events. The fluctuations in the polarity of the field for the past 4.5 million years are shown in Fig. 6.11. The fine structure in the pattern of reversals provides an aid to geological correlation on a finer scale than is possible by direct (radioactive) dating, as has been demonstrated by comparisons between ocean bottom cores (see, for example, Hays and Opdyke, 1967). It appears likely that closer scrutiny will reveal even more short events.

TABLE 6.1: PALEOMAGNETIC POLARITIES OF IGNEOUS ROCKS AND THEIR BAKED CONTACTS. (The original analysis was by R. L. Wilson and has been extended by Irving, 1964, to include additional data. Irving's figures are quoted here. N = normal, R = reversed.)

Igneous Polarity	Baked Contact Polarity	Number of Cases
N	N	34
R	R	49
Oblique	Oblique	2
N	R	2
R	N	0

With the information in Fig. 6.11, we can examine the significance of the numbers in Table 6.1. Reversals appear to have occurred, on average, once every 200,000 years. Since 2 out of 87 baked contacts had oblique polarities and therefore were apparently magnetized during reversals, it would appear that reversals are in process for about 2/87 of the time for the past few million years, and that each reversal therefore takes 2/87 of 200,000 years, i.e., 4600 years. The statistics of this estimate are, of course, very poor, but it coincides exactly with the estimate by Cox and Dalrymple (1967), which was based upon a much larger number of individual specimens (i.e., not baked contacts). It therefore appears that the field has been in the process of reversing for about 1.5% of the time during the past few million years. It is also noted that two of the baked contacts disagree in polarity with the igneous rocks which heated them. Again this is reasonable; ilmenite-hematite is sufficiently common that we should expect some self-reversals in baked sediments.

An excellent example of the evidence for progressive reversal of the field over a few thousand years is reproduced in Fig. 6.12. The actual process of

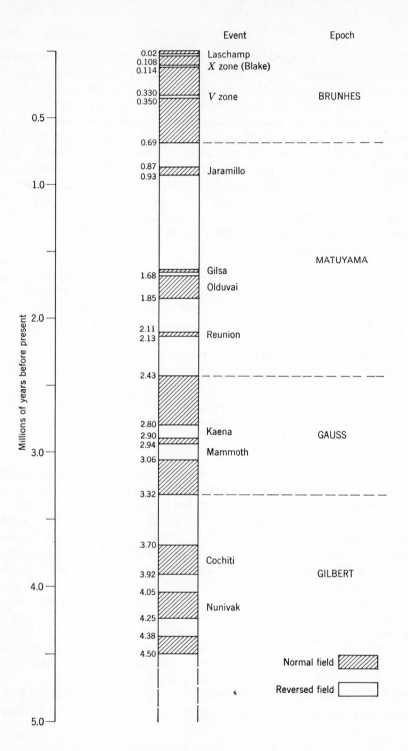

change cannot be made clear from measurements in a single locality, but the field strength is reduced by a factor of 4 or 5 during a reversal, so its seems likely that the nondipole field and possibly also an equatorial component of the dipole field remain. This is consistent with the example in Fig. 6.12.

Overall, the numbers of normal and reversely magnetized rocks are about equal and, within the experimental uncertainties, the reversals are true 180°

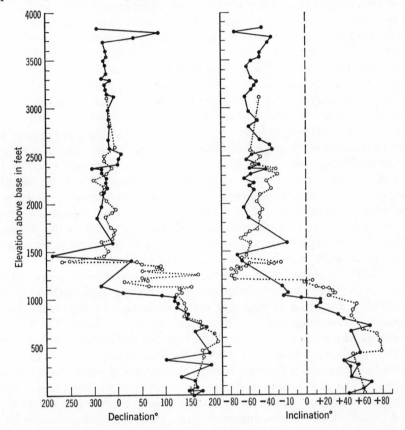

Figure 6.12: Details of paleomagnetic directions through a geomagnetic reversal, obtained by Van Zijl, Graham, and Hales (1962) and reproduced from the redrawn figure of Irving (1964). The measurements were on rocks from two localities, indicated by the continuous and dotted lines, from an extensive series of flat-lying lavas. The intensity decreased by a factor of 4 to 5 during the reversal.

Figure 6.11: Polarity of the geomagnetic field for the past 4.5 million years, based on summaries of the data by Cox and Dalrymple (1967b), Ninkovitch et al. (1966), Hays and Opdyke (1967), and on data still to be published which were communicated to the author by N. D. Opdyke, J. D. Smith, and J. H. Foster.

changes. In terms of the symmetry principles discussed in Section 6.5, we may therefore say that the basic causes of the geomagnetic field appear to have axial symmetry without polar asymmetry, i.e., there is no distinction between the opposite axial directions, because the field may have either polarity with equal probability. This is in accord with the geomagnetic dynamo theory; a field of either polarity may be self-maintained.

Uffen (1963) suggested that weakening or disappearance of the field during a reversal removes the magnetic screen which shields the Earth from cosmic radiation and that the greatly enhanced mutation-producing radiation causes a dramatic acceleration of evolution. Opdyke et al. (1966), Watkins and Goodell (1967), and others have reported evidence of faunal extinctions coincident with geomagnetic reversals in sedimentary cores.

However, Harrison (1958) has pointed out that one would not expect striking changes in mutation rates to occur in deep sea fauna, which are protected from cosmic radiation by the depth of sea water, and even at sea level significant enhancement of cosmic radiation during reversals has been doubted (Waddington, 1967; Black, 1967), It is possible that reversals are correlated with other effects such as climatic changes, which are actually responsible for faunal extinctions, although even this assumption appears to be incapable of explaining all of the observations. Useful evidence may come from comparisons of mutation rates during periods such as the Permian, when the geomagnetic field apparently had a constant polarity for tens of millions of years, with more recent periods of frequent reversals.

Irving (1966) suggested that reversals occurred frequently during periods of rapid polar wander and/or continental drift. This is a particularly intriguing suggestion because it implies that the configuration of the mantle has a direct influence upon the core motions which are responsible for the geomagnetic field and we must inquire just how this is possible. It is difficult to explain on the convective dynamo theory but appears to follow more naturally from the precessional dynamo, because the process of polar wander would cause the axis of symmetry of the core to depart from the axis of rotation of the Earth and thus modify the differential precessional torques on the core and mantle. It nevertheless appears surprising that the geomagnetic dynamo, with a time constant of order 10^3 years, could be sensitive to axial changes on a time scale of order millions of years. Evidently a further search must be made for correlations with other possibly relevant factors, such as the obliquity of the ecliptic, to indicate any common causal connection.

Geomagnetic reversals have provided striking evidence of the spreading of ocean floors outward from the ocean ridges. Magnetic surveys have revealed linear magnetic anomalies parallel to the ridges in the Atlantic, Pacific, and Indian oceans and they are interpreted as being due to alternating magnetic polarities of crustal igneous rocks which have been extruded more or less continuously from the ridges and spread outward. The similarities of the patterns in widely separated areas, revealed by the extensive work of Heirtzler et al. (1968), leave little doubt as to the worldwide significance of the phenomenon. Implications concerning the mechanical properties of the mantle are considered in Chapter 7.

6.7   Polar Wander and Continental Drift

We turn now to a time scale of many millions of years to obtain evidence of very slow changes which become apparent when the relatively transient secular variation is averaged out by extensive sampling of rocks whose ages may span several million years. On this time scale even reversals may appear frequent and the polarity of the field is neglected. By the axial dipole hypothesis the averaged paleomagnetic poles are poles of rotation and we are simply considering slow movements of the pole with respect to sections of the crust in which we find rocks suitable for paleomagnetic work.

Poles calculated from measurements on rocks younger than about 20 million years do not depart from the present geographic pole by distances greater than the experimental uncertainties. Substantial deviations appear only in data from rocks which are more than about 30 million years old (lower Tertiary or earlier).

Since there are disparities between data from different continents we must now consider separately the paleomagnetic results from each of several large land masses. The term "continents" is commonly used rather loosely for these land masses, whose boundaries are not in fact everwhere coincident with continental divisions. The term "region" will be used where this is so. The best sampled region is Europe-North Asia, represented by the stippled area in Fig. 6.13. Measurements have been made on rocks from sites indicated by heavy crosses, except that some crosses are necessarily omitted from areas such as the British Isles which have been sampled particularly intensively. Paleomagnetic data for rocks of late paleozoic age or younger (i.e., for the past 300 million years) from all parts of this region are mutually consistent. Rocks from Siberia give the same poles as rocks of the same ages from Western Europe. In view of the great geographical separation of Eastern Siberia and Western Europe this agreement, which is particularly well documented for the Permian period (230 to 280 million years ago), gives strong support to the hypothesis that the geomagnetic field has been predominantly dipolar for at least 300 million years. Plotted on Fig. 6.13 is a series of positions of the pole relative to Europe-North Asia for the geological periods indicated by the initial letters. The poles are averages obtained by Irving (1964) for each of the periods and the arrows joining them constitute a polar wander curve, a dated path which the pole has followed. The mean pole positions are subject to errors of order 10°, so that the detailed

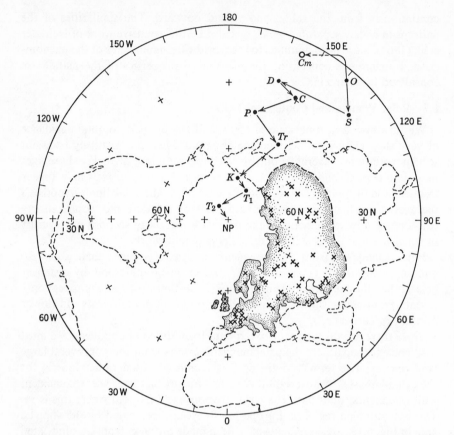

Figure 6.13: Polar movement relative to Europe-North Asian region. The paleomagnetic poles are means for the geological periods Cambrian (*Cm*), Ordovician (*O*), Silurian (*S*), Devonian (*D*), Carboniferous (*C*), Permian (*P*), Triassic (*Tr*), Jurassic (*J*), Cretaceous (*K*) Lower Tertiary (T_1) and Upper Tertiary (T_2). Figure reproduced, by permission, from Irving (1964).

irregularities of the polar path cannot be regarded as certain but the general trend of polar movement is clear; it exceeds 90° in 500 million years, an average rate of 1/5° per million years or 2 cm year^{-1} if the irregularities are neglected.

Other evidence, considered in Chapter 7, indicates that large-scale tectonic processes occur at rates of a few centimeters per year and it is therefore entirely reasonable to regard polar wander as a result of such a process. In Section 7.3 the conclusion is reached that the stresses which are supported by the lower mantle are too large to admit the possibility of lower-mantle convection at a rate of a few centimeters per year, i.e., a mantle creep rate of order 10^{-15} to 10^{-16} sec^{-1}. However, this does not preclude lower-mantle adjustment of the equatorial bulge to a changing axis of rotation. Polar wander at an angular rate $\dot{\alpha}$ in the Earth, of ellipticity $\epsilon = 3 \times 10^{-3}$, requires a deformation rate of only $\epsilon\dot{\alpha}$, which is therefore only 0.3 % of the deformation rate required to cause continental drift. We may of course suppose that polar wander and continental drift are caused by the same general process of mass redistribution. The important point here is simply that polar wander need not be inhibited by a stable equatorial bulge.

The comparison of pole paths for different regions reveals a major disagreement, as in Fig. 6.14. These polar wander curves are compatible with the hypothesis of a dipole field only if the regions have moved with respect to one another, i.e., if continental drift has occurred. Each of the curves demonstrates polar wander, but if the axial dipole hypothesis is accepted, the differences between the polar wander curves can only be interpreted as continental drift. The first polar wander curves to be compared were for Europe

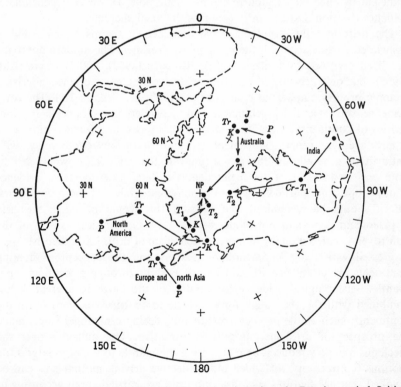

Figure 6.14: Comparison of pole paths for four regions since the Permian period. Initials for the geological periods are the same as in Fig. 6.13. Figure reproduced, by permission, from Irving (1964).

and North America, and these were found to be reconcileable if Europe and North America were adjacent, with the Atlantic closed, 300 million years ago. It may still be possible to quarrel with the significance of the discrepancy between European and North American data, but there is no doubt at all that both differ from the Australian results to an extent which rules out experimental uncertainty. A relative movement exceeding 90° since the Permian (250 million years) is indicated, corresponding to an average rate of nearly 4 cm year^{-1}. Confirmation of the dipole hypothesis by the data from widely scattered parts of Eurasia leaves very little scope for antidrift theorists. The hypothesis of continental drift, according to which continent-sized blocks of the Earth's crust have retained their form but have moved relative to one another, is now a virtually inescapable conclusion of paleomagnetism. It should be noted that although the axial dipole hypothesis is basic to paleo-magnetism, this particular conclusion depends only upon the dipole character of the geomagnetic field and does not require the field to be axial, since it is sufficient to refer each region to the magnetic pole. However, paleoclimatic evidence (Section 6.8) generally confirms the axial character.

The drift hypothesis antedates paleomagnetism by many years and is associated particularly with the name of A. Wegener. The modern approach has been reviewed by Bullard (1964), Runcorn (1962), and Blackett et al. (1965), but only recently has a comprehensive statement of the underlying tectonic processes appeared (Isacks et al., 1968; see Section 7.1). The early evidence was essentially geological and depended upon appeals to similarities of continental margins in shape, rock types, and fossil occurrences. It produced no general agreement but only a furious controversy, with antidrifters generally in the majority until the mid-1950s, when serious note was first taken of the paleomagnetic data. Corroborative evidence is now appearing, particularly from geophysical studies of the ocean floors, but it should be recognized that little of this would be taken seriously if paleomagnetism had not pioneered the way. Geophysical aspects of the controversy centered on the supposed mechanism of drift. Early theorists were impressed with the phenomenon of isostasy and the concept of floating continents, which they envisaged to move as ships through a yielding upper mantle. Seismological evidence for solidity of the Earth to a great depth, combined with the very weak forces which could be invoked to move the continents, such as the *polfluchtkraft* or pole-fleeing centrifugal force, made the prospect of drift mechanically unattractive. The difficulty over the mechanism is now reduced by considering the mantle to undergo large-scale motions (convection), for which more effective driving mechanisms can be invoked; surface features such as continents are carried about according to the convective pattern. The mechanics of this problem are discussed further in Chapter 7.

An alternative to the convection hypothesis, which has been suggested, is that the Earth has undergone a gross radial expansion during geological time. Without a violation of accepted physical principles, this suggestion must be discounted as incompatible with any plausible available energy source (Beck, 1961). One explanation is that the gravitational constant G has decreased with time but this would have affected the Sun similarly and the energy of solar radiation would have changed dramatically. Since past climates were generally similar to those at present, we must conclude that there has been no major change in G and that the hypothesis of substantial Earth expansion must be discounted.

The detailed pattern of continental drift is still rather rudimentary. Historically two alternative hypotheses have been used as bases for theories: (1) a single supercontinent (Pangaea) was formed initially and remained intact during the paleozoic era but broke up subsequently; (2) two primeval continents (Laurasia and Gondwanaland) formed separately in the northern and southern hemispheres and broke up at the end of the paleozoic period. No unique solution to the problem is possible because, apart from observational inaccuracies, in principle paleomagnetism can provide only part of the evidence necessary for continental reconstructions. The latitude of a land mass is determined unambiguously, as is its geographic orientation with respect to the poles, but its longtitude is arbitrary. The indeterminacy is reduced somewhat by imposing the obvious conditions that two land masses do not overlap at any stage or "cross over." A particularly interesting feature of the drift hypothesis, which is substantiated by paleomagnetic observations, is the separation of the southern continents, Africa, Australia, South America and also India, which were apparently closely grouped around Antarctica during the Jurassic period 150 million years ago (Fig. 6.15). The mutual proximity of Africa and South America during the Paleozoic is particularly well demonstrated by the data of M. W. McElhinny, plotted in Fig. 6.16. It nevertheless appears likely that the concept of primeval supercontinents is too simple and that if a complete record were available for 2000–3000 million years we would find that continental masses had been repeatedly broken up and rejoined in different ways. Particularly suggestive of this is the movement of the Indian peninsula, which was evidently once part of or very close to a composite land mass with the southern continents, but has pushed up against Asia; the massive folding of the Himalayas suggests that it is still pushing.

It is possible that polar wander and continental drift are not steady processes, but are intermittent with long quasistatic periods, punctuated by shorter drift episodes. Irving (1966) found that almost all of the polar movement relative to Australia from the upper Silurian (400 million years ago) to mid-Cretaceous (100 million years ago) occurred during a 20 million year episode in the Carboniferous period (about 300 million years ago). The drift episode

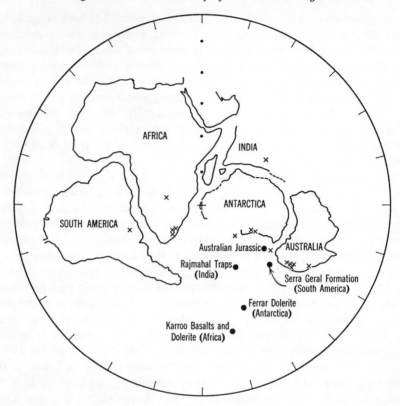

Figure 6.15: A reconstruction of Gondwanaland, made in 1937 by A. L. du Toit on the basis of continental fitting and geological similarities, with Jurassic pole positions plotted by E. Irving. Although even better agreement between the poles can be produced by a slightly different arrangement of the continents, the approximate agreement justifies the approach of the early "drifters." Figure reproduced, by permission, from Irving (1964).

coincided with strong orogenic (tectonic) activity in eastern Australia. Before demanding an intermittent mechanism for drift, it will be vital to know whether similar drift episodes were really synchronous all over the Earth, but it is of interest to note that Gastil (1960) found the radiometric ages of rocks to be grouped in a manner suggestive of a more-or-less cyclic sequence of tectonically active and quiescent periods, with a cycle of 250 to 500 million years.

Longitude and latitude observations are capable in principle of observing continental drift as a disparity in polar movements as seen from different continents, but no unambiguous evidence has yet been reported. The cyclic (annual and Chandler period) variations in latitude correspond to polar movements of about 10 m, so that, with the irregularity in wobble, secular

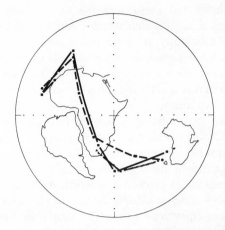

Figure 6.16: Polar wander curves for three southern continents, Africa (solid line), South America (dashed line) and Australia (hatched line) for the Paleozoic period, drawn for the continents in fixed relative positions, but different from the present. This is the most convincing demonstration that Africa and South America were a single continent for several hundred million years during the Paleozoic. Figure courtesy of M. W. McElhinny, who presented it to the International Union of Geological Sciences Symposium on Continental Drift, Montevideo, October 1967.

movements of a few centimeters per year are obviously difficult to distinguish with certainty. Astronomical determinations of longitude are too imprecise to indicate relative movements, but latitude has been reliably measured for a sufficient number of years to indicate a steady movement of the pole toward about 70° W at about 0.003 sec year^{-1} or 10 cm year^{-1} (Markowitz et al., 1964, and several papers in the conference proceedings edited by Markowitz and Guinot, 1968). The significance of this result is not entirely clear because data from only two of the standard latitude stations are emphasised; there is no question of distinguishing polar wander from continental drift. It is possible that in the foreseeable future observations of artificial satellites will give less ambiguous evidence of relative movements of continents.

6.8   Paleoclimates

That ancient climates have differed remarkably from the present is attested by occurrences of fossil organisms and rock formations in regions where they could not possibly develop now. The discovery of coal seams in Antarctica is probably the best known of these observations. The study of paleoclimatology is deeply rooted in geology and there is an impressive accumulation of data, reviewed in collections of papers edited by Nairn (1961, 1964). As Blackett (1961) and Irving (1964, Chapter 9) point out, a correlation between the ancient climates and paleomagnetic latitudes is essential to our acceptance of the axial dipole hypothesis and constitutes the most direct evidence for its validity over geological time. Paleolatitudes could be determined approximately, but without ambiguity, from statistical studies of fossil occurrences and rock types if the Earth had a constant pattern of climatic zones. However, there are changes in climate which occur more rapidly than polar wander or continental drift, so that it is only the gross features of climatic change that test the axial dipole hypothesis.

The ice ages are the most striking features of the geologically rapid, world-wide changes in climate. Coincidence of Pleistocene ice ages in different sectors of the northern hemisphere excludes the possibility of extremely rapid polar wander, which might have caused alternating glacial advances and retreats, in opposite phase in Eurasia and North America; we can also note that such rapid polar wander is incompatible with the axial stability implicit in the nonequilibrium equatorial bulge. The term "ice age" is commonly used to mean the climatic changes of the past few hundred thousand years. Major glaciations have occurred at earlier geological times but cannot be studied in the same detail. Nevertheless, it is apparent that there is a very wide spectrum of frequencies in the climatic fluctuations.

The most directly quantitative estimates of ancient climatic features are the paleotemperatures derived from analyses of oxygen isotopes in the fossil shells of certain marine organisms (Bowen, 1966). The measurements depend upon the slight isotopic fractionation which occurs in the very slow deposition of carbonate from water and dissolved carbon dioxide and calcium; thermodynamic equilibrium is evidently maintained during the process and the minimum of free energy occurs when O^{18} is slightly more concentrated in $CaCO_3$ than in H_2O molecules. The excess is greater at low temperatures so that the abundance of O^{18} in the carbonate is dependent upon the temperature at which it was formed, always assuming, of course, a constant isotopic composition for the sea water.

The relevant thermodynamic analysis was developed by Urey (1947), who showed that a knowledge of the vibrational energy states* E_i of simple molecules such as CO_2, with alternative isotopic compositions $CO^{16}O^{18}$, CO_2^{16}, allowed the partition functions Z to be calculated for each composition:

$$Z = \sum_i \exp\left(-\frac{E_i}{kT}\right) \qquad (6.10)$$

The free energy of a molecular mixture in equilibrium, such as

$$CO_2 + H_2O \rightleftharpoons H_2CO_3 \qquad (6.11)$$

can then be calculated, for arbitrary distribution of available O^{18} and O^{16}

---

* Only vibrational states are significantly excited at normal temperatures.

atoms, and minimized with respect to the isotopic distribution. The carbonate molecules are favored by the heavier O^{18} atoms but only slightly so. As the temperature is raised, the total energy of the system increases and the isotopic distribution has proportionately less effect upon the free energy; the isotopic concentrations in the reacting materials thus approach equality at high temperatures. Calculations for the carbonate reaction are not exact but paleotemperature estimates are "calibrated" by measurements on specimens grown in controlled environments and in well-observed natural environments. Emiliani (1955) gives the O^{18}/O^{16} fractionation factor between $CaCO_3$ and H_2O as 1.025 at 0°C and 1.021 at 25°C.

The natural O^{18}/O^{16} abundance ratio is approximately 1/500. Measurement of this ratio to 1 part in 6000, corresponding to temperature variations of ±1°C, demands a very sophisticated mass spectrometric technique, such as that used by Emiliani (1955) on deep sea cores from the Pacific, Atlantic, and Caribbean. These are tropical regions in which the annual temperature variation is relatively slight, so that no uncertainty arises from selective seasonal growth of carbonate and the observed variations reflect temperatures over the Earth as a whole. Temperature fluctuations over the past 300,000 years, as obtained by Emiliani, are shown in Fig. 6.17. These results confirm that the climatic fluctuations responsible for the Pleistocene glaciations were worldwide phenomena.

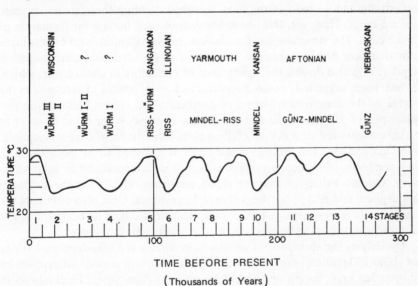

Figure 6.17: Tropical, deep-ocean temperatures for the past 300,000 years, obtained from ratios of 0^{18} and 0^{16} isotopes, with names of the recognized glacial and interglacial periods. Reproduced from Emiliani (1955) by permission of the University of Chicago Press.

Theories of the causes of ice ages have been summarized by Scheidegger (1961). It is probable that no single mechanism can be found to explain all of the observations. Variations in insolation, or solar radiation falling on the Earth, due to its rotational and orbital characteristics, were studied by M. Milankovitch, who explained the periodicity of Pleistocene glaciations in this way. Van den Heuvel (1966) extended this work by making a Fourier analysis of Emiliani's (1955) O^{18}/O^{16} temperatures; he found periods of 40,000 years and 12,825 years, which are in satisfactory agreement with the period of oscillation of the obliquity of the ecliptic and with half the precession period. In a comprehensive discussion of the whole problem, Öpik (1967) discounted the orbital variations as quite inadequate to account for the gross, longer-term variations in climate and even doubted their effectiveness in short-term glacial advances; he favored an intrinsic variability of the Sun as the ultimate cause of climatic changes.

In considering the effectiveness of orbital variations in modulating the insolation, we can note that if the ellipticity of the Earth's orbit, the mean Earth-Sun distance and the solar emission do not vary, then the total radiation falling on the Earth is constant. The orbital variations in insolation which may occur must be opposite in the two hemispheres, but glaciations occur simultaneously in both hemispheres. To some extent the formation of polar caps may be self-catalyzing; the presence of ice increases the Earth's reflectivity and so reduces the heat absorbed, tending therefore to increase the ice cover. However, this instability cannot itself initiate the formation of polar caps. The simultaneous fluctuations in glaciation in both hemispheres can therefore not be explained entirely in terms of orbital characteristics. Öpik (1967) also doubts the significance of variability in cloud cover, which, it has been suggested, could be correlated with orbital characteristics by virtue of the uneven distribution of continents and oceans between the two hemispheres. He points out that the general tendency is for cloud to occur in rising currents and clear sky over falling currents in the convective atmospheric motions and that the average cloud cover is expected to be close to 50%, independently of moderate variations in surface temperatures or solar emission. Quoted values of average cloud cover are 47.3% for the northern hemisphere and 53.1% for the southern hemisphere. Öpik also discounts the variation in atmospheric reflectivity due to volcanic dust as being too slight and too short-lived to be effective in producing climatic changes.

Underlying the shorter-term variations in climate is a long-term periodicity of about 250 million years, with long "normal" warm periods interrupted by shorter ice ages, lasting no more than a few million years. Fluctuations in solar output by a few percent appear to provide the only plausible explanation and since the solar constant appears to have varied by as much as 0.5% since the beginning of the century, there is no difficulty in accepting solar variability. On this basis we may suppose that the astronomical cycles, with periods of 10^4 to 10^5 years, superimpose their own relatively minor effects upon the climatic changes caused by variability in the Sun.

Eventually the oxygen isotope method may be applied to give quantitative data about temperatures in remote geological periods, but there are difficulties. We cannot expect the isotopic composition of sea water to be constant on this time scale; presuming the water to have been derived largely from the mantle, we note that basic igneous rocks are 0.65% richer in O^{18} than is sea water, which must therefore have received progressive O^{18} enrichment in the course of geological time. The selection of fossil types which resist chemical changes for more than 10^8 years is also a considerable problem. Thus we note the limitation of the paleoclimatic method of estimating ancient latitudes and, for the purpose of comparison with paleomagnetic data, restrict consideration to climatic changes which were very prolonged and therefore would certainly be observed in all continents if they were worldwide.

On this basis the most striking paleoclimatic evidence is from the Permian and Carboniferous periods when there were very extensive glaciations in Antarctica, Australia, Africa, South America, and India, while European and North American rocks from the same period indicate tropical climates, with corals, salt deposits, and desert sandstones (see Blackett, 1961; Irving, 1964). It is further noted that the Indian ice moved northward, although India is now in the northern hemisphere. These observations are basic to the hypothesis that the southern continents were grouped around Antarctica as a single supercontinent, Gondwanaland, during the Paleozoic and have subsequently moved apart to their present positions. Blackett (1961) particularly emphasized that this hypothesis is compatible with the available paleomagnetic data (see also Fig. 6.15), which indicate high southerly latitudes for the southern continents but low latitudes for the northern continents. Alternative explanations are so devious that we can hardly avoid the conclusion that the axial dipole hypothesis and continental drift are confirmed.

Comparisons of paleomagnetic data with other paleoclimatic indicators are made by Briden and Irving (1964). They point out that the latitude distribution of modern corals is symmetrical about the equator, being concentrated in low latitudes, in almost all cases less than 30°, whereas fossil carbonates appear predominantly in the continents of the northern hemisphere, at latitudes higher than 30°. The paleolatitudes of these carbonates—the latitudes in which they were formed—as determined from paleomagnetic measurements, are, however, grouped about the equator in roughly the same way as the modern corals. The distribution of carbonates is thus consistent with a more or less constant climatic zoning of the Earth, but movement of the continents relative to the climatic zones.

6.9 Intensity of the Paleomagnetic Field

The variations in the strength of the geomagnetic field over the past several thousand years are discussed in Section 6.4. Essentially the same method of comparing natural remanence and laboratory thermoremanence has been used on numerous rocks, dated back to the Silurian period, about 400 million years ago. The results have been summarized by Briden (1966) and P. J. Smith (1967) and are represented in Fig. 6.18.

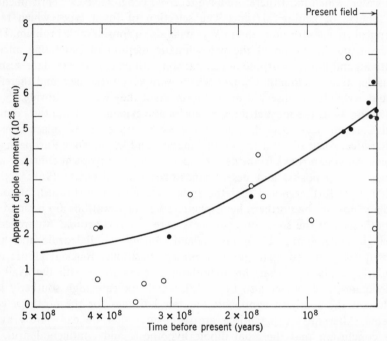

Figure 6.18: Estimated strength of the Earth's magnetic dipole moment for the past 400 million years, drawn from data by P. J. Smith (1967) (solid circles) and Briden (1966) (open circles). The curve represents the trend of Smith's values. Each point represents an average of several determinations and thus averages at least partially the short-term variability of the field.

On a much shorter time scale than that represented here, the field is highly variable, so that a wide scatter of results is not surprising. Nevertheless, two tentative conclusions can be drawn. The present field is substantially stronger than the average over any prolonged interval in the past 500 million years, even in the last million years or so; and the dipole moment has generally increased since the Silurian period. Taking the data of P. J. Smith (1967) alone, a smooth curve fits the data reasonably well; these values are mostly averages of larger numbers of independent determinations and therefore appear to represent the trend accurately. A gradual decay of the natural remanence would of course explain these observations. Care was taken to select from the remanence only the part with high blocking temperature, which is generally the most stable magnetically, but further work on the magnetic viscosity of rocks is needed to ensure that this technique avoids completely the problem of decayed natural remanence.

There is a particular geophysical interest in extending these observations back 2×10^9 to 3×10^9 years. If, as Runcorn (1962) has supposed, the core developed since that time, then rocks of that age should indicate no geomagnetic field. Measurements reported by Evans and McElhinny (1966) and Carmichael (1967) have extended the method back to 2.6×10^9 years. Carmichael showed that the average field strength went through a minimum about 5×10^8 years ago and that the curve of Fig. 6.17 does not extrapolate back to give a negligible pre-Cambrian field, but that between 1×10^9 and 2×10^9 years ago the field was stronger than at present. Evans and McElhinny found the magnetization of a 2.6×10^9 year-old gabbro from South Africa to indicate a field at that time comparable in strength to the present field. This implies that the core was essentially complete and producing a substantial field at an early stage in the Earth's history (see also Section 1.5).

Chapter 7

CREEP AND ANELASTICITY
OF THE MANTLE

"Conclusions and hypotheses concerning the earth's interior are in a state of flux."

GUTENBERG, 1959, p. v.

7.1 Evidence of Motion within the Mantle (Global Tectonics)

In Chapter 6 we obtained rates of 2 cm yr^{-1} for polar wander and 4 cm yr^{-1} for relative drift of continents. It may be doubtful whether polar wander demands complex motions within the mantle, but continental drift certainly does and the rate must be at least several centimeters per year. This appears to be a characteristic rate, as indicated by other observations. However, we must be careful not to regard the merely corroborative evidence as actual proof of prolonged, large-scale movements. Paleomagnetism provides proof as nearly as we are ever likely to obtain it; most of the other observations, if taken alone, would leave much more room for doubt, but taken together there is an impressive concordance of evidence.

Major horizontal movements between sections of the crust are noticeable on transcurrent faults (Benioff, 1962) of which four of the best known examples are shown in Fig. 7.1. In several cases cumulative movements of hundreds of kilometers can be recognized. Of the currently active faults, the most intensively studied is the San Andreas system in California, where geodetic surveys (Whitten, 1956) indicate zones of shear with northward movement of the west side of the fault relative to the east side amounting in places to as much as 4 cm year^{-1}. At several points progressive, but irregular movement of order 1 cm yr^{-1} is occurring without earthquakes across shear zones of widths only 1 ft to a few tens of feet (Steinbrugge et al., 1960; Tocher, 1960; Radbruch et al., 1966). Sudden movements, in places exceeding 5 m, occurred over hundreds of kilometers of the San Andreas fault in the earthquakes of 1857 in Southern California and in 1906 in the San Francisco

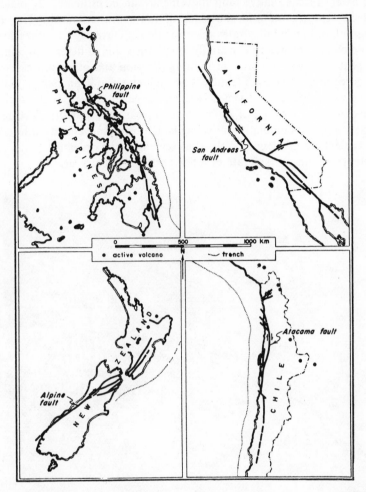

Figure 7.1: Maps to the same scale of four of the world's major transcurrent faults. Reproduced, by permission, from Allen (1965).

area (see Section 4.2), but the intervals between such events are too long to allow us to make a statistical estimate of the average rate of movement due to earthquakes. One such jump every century or so would give an average movement of a few centimeters per year. A striking example of such sudden transcurrent displacement is shown in Fig. 4.6. Apart from secondary faults normal to the trend of the San Andreas system, all of the Californian movements are in the same sense and indicate a general northward movement of the Pacific side relative to the continental side at an average rate of a few centimeters per year.

The extent of cumulative fault movements can be estimated by matching geological features on opposite sides of a transcurrent fault. The aerial photograph in Fig. 7.2 illustrates the mismatch of currently adjacent features. The matching process is essentially qualitative and different amounts of movement are given by different authors. Erosion and other processes tend to obscure the similarities of features once in juxtaposition and some disagreement between different estimates is to be expected also from the fact that matched features may be of different geological ages and therefore record genuinely different cumulative movements. However, movements of hundreds of kilometers are involved and the time scale is of the order 10^7 years, again consistent with an average rate of a few centimeters per year.

Figure 7.2: Mismatch of geological features across a branch of the San Andreas Fault near Indio, California. Photograph by Spence Air Photos.

Transcurrent faulting has frequently been cited as evidence of large-scale mantle movements, but many earthquakes are not of the transcurrent or strike-slip type. Dip-slip, involving vertical movement between blocks, is dominant along large parts of the most active seismic belts which are delineated in Fig. 4.1. Directions of movement determined by fault plane studies in seismology (Section 4.2) are mutually consistent for the earthquakes in any one region and indicate an underlying global pattern of movement. Wilson (1965b) postulated such a global pattern, in which transcurrent movements appear as "transform" faults, preserving continuity of the lithosphere (crust plus rigid part of the upper mantle) between the rising and sinking limbs of mantle convection, and Morgan (1968) has elaborated this idea. It is now widely accepted and is basic to our understanding of the global tectonics (Isacks et al., 1968). However, although transcurrent faults are relegated to a rather secondary role in the tectonic pattern, they are geologically the most obvious type of fault and so are very important indicators of large-scale, prolonged mantle motion.

The global tectonic pattern, illustrated in Figs. 7.3 and 7.4, is based on the sea floor-spreading hypothesis of Hess (1962) and Dietz (1961), according to which new oceanic lithosphere (crust plus some cool, uppermost mantle material) is formed more or less continuously at the ridges and spreads outward. Various estimates of the rate of spreading all indicate several centimeters per year. The most direct estimates are based on the observation of Vine and Matthews (1963) that linear magnetic anomalies parallel to the ridges correspond to reversals of the geomagnetic field over the past several million years, each strip of ocean floor acquiring the magnetic polarity of the field at the time it cooled. Since the history of the field polarity is known for the past several million years (Section 6.6), it provides a time scale for the spreading. Taking the widths of the anomalies to be 20 km and the average interval between reversals to be 5×10^5 years, we obtain a spreading rate of 4 cm year^{-1}. Heirtzler et al. (1968) identified corresponding anomalies flanking ridges in the Atlantic, Pacific, and Indian Oceans and deduced spreading rates of 2 to 4 cm year^{-1}. LePichon (1968) made some higher estimates and also used these to deduce rates of convergence of lithospheric blocks at island arcs. A value as high as 10 cm year^{-1} appears to be reasonable for Japan.

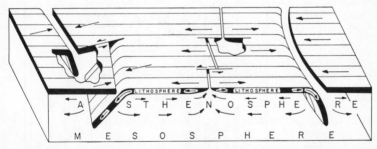

Figure 7.3: Schematic illustration of global tectonic movements. The motion is supposed, to be of a convective type and to originate in the asthenosphere or "soft" layer of the upper mantle; the harder lithosphere is carried along from line sources (ocean ridges) to line sinks (island arcs and similar structures). Planes of earthquake foci mark the plunging slabs of lithosphere under island arcs (Fig. 4.2). Earthquakes along ridges and along transform faults between sections of ridges or island arcs are all shallow. The distribution of the lines of rising and sinking material over the Earth is indicated in Fig. 7.4. Figure reproduced, by permission, from Isacks et al. (1968).

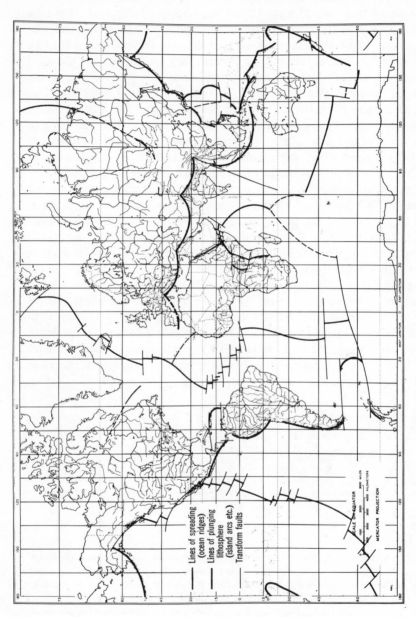

Figure 7.4: Tectonic pattern of the Earth, showing the centers of sea floor spreading (ocean ridges) and island arcs and similar structures, in which slabs of lithosphere appear to be sinking, and some of the interconnecting transform faults. Many of the details are still unclear and will be subject to revision, but the general features are now almost universally accepted after a 15-year rearguard action by opponents of continental drift. The accumulated evidence is summarized by Isacks et al. (1968).

The first of the magnetic lineations to be well documented were those off the coast of California (Fig. 7.5), where R. G. Mason (1958) and Vacquier (1962, 1965) noted breaks in the linear anomalies, very suggestive of faulting and displacement, and apparent as topographic features on the ocean floor, although not associated with seismic activity at the present time. These can now be seen as examples of the transform fault structure, illustrated in the inset to Fig. 7.5.

The persistence of sea-floor spreading for at least 100 million years is attested by several geological observations, although none of them give evidence as clear as that obtained for the past few million years from the magnetic anomalies and the movement may not have been continuous. Heezen (1962) noted the general youthfulness of ocean basin sediments; although at least some parts of the North Atlantic are now known to be much older than Heezen realized, most of the ocean floors are evidently much younger (of order 100 million years) than the average for continents (of order 1000 million years) and they must have been generated by some process such as sea-floor spreading. Wilson (1965a) reported evidence that the ages of sea mounts (extinct submarine volcanoes) increased with their distances from the nearest ridges, and suggested that they were merely ridge volcanoes displaced by spreading of the sea floors away from the ridges. Again we must expect to find exceptions, if only because we cannot expect the tectonic pattern to be constant in time. For evidence of tectonic activity on a time scale longer than a few tens of millions of years, we still have little more than the paleomagnetic data. As noted in Section 6.7, continental drift may not be a continuous process on this time scale, but rather spasmodic with quasi-static periods interrupted by drift "episodes."

Lee and MacDonald (1963) attempted to obtain evidence of a convective pattern by comparing spherical harmonic analyses of gravitational potential and surface heat flow, although the world coverage of heat flow data is probably inadequate for the analysis to be significant. The supposition was that a correlation between geoidal lows and high heat flow represented rising limbs in the convective pattern. However, gravity anomalies associated with tectonic features such as island arcs would only be apparent as very high-order harmonics in such an analysis and it is much more likely that the low-order harmonics in the geoid are associated with features of the lower mantle, but not the core-mantle boundary. On a much smaller scale, localized high heat flows are associated with ocean ridges and this has been supposed to confirm the idea of rising convective currents under ridges. However, the argument is unconvincing because local high heat flow merely requires volcanism and this occurs also in island arcs, where lithospheric material is believed to be sinking. Thus heat flow data do not help to clarify the convective pattern.

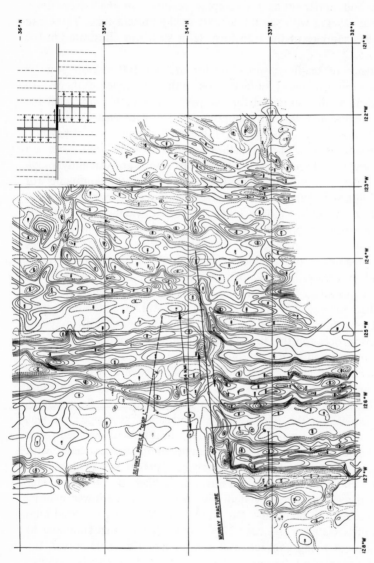

Figure 7.5: Pattern of geomagnetic anomalies off the Californian coast. The linear features are presumed to indicate alternately normal and reversed magnetic polarities of igneous rocks which have spread outwards from a line source (ocean ridge). The breaks in the lineations correspond to offsets of sections of the ridge, as illustrated in the inset. The transform fault, across which transcurrent displacement is occurring, is limited to the heavily shaded line between the sections of the ridge; the extensions to this fault are "welds" of materials of different ages, but are not faults in the seismic sense. Reproduced, by permission, from Vacquier (1962).

Possibly convection is at least a slight misnomer; it is not independent of the process of continental accretion. As pointed out in Section 7.3, the differentiation of the crust from the mantle is an important contributor to the energy budget of the tectonic engine. Some compromise between the thermally driven (convective) tectonic engine and the chemical engine of Ringwood and Green (1966) (see Fig. 4.3) is required. However, the fundamental fact that the mantle is undergoing large-scale motion can hardly be in doubt and little more than this need be assumed in order-of-magnitude estimates of its rheological properties. In Section 7.3 convective cells of average radii 2000 km are assumed, in accordance with Menard's (1965) estimate from the pattern of ocean ridges, but this figure can be varied without significantly affecting the conclusions.

Movement of an apparently different type is occurring in the Baltic and in the Great Lakes region of Canada, which were heavily loaded with ice during the most recent glaciation and are now apparently springing back toward isostatic balance. Contoured rates of rise for Fenno-Scandia are shown in Fig. 7.6. These observations have been widely used to estimate the anelastic response of the upper mantle to applied stresses, usually in terms of a New-tonian, fluid viscosity, for which a value of 10^{21} to 10^{22} poises is required.

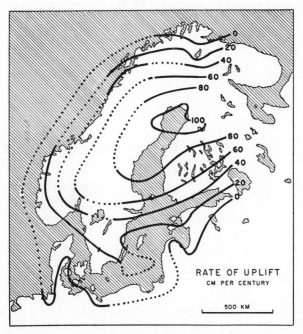

RATE OF UPLIFT
CM PER CENTURY

500 KM

Figure 7.6: Postglacial rebound of Fenno-Scandia. Reproduced, by permission, from Gutenberg (1958b). (Copyright Academic Press.)

7.2   Crystal Dislocations and Creep

Figure 7.7 shows three simple models which illustrate the elastic and anelastic responses of a solid to applied stresses. All of them involve, as a mechanical element, a dashpot with viscous fluid and they are in fact attempts to generalize to solids the Newtonian viscous behavior of liquids. These models have featured prominently in discussions of the mechanical properties of the Earth (e.g., Jeffreys, 1962; Scheidegger, 1963), the choice of model depending upon which properties are to be emphasized. They may, with suitable adjustment of available parameters, be made to fit reasonably well to laboratory observations of the variation of creep rate with time at constant stress, but they do not then extrapolate well to other observations and so cannot be applied in any realistic way to the relatively unknown conditions in the interior of the Earth. The anelastic response of a crystalline solid is normally non-Newtonian (i.e., nonlinear in stress) and we can expect to comprehend the rheological behavior of the Earth only by taking account of the fundamental atomic processes involved. A useful introduction to the mechanical properties of solids is by Cottrell (1964) and an application to the Earth by Orowan (1967b).

There are still few fundamental measurements of creep in rocks. The most comprehensive experiments are those of D. Griggs and co-workers who

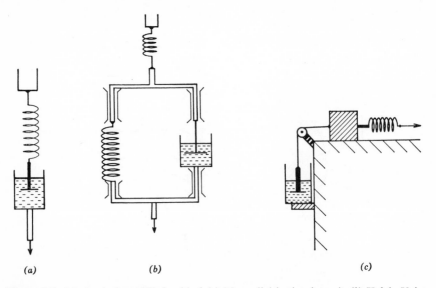

(a)   (b)   (c)

Figure 7.7: Mechanical models for ideal (*a*) Maxwell (elasticoviscous), (*b*) Kelvin-Voigt (firmoviscous), and (*c*) Bingham (plasticoviscous) solids. Firmoviscosity is often represented without the immediate elastic response of the top spring. Note that to move the block in (*c*) friction must be overcome and this represents a finite yield point.

conclude (Griggs et al., 1960): "Our results with rocks and rock-forming minerals without exception follow the empirical laws developed in the study of metals." We are therefore justified in making use of the more extensive literature on creep in metals and applying the basic conclusions to our consideration of creep in nonmetallic *crystalline* materials. Fundamental solid state concepts of creep are applied to rocks and to the mantle by Murrell and Misra (1962; see also Misra and Murrell, 1965), Stacey (1963b), and by several authors in the conference proceedings edited by Tozer (1967).

Probably the most important creep process depends upon movements of extended linear crystal defects, known as dislocations, which occur in all crystals and are treated comprehensively by Cottrell (1953) and Friedel (1964); a more elementary discussion is by Kittel (1966). There are two basic types of dislocation, represented in Fig. 7.8. Both types can, in principle, be produced in a perfect single crystal by making a half-cut through it and displacing atoms on opposite sides of the cut by a single lattice spacing. The displacement is known as the Burgers vector, which is normal to the dislocation axis in the case of an edge dislocation and parallel to the axis of a screw dislocation.

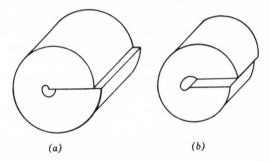

(a)   (b)

Figure 7.8: Forms of the atomic displacements in (a) edge and (b) screw dislocations. The dislocations are at the centers of the cylindrical crystals which are drawn as hollow only for convenience in representation. Actual atomic displacements very close to the dislocation are non-Hookeian and are not quite as simple as represented here. Leftward movements of the dislocations in this figure cause increasing distortions of the originally cyclindrical bodies. The cumulative effect of movements of several dislocations is indicated in Fig. 7.9.

To proceed from state (a) to state (b) in Fig. 7.9 by displacing two planes of atoms coherently would require a stress of about $q/30 \approx 3 \times 10^4$ kg cm^{-2}, where q is Young's modulus. This is much greater than the strength of all but a few specially prepared materials, such as iron whiskers, which are free of defects. However, the same displacement can occur relatively freely by the progressive movement of a dislocation. For this reason the creep behavior of most materials is determined by the defects. Dislocations are, however, not

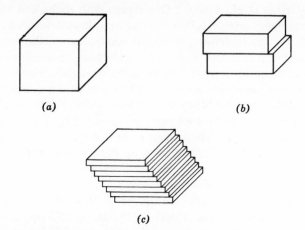

(a)   **(b)**

(c)

Figure 7.9: Cubical sample (*a*) before and after the movement through it of (*b*) one and (*c*) several dislocations. The deformation of single crystals commonly occurs on preferred slip bands, on each of which there are many atomic displacements, so that a sheared specimen may have the macroscopic appearance of (*c*).

completely free to move. They are impeded by interactions with one another and with other crystal defects (lattice vacancies, etc.) and the rate-controlling process, which determines how fast creep proceeds, is the overcoming of the obstacles (dislocation climb; see Weertman, 1955). Depending upon the nature of the obstacles, there are alternative mechanisms, but basically they depend upon diffusion processes, so that it is self-diffusion which is ultimately the rate-controlling process in creep, particularly at the high temperature with which we are concerned in the Earth.

We can therefore represent the probability dP of a particular dislocation climb event occurring in time dt by the equation for a rate process:

$$dP = C \exp\left(-\frac{U}{kT}\right) dt \tag{7.1}$$

where U is the activation energy for self-diffusion and kT is the thermal energy at temperature T. C is a numerically large frequency factor, generally referred to as the attempt frequency, which represents the frequency with which the dislocation may attempt to climb the barrier and may be related to the frequency spectrum of thermal lattice vibrations (phonons) but is generally smaller. Thus the rate of increase in the deformation ϵ of a crystal by creep is

$$\dot{\epsilon} = A \exp\left(-\frac{U}{kT}\right) \tag{7.2}$$

where A includes C as well as the number density of dislocations available to move and the contribution to creep deformation which occurs each time one does so.

In general the activation energy U is a function of the physical state of the body, depending upon confining pressure p, temperature T, applied stress σ, and thermal and mechanical history, in particular the integrated deformation ϵ. Now consider a body creeping in a physically steady state, i.e., it is macroscopically unaffected (except as regards its shape) by the progressive deformation and U is therefore independent of ϵ; p and T remain constant and we are interested in the dependence of $\dot\epsilon$ upon σ. The activation energy for atomic movements in the direction aided by the stress is reduced and for small or moderate stresses a linear relationship

$$U = U_0 - \beta\sigma \tag{7.3}$$

suffices. This gives the exponential law of steady state creep:

$$\dot\epsilon = \left[A \exp\left(- \frac{U_0}{kT} \right) \right] \exp\left(\frac{\beta\sigma}{kT} \right) \tag{7.4}$$

At very small stresses Eq. 7.4 is insufficient because it neglects the possibility of atomic displacements occurring in the direction opposed by the stress, for which the activation energy is

$$U' = U_0 + \beta\sigma \tag{7.5}$$

Thus, subtracting from Eq. 7.4 a reversed creep with activation energy U', we obtain the more general expression for steady state creep:

$$\dot\epsilon = \left[2A \exp\left(- \frac{U_0}{kT} \right) \right] \sinh\left(\frac{\beta\sigma}{kT} \right) \tag{7.6}$$

This is commonly referred to as the Eyring equation, which includes as a special case $(\beta\sigma/kT \ll 1)$, the behavior of viscous fluids. If the frequency factor C in Eq. 7.1 is independent of σ, then the square-bracketed factors in Eqs. 7.4 and 7.6 are constants, so long as the material remains in a steady state. However, an analysis of dislocation climb by Weertman (1955, 1957) led to a rather different steady state creep law, in which it is the pre-exponential factor which is stress-dependent:

$$\dot\epsilon = A'\sigma^n \exp\left(- \frac{U}{kT} \right) \tag{7.7}$$

where U is the energy of self-diffusion, independent of σ, and $n \approx 3$ to 4. There is no certainty which of the equations (7.6, 7.7) is more applicable to the high temperatures and generally low creep rates in the interior of the Earth and it appears likely that some compromise may be better than either of them, i.e., that both the activation energy and the frequency factor in Eq. 7.1 are stress-dependent. Experiments on dislocation velocities in lithium fluoride crystals by Johnston and Gilman (1959) indicate the validity of Eq. 7.6 over a wide range of creep rates and Heard (1963) found its application to geological materials satisfactory, but more recently Heard (1968) favored Eq. 7.7.

Steady state creep is normally observed in solids which are deformed slowly at absolute temperatures above half their melting points, so that we must expect a steady state law to apply to all but the outer 20 km of the Earth. Physically, the requirement for steady state creep is that the temperature be high enough to cause spontaneous recovery from the work-hardening produced by deformation.

Other creep processes which have been considered in connection with mantle convection include grain boundary sliding and high temperature creep by pure diffusion (Nabarro-Herring creep). The latter is only effective with small crystal sizes, which are unlikely to be found at appreciable depth within the Earth, but the grain boundary sliding process must necessarily occur where individual crystals are deformed. A boundary between grains, whether of the same or different materials, is a confused termination of two crystal lattices, which can only lock together in a very irregular fashion. If the materials are the same and the angle between the crystallographic axes is small, then the boundary becomes simply a plane of spaced dislocations. Then by increasing the angle, or changing the lattice spacings of the crystals, we multiply the dislocations until they become indistinguishable. The movement of a crystal boundary in creep is a result of atomic displacements, as in dislocation movements. We therefore expect a similar creep law to apply and since both processes proceed together, whichever is the more difficult will become the rate-controlling process. It is noted, however, that flow processes which involve simple frictional sliding between grains are very strongly inhibited by pressure (Paterson, 1967) and are therefore discounted in the Earth, except possibly in the upper part of the crust.

The law of steady state creep appropriate to flow under prolonged, steady stress does not apply to oscillatory stresses. To see why this is so we must consider the motion of dislocations in more detail. Dislocations do not necessarily extend to the boundaries of a crystal; they may terminate instead on a lattice defect, such as an impurity. Even if dislocations extend past such point defects, they are pinned to them and even more strongly pinned to the points of intersection of dislocations with different directions. Crystals are interlaced with a tangle of dislocations resembling the tinsel on a well-decorated Christmas tree, the dislocations being locally pinned at their intersections but free to move on the untangled sections between. The effect of the pinning is to restrict movement of a dislocation so that instead of advancing uniformly under the influence of a stress field it forms loops in the manner of Fig. 7.10. The stress field of a dislocation is associated with a definite strain energy and, as the length of the dislocation increases, energy must be supplied to it by a favorably oriented stress field. If the stress is very small the dislocation can only extend to a small loop as in Fig. 7.10*b* and this is the situation of interest in the damping of small amplitude stress waves (Section 7.4). Orowan (1967a) disputed the applicability of this "vibrating string" model of the anelastic response of a dislocation to small oscillatory stresses. There are other mechanisms of anelastic damping and this problem is considered further in Section 7.4.

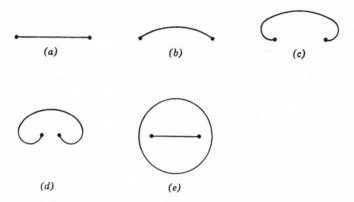

(a) (b) (c)

(d) (e)

Figure 7.10: Progressive movement of a length of dislocation pinned at its ends, showing expansion to form a loop which breaks away (the Frank-Read mechanism of dislocation multiplication).

Under a sufficiently large stress a dislocation loop may expand indefinitely and eventually break away from its pinning points, allowing a new loop to form. The loop expansion is opposed by barriers due to interactions with other loops from different sources and by lattice vacancies, etc., and so prevents further loop multiplication by repelling further dislocations of the same sign. (This occurs because a double dislocation would have twice the strain energy of two well separated single dislocations.) The material is then work-hardened. The work-hardening may be removed by heating (annealing) but at low temperatures it makes creep progressively more difficult. A detailed review of the work-hardening process is given by Nabarro et al. (1964).

Many laboratory observations are of *transient creep*, in which the creep rate at constant stress decreases with time by virtue of work-hardening. The two most important laws of transient creep may be derived very simply from Eq. 7.4 (Stacey, 1963b). If work-hardening occurs such that the increasing difficulty in causing activation of creep increments can be represented by a linear increase of U_0 with integrated creep ϵ,

$$U = U_0 - \beta\sigma + \alpha\epsilon \tag{7.8}$$

then by integrating (7.2) we obtain the logarithmic creep law:

$$\epsilon = \frac{kT}{\alpha} \ln\left\{\frac{\alpha A}{kT} \exp\left[-\frac{(U_0 - \beta\sigma)}{kT}\right] \cdot t + 1\right\} \tag{7.9}$$

This is observed at low temperatures in the initial stages of work-hardening, which cannot, however, increase indefinitely but tends to saturate, in which case we obtain Andrade's Law:

$$\epsilon \propto t^n, \qquad n \approx \tfrac{1}{3} \tag{7.10}$$

The transient creep laws must be presumed to apply to the crust, but the bulk of the Earth satisfies the conditions necessary for the observation of steady state creep—temperature above half of the melting point and slow creep rates. A steady state law must therefore be used to appraise the mechanics of convection in the mantle.

7.3 Strength of the Mantle and Convection

The activation energy U in Eq. 7.2 is, as already noted, a function of several parameters which represent the physical state of a body and a completely general treatment of the behavior of rock under the conditions of temperature and pressure in the interior of the Earth is not possible. Nevertheless, a very simple argument shows what to expect. The complexity of the problem is reduced greatly by restricting consideration to steady state creep, so that U is independent of ϵ, and in Section 7.2 a simple linear dependence of U upon σ was shown to lead to a creep law in excellent accord with observations. Stacey (1963b) further assumed that U was linear also in pressure p and temperature T:

$$U(\sigma, p, T) = U_0 - \beta\sigma + \gamma p - \delta T \tag{7.11}$$

where β, γ, δ are constants. This generalizes Eq. 7.3 to situations in which p and T are not constant, but U is independent of ϵ, i.e., the creep is a steady state effect; roughly speaking, Eq. 7.11 applies to the mantle but not to the crust, where we expect a transient creep law to apply. There is no justification for supposing that the dependences of U upon p, T, and σ are linear and independent at the extreme pressures and temperatures of the lower mantle, but we can reasonably expect (7.11) to apply approximately to the upper mantle. Making U the subject of Eq. 7.2, equating to (7.11) and then rewriting with σ as subject, we obtain

$$\sigma = \left[\frac{U_0}{\beta}\right] + \left[\frac{\gamma}{\beta}\right]p - \left[\frac{\delta}{\beta} - \frac{k}{\beta}\ln\left(\frac{\dot{\epsilon}}{A}\right)\right]T \tag{7.12}$$

This gives the (deviatoric) stress required to cause creep at a rate $\dot{\epsilon}$ at pressure p and temperature T. Owing to the logarithmic dependence of σ upon $\dot{\epsilon}$, there is in effect a distinct value of stress below which the creep rate is insignificant and above which it increases exponentially, as is apparent from Eq. 7.2, which has been assumed. Thus by putting the creep rate equal to a just significant value, the precise selection of which is not important because it occurs as a logarithm, we obtain the creep strength σ_s. The square bracketed factors in Eq. 7.12 being constants, we have, in terms of new constants λ and μ,

$$\sigma_s = \sigma_{s0}[1 + \lambda p - \mu(T - T_0)] \qquad (7.13)$$

where $\sigma_{s0} = U_0/\beta$ is the creep strength at zero pressure and standard temperature T_0. The retention of finite T_0 is not necessary but serves as a reminder that the analysis only applies at temperatures above half of the melting point. For greater generality Eq. 7.6 should have been used instead of (7.2) to obtain an equivalent of (7 13), which is an approximation valid under the condition $\beta\sigma \gg kT$. However, the opposite extreme, $\beta\sigma \ll kT$, is the condition for a Newtonian viscous liquid ($\dot{\epsilon} \propto \sigma$ in Eq. 7.6), which is sufficiently inappropriate that we may take Eq. 7.13 as a reasonable approximation.

The variation of pressure with depth within the Earth is known quite reliably and reasonable estimates of internal temperature are available (Chapter 9). By using these in Eq. 7.13, it is possible to derive two important conclusions about the variation of strength with depth, without having an explicit knowledge of the constants in the equation. The most important condition to be satisfied is that the mantle has finite strength at all depths, so that a lower limit can be assigned to the value of λ. The temperature at which the strength vanishes at zero pressure can be referred to crudely as the zero pressure melting point, T_{m0}, although complete melting of the mixed phases in a rock would occur only at a substantially higher temperature. Then

$$1 = \mu(T_{m0} - T_0) \qquad (7.14)$$

and substituting for μ in (7.13) we obtain

$$\frac{\sigma_s}{\sigma_{s0}} = \left(\frac{T_{m0} - T}{T_{m0} - T_0}\right) + \lambda p \qquad (7.15)$$

The approximately linear variation of p with depth z and the strongly convex curve of $T(z)$ compel the inference that the upper mantle has a region of greatest weakness at a depth of 100–150 km. Knowing the variation of melting point with depth (see Eq. 9.30 and Fig. 9.4), we can obtain a quantitative estimate of the variation of strength with depth, since strength is zero at the melting point ($T = T_m$):

$$0 = \frac{T_{m0} - T_m}{T_{m0} - T_0} + \lambda p \qquad (7.16)$$

Thus Eq. 7.15 can be written in a form convenient for plotting strength versus depth directly from the data in Fig. 9.4:

$$\frac{\sigma_s}{\sigma_{s_0}} = \frac{T_{m0} - T}{T_{m0} - T_0} + \frac{T_m - T_{m0}}{T_{m0} - T_0} = \frac{T_m - T}{T_{m0} - T_0} \tag{7.17}$$

Although Fig. 7.11 can be extended into the lower mantle, its validity there becomes increasingly doubtful because the pressure is so high that we must expect the pressure and temperature dependences of σ_s to become nonlinear and to involve cross-terms. Moreover, we can hardly extrapolate Eq. 7.17 with confidence through the mineral phase changes which begin below 350 km. Nevertheless, it is difficult to see any reason why the creep strength should not continue to increase with depth.

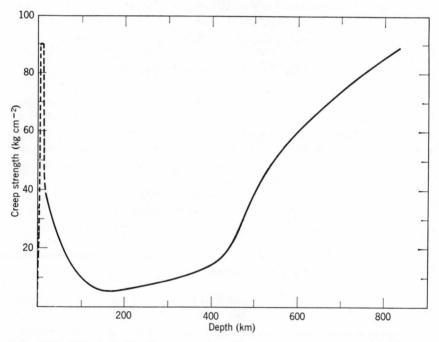

Figure 7.11: Variation of creep strength with depth in the upper mantle according to Eq. 7.17, with temperature T and melting point T_m from Fig. 9.4 (solid curves). T_{m0} is assumed to be 1400°K. The scale is normalized somewhat arbitrarily to a strength of 10 kg cm^{-2} at 100 km depth. Above 20 km, $T < \frac{1}{2}T_m$ and the equation is not strictly applicable; transient creep occurs in this range and the curve has therefore been extrapolated to strengths compatible with the dislocation theory of earthquakes. At very small depths the strength is presumably limited more by friction across joints and cracks (which may be greatly reduced by fluid pore pressure; see Section 7.5) than by the intrinsic properties of the crystals and diminishes rapidly towards the surface.

The forces causing convective motion in the mantle must be sufficient to overcome its strength. It is possible to postulate adequate forces in terms of slight density variations because of the enormous scale of the convection, but the energy requirement imposes a severe restraint on possible models (Stacey, 1963b, 1967b). We have to explain motion of about 4 cm year^{-1} in " cells " of order 2000 km $= 2 \times 10^8$ cm radius and this must involve shearing of the mantle at a rate of about 2×10^{-8} year^{-1} or rather more if the upper mantle only is involved. Meinesz (1962) suggested that the effect of a finite yield point was to make the convection erratic and that each overturn of a convective cell was followed by a quiescent period, but compatibility of the rate of observed currently active deformation and the average rate over hundreds of millions of years is evidence against this hypothesis. The energy dissipation caused by the convective motion can be estimated from the simple equation

$$\text{Work} = \text{Force} \times \text{Distance}$$

or its time-derivative

$$\text{Power} = \text{Force} \times \text{Velocity}$$

which, applied to the deformation of solid material, becomes

$$\text{Power dissipated} \left(\frac{dE}{dt} \right) = \text{Stress } (\sigma) \times \text{Rate of deformation } (\dot{\epsilon}) \quad (7.18)$$
$$\times \text{Volume } (V)$$

In the present case we equate the stress to the creep strength σ_s of the convecting part of the mantle. The rate of deformation is about 2×10^{-8} year^{-1} if the whole mantle (9×10^{26} cm^3) participates in the convection or proportionately more if a relatively thin layer in the upper mantle only is convecting, so that Eq. 7.18 gives

$$\frac{dE}{dt} = 1.8 \times 10^{19} \, \sigma_s \, \text{ergs year}^{-1} \quad (7.19)$$

This energy dissipation may be compared directly to the maximum energy available from the convecting engine, to obtain a maximum value of σ_s which is thermodynamically compatible with convection:

$$\frac{dE}{dt} = \eta \frac{dQ}{dt} \quad (7.20)$$

where η is the thermodynamic efficiency of the convective engine in producing (and dissipating) mechanical energy in the process of transferring heat to the surface, and $dQ/dt = 10^{28}$ ergs year^{-1} is the total geothermal flux. Assuming a thermodynamically ideal convective cycle in an upper mantle with an adiabatic temperature gradient, Stacey (1967c) found the efficiency to be

$\eta = 0.13$. Allowing that at least 50% of the heat flux originates in crustal radioactivity, which cannot contribute to the convective energy, that the mantle heat sources are distributed and not concentrated at the bottom of the convective layer, and that the temperature gradient must exceed the adiabatic gradient in a substantial part of the upper mantle, further reducing η, we find that a generous upper limit is $\eta = 0.025$. Using this value in Eqs. 7.19 and 7.18 we obtain the upper limit for the strength of the convecting part of the mantle, $\sigma_s = 1.4 \times 10^7$ dynes cm^{-2} (14 kg cm^{-2}). But the stresses in the lower mantle must be 10^8 dynes cm^{-2} to support the nonequilibrium ellipticity (Section 2.1), so that convection in the lower mantle must be discounted. The same conclusion is reached if the lower mantle is assumed to be a liquid with viscosity of 10^{26} poise, as has been postulated to explain the ellipticity.

Having restricted convection to the upper mantle, we must examine the role which chemical differentiation may play (Stacey, 1968; Ringwood and Green, 1966; see Fig. 4.3). Following the conclusions of radioactive age measurements, discussed in Section 8.3, we can suppose that the crust has been derived from the upper mantle, whose depth d we can assume to be 1000 km, at a uniform rate over a period $\tau = 3 \times 10^9$ years. A total volume $V = 10^{10}$ km^3 of crustal material has appeared, with a density contrast of $\Delta\rho = 0.5$ gm cm^{-3} relative to the ultrabasic mantle. Since the average depth from which this material was derived is assumed to be $d/2 = 500$ km, the average annual release of gravitational energy is

$$\frac{dE}{dt} = \frac{V\Delta\rho g d}{2\tau} = 0.8 \times 10^{26} \text{ ergs year}^{-1} \tag{7.21}$$

Since this is smaller by a factor of only 3 than the estimated upper limit of mechanical energy available from thermal convection, it is apparent that chemical differentiation is comparable in importance to convection as a driving mechanism for mantle motions.

The role of chemical differentiation does not modify the conclusion that the mean strength of the convecting upper mantle cannot exceed about 14 kg cm^{-2}. This limit is appreciably lower than the strength of the crust deduced from dislocation models of large earthquakes (Section 4.2). The elastic shear strain released during the very shallow San Francisco earthquake of 1906 is estimated from Fig. 4.9 to be about 3×10^{-4}, which corresponds to a stress of about 90 kg cm^{-2} in rock of rigidity 3×10^5 kg cm^{-2}; a similar stress is deduced from the model of the Alaskan earthquake represented in Fig. 4.12. Assuming validity of the convection hypothesis, this figure cannot be representative of the upper mantle; the strength must decrease with depth. Dislocation models are not yet sufficiently sophisticated to accommodate a strength varying with depth, but the relevant equations can be handled by digital computer if geodetic observations of surface displacements can be made accurate enough to justify this complication.

A reason for supposing convection to be confined to the upper mantle, independently of the consideration of strength, is that the convective tendency is greatest where the temperature gradient is strongest. The temperature gradient in the lower mantle is certainly much lower than in the top few hundred kilometers (see Section 9.3, especially Fig. 9.4) and may be less than the adiabatic gradient, below which convection cannot occur even in a liquid. Elsasser (1963) discussed the restriction of convection to the upper mantle and pointed out that it must be of anisometric form, i.e., with short vertical limbs and long horizontal ones, instead of the neatly symmetrical and equi-dimensional (isometric) cells which are envisaged for whole-mantle convection. Anisometric convection is a natural consequence of the variation of strength with depth, and particularly of the existence of a weaker layer in the upper mantle.

7.4 Damping of Stress Waves

Vibrating solids dissipate mechanical energy by a number of processes, known collectively as *internal friction*. The standard laboratory techniques for measuring internal friction are to observe the damping of oscillatory motions which apply cyclic stresses to a solid under investigation, commonly a resonant oscillation of the solid itself, or to measure the attenuation of acoustic or ultrasonic waves. For the Earth the equivalent observations are damping of free oscillations and seismic waves. We also consider the extent to which the mantle may be responsible for tidal friction and damping of the Chandler wobble. There are numerous mechanisms of internal friction and we have no certain knowledge which of them is most important in the Earth, although three now appear most likely to be important: dislocation damping, grain-boundary "sliding," and stress-induced ordering.

Dislocation damping in metals is reviewed by Niblett and Wilks (1960) and its application to the mantle by Gordon and Nelson (1966). A simple theory which accords well with many observations is that lengths of disloca-tions between points of pinning are bowed out by a stress field as in Fig. 7.10b and oscillate with the stress in the manner of a stretched string, but lag in phase. The consequent lag in strain causes mechanical dissipation. Orowan (1967a) doubted the effectiveness of dislocation damping in the mantle and prefers grain boundary movement. The boundary between two crystals must be irregular on the atomic scale and many of the atoms in the boundary will have alternative positions of approximately equal energy, separated by low energy barriers, so that they may readily jump over under the influence of thermal activation. Application of a stress will favor occupation of certain positions and reversal of the stress will favor others, so that the atoms jump to and fro in an oscillatory stress field. Again the lag in phase is responsible for energy dissipation. Anderson (1967a) favored stress-induced ordering of small interstitial or substitutional atoms, particularly in the lower mantle. In a stressed crystal lattice the occupation of certain of the lattice (or interstitial) sites may be energetically favored. For example, the centers of the cube edges in a body-centered cubic crystal are interstitial sites forming three sublattices; one set of these sites is dilated more by extension of the crystal than the others but contracts more under compression. As with the grain boundary movements, the atoms jump to and fro, but with a phase lag relative to a cyclic stress.

None of these processes appears anelastic damping adequate to explain the observations on rocks. All three processes are basically quite similar relaxation phenomena and give attenuation which is strongly dependent upon frequency of the oscillations, being greatest for frequencies corresponding to the relaxation times for the processes of atomic diffusion. Damping independent of frequency and amplitude, which is normally observed in rocks and is presumed to occur in the mantle, requires that there be a number of diffusion processes with a wide spectrum of activation energies. The activation energies are very much smaller than those responsible for steady creep and the superposition of a substantial steady stress has no significant effect upon the damping of a small oscillatory stress (Baker, 1957). The extent of anelastic deformation can be seen from the fact that 2 to 3 % of the elastic strain energy in a rock is dissipated by internal friction in each cycle of a stress wave, so that the anelastic yielding must be at least this fraction of the elastic deformation, and is enormously greater than the creep rate produced by a steady stress of similar magnitude. The attenuation of seismic waves by internal friction is unaffected by convection in the mantle and therefore gives no evidence of it (Stacey, 1963b, 1967b). Arguments used by opponents of continental drift who maintained that convection is incompatible with the seismic behavior of the mantle as an elastic solid are incorrect.

The magnitude of internal friction is expressed in terms of the parameter Q, which may be defined by the loss ΔE of strain energy E per cycle of a stress wave:

$$\frac{\Delta E}{E} = \frac{2\pi}{Q} \tag{7.22}$$

The mechanical Q considered here is mathematically equivalent to the Q of an oscillatory electrical circuit; the relationship is made more obvious by an alternative definition which refers to the forced vibration of a specimen with resonance frequency f, the amplitude of vibration being reduced by $1\sqrt{2}$ at frequencies $\Delta f/2$ on either side of resonance. Then

$$Q = \frac{f}{\Delta f} \tag{7.23}$$

Equations 7.22 and 7.23 are valid for $Q \gg 1$. For cyclically stressed material the energy loss per cycle is the area of the mechanical hysteresis loop (Fig. 7.12), in which the phase of the strain lags behind that of the stress. Assuming the lag to be represented by a constant phase angle ϕ, then

$$\tan \phi = \frac{1}{Q} \tag{7.24}$$

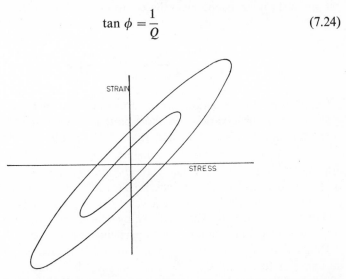

Figure 7.12: Mechanical hysteresis due to internal friction. The area of the loop represents the energy dissipation per cycle of oscillatory stress, the width of the loop being here greatly exaggerated. Note that the energy is proportional to the *square* of strain amplitude, so that for Q to be amplitude-independent the loop area must be proportional to the square of its length, i.e., width proportional to length.

Knopoff (1964a) has reviewed the evidence for frequency-independence of Q in rocks, including field evidence of attenuation of local seismic signals; although there is no compelling reason for frequency-independent Q under all conditions from solid state considerations, we suppose that it applies to all depths in the mantle and for all seismic frequencies, including the very low frequencies of free oscillations. With this assumption there are several ways of determining the variation of Q with depth in the mantle from the damping of free oscillations and frequency spectra of body waves (Anderson, 1967a). There is sufficient agreement between the results of different methods to make frequency-independence of Q a reasonable assumption, but it is nevertheless an assumption that needs careful examination since the conclusions about Q within the Earth are otherwise quite invalid.

Values of Q are higher for compressional waves than for shear waves. In liquids the propagation of sound waves is purely compressional and Q is very high indeed. Also the $_0S_0$ radial mode of free oscillation, in which the strain is purely compressive with no shear component, has an exceedingly high Q

(Slichter et al., 1966.)* These observations invite the conclusion that it is only the shear component of strain which is responsible for the attenuation of compressional waves in solids. If this is so then we expect the ratio of the Qs for compressional and shear waves to be simply the ratio of total strain energy to shear strain energy in the compressional wave (see also Anderson, 1967b):

$$\frac{Q_P}{Q_S} = \frac{k + \frac{4}{3}\mu}{\frac{4}{3}\mu} = \frac{3}{2}\left(\frac{1 - v}{1 - 2v}\right) \approx 2.38 \tag{7.25}$$

$v \approx 0.27$ being Poisson's ratio for the upper mantle, where anelastic losses are greatest. From observations on short-period waves propagated through the core, Kanamori (1967) estimated that $Q_P/Q_S = 1.90$ for the mantle, with $\overline{Q_S} = 230$, averaged for the whole mantle. This result indicates that P wave attenuation is not simply shear strain dissipation; theoretically the situation is unclear and further observations are desirable. However, for P and S waves of similar periods, having different velocities ($V_P/V_S \approx 1.8$), the S waves are attenuated more rapidly with distance, by a factor of about 4. To be meaningful, values of Q must be referred to particular wave types.

Detailed information on Q in the mantle has been obtained from the damping of free oscillations of different periods. These oscillations are, in effect, standing surface waves which penetrate to different depths within the mantle, the waves of longer periods sampling greater depths. The frequency dependence of Q for the oscillations is attributed to the depth-dependence of Q, the greater persistence of the lower modes being due to the fact that they stress the lower mantle in which Q is very high. By comparing the decays of torsional modes, Anderson and Archambeau (1964) obtained a profile of Q_S for the mantle (Fig. 7.13), which is quite similar to the variation of creep strength with depth (Fig. 7.11). Materials with high yield points also have high Qs, but there is no known quantitative correspondence between the two.

The observed increase in Q with period for free oscillations, which is naturally explained by the Q versus depth curve, allows us to discount as most improbable the supposition that the intrinsic Q of the mantle actually decreases with period of oscillation. This is important in discussing the role of the mantle in tidal friction and damping of the Chandler wobble. However, Press (1966) has pointed out that aftershocks may reinforce the free oscillations generated by a large earthquake, and thus lead to a systematic overestimate of Q. It is therefore important that free oscillation data are supplemented by body wave attenuation measurements.

---

* It is possible that the slight damping of the $_0S_0$ mode is due primarily to the development of shear strains across horizontal inhomogeneities in the upper mantle.

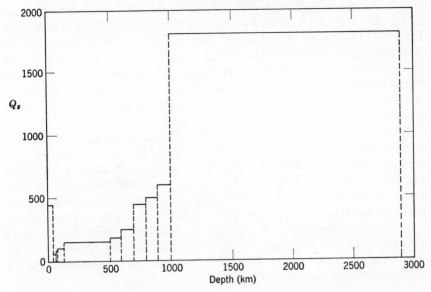

Figure 7.13: Q_S for shear waves in the mantle as a function of depth, from data quoted by Anderson (1967a), on the basis of the analysis by Anderson and Archambeau (1964). Model MM8 down to 1000 km depth and the mean of models F, G, H from 1000 to 2900 km.

The lunar tide causes a cyclic strain of the mantle with a period $\tau = 4.46 \times 10^4$ sec (13 hours) and amplitude ϵ about 5×10^{-8}. For the whole mantle, volume $V = 9 \times 10^{26}$ cm^3, the tidal strain energy is

$$E = \tfrac{1}{4}\mu\epsilon^2 V \tag{7.26}$$

By Eq. 7.22 the loss of energy per cycle is

$$\Delta E = \frac{2\pi}{Q} E \tag{7.27}$$

and the rate of dissipation of energy is therefore given by

$$\frac{dE}{dt} = -\frac{2\pi}{Q} \cdot \frac{E}{\tau} = -\frac{\pi\mu\epsilon^2 V}{2Q\tau} \tag{7.28}$$

where μ is the rigidity of the mantle. Taking as average values for the mantle $Q_S = 250$, $\mu = 2 \times 10^{12}$ dynes cm^{-2}, we obtain a generous estimate of tidal dissipation:

$$\frac{dE}{dt} = -6 \times 10^{17} \text{ ergs sec}^{-1} \tag{7.29}$$

This is hardly a significant contribution to the total tidal dissipation of 2.74×10^{19} ergs sec^{-1}, which was obtained in Section 2.4. It is too small by a factor of 15 to explain the "missing" 10^{19} ergs sec^{-1}, tentatively ascribed to core dissipation. Since the tidal period is only about 15 times that of the slowest of the free oscillations we can discount the possibility of a sufficiently strong frequency dependence of internal friction to give an average Q 15 times smaller at the tidal period than at free oscillation periods. The mantle is therefore not an appreciable sink of tidal energy.

We can consider similarly the dissipation of wobble energy in the mantle. It is convenient to calculate the deformation of the mantle associated with the wobble in terms of the difference between the observed Chandler period τ_0 and the free Eulerian period for a rigid Earth τ_R. For wobble of amplitude α the elastic deformation is such that the effective equatorial bulge deviates by a smaller angle $[(\tau_R/\tau_0)\alpha]$ from the state of symmetrical rotation. The deformation thus has the effect of shifting the equatorial bulge by an angle $[1 - (\tau_R/\tau_0)]\alpha$. For ellipticity H [the dynamical ellipticity, $H = (C - A)/C$, will be used here since the whole mantle is involved and not merely the surface equatorial bulge] the strain of the mantle is

$$\epsilon = H\left(1 - \frac{\tau_R}{\tau_0}\right)\alpha \tag{7.30}$$

and the strain energy of the whole mantle is therefore

$$E = \tfrac{1}{2}\mu\epsilon^2 V = \tfrac{1}{2}\mu H^2\left(1 - \frac{\tau_R}{\tau_0}\right)^2\alpha^2 V \tag{7.31}$$

By Eqs. 7.22 and 7.31 the energy dissipation per cycle is

$$\Delta E = \frac{2\pi}{Q} E = \frac{\pi\mu}{Q} H^2\left(1 - \frac{\tau_R}{\tau_0}\right)^2\alpha^2 V \tag{7.32}$$

This may now be related to the total energy of the wobble E_w, which has to be dissipated to damp the wobble. E_w is much larger than the strain energy associated with the wobble and must not be confused with it. From Eq. 2.43,

$$E_w = \tfrac{1}{2}AH\omega^2\alpha^2 \tag{7.33}$$

and the time constant for decay of the wobble, due only to internal friction in the mantle, is n wobble periods where

$$n = \frac{2E_w}{\Delta E} = \frac{A\omega^2 Q}{\pi\mu H[1 - (\tau_R/\tau_0)]^2 V} \approx 600 \tag{7.34}$$

The factor 2 arises here because energy is proportional the square of wobble amplitude so that the time constant for decay of wobble energy is only half of the time constant for decay of the amplitude.

The spectrum of the observed wobble frequency gives a Q_w for the wobble (which must not be confused with the intrinsic Q for the mantle material) of about 30, corresponding to a damping time constant of about 10 periods. With a mantle Q of 250, internal friction is therefore deficient by a factor of 60 to explain "damping" of the wobble. A mantle Q as low as 4 at the wobble frequency can hardly be expected, although there are no other observations which compel us to rule out this possibility, and the mantle has been examined as a possible sink of wobble energy by numerous authors. However, as pointed out in Section 2.5, core-mantle coupling provides a mechanism for both generation and damping of the wobble and there is no need to postulate extraordinary anelastic damping.

7.5   Origin of Seismic Stresses and the Earthquake Mechanism

Theories of the focal mechanism(s) of strain release in earthquakes are discussed in Section 4.2. These theories are based on the strain rebound hypothesis, which assumes that elastic strain released during an earthquake was previously built up in the focal region over a prolonged period. Here we are concerned with the processes by which the elastic strain is built up and with the mechanism of earthquake triggering. These are major unsolved problems, but observations impose some boundary conditions on conjecture.

The simplest supposition, that stress builds up across a fault linearly with time until a definite breaking point is reached, is readily dismissed. There is no regularity in the intervals between major earthquakes at any point on a fault, although the intervals are sufficiently long that statistical evidence on this point is not good; perhaps a hundred years would be a typical interval in California, but shorter periods in Japan and Chile. Taking the magnitude of the stress release in an earthquake to be 10 kg cm^{-2} ($10^7 \text{ dynes cm}^{-2}$), the required rate for linear build-up would be about $0.1 \text{ kg cm}^{-2} \text{ yr}^{-1}$ or 3×10^{-3} dynes $\text{cm}^{-2} \text{ sec}^{-1}$. This is slower by a factor of 1000 than the rate of change of lunar tidal stress in the crust, which has an amplitude of about 5×10^4 dynes cm^{-2} cycled every 13 hours and therefore a peak rate of change of about 7 dynes $\text{cm}^{-2} \text{ sec}^{-1}$. If the linear stress buildup were a valid hypothesis, we would expect earthquakes on a particular fault, having a common orientation of stress release, to occur always at a particular phase of the lunar tide, when the tidal stress increases the total stress and triggers the stress release. It is usually considered that earthquake occurrences show no correlation with lunar tides (Knopoff, 1964b; Simpson, 1967).[*] Two conclusions are possible. Either the rate of stress buildup before an earthquake substantially exceeds 7 dynes $\text{cm}^{-2} \text{ sec}^{-1}$, i.e., it is highly nonlinear over the interval

---

[*] There are, however, contrary opinions (see, for example, Berg, 1966), and the above argument may not apply to rapid sequences of aftershocks.

between earthquakes and occurs almost entirely in the last two or three months or less, or, perhaps more likely, the stress itself is not directly responsible for triggering earthquakes at all. Quite possibly both conclusions are partially correct.

Except for shallow earthquakes, we must also dismiss the supposition that fault movement is a simple sliding friction across the fault face. The coefficient of friction between rock surfaces is of the order unity, so that, for sliding friction to occur, the shearing stress must exceed the normal stress across the fault. At depth, the normal stress may be equated approximately to the overburden pressure, nearly 300 kg cm^{-2} per km for rock of average density. If the creep strength of rock is σ_S kg cm^{-2}, sliding friction is only possible to a depth of about ($\sigma_S/300$) km, because at greater depths a shear stress would deform the rock bodily before causing frictional sliding. Several considerations indicate strengths of the order 100 kg cm^{-2} for the crust and less for the mantle. This limits the depth of sliding friction to about 300 m. Thus even a surface shock, such as San Francisco 1906, in which fault movement extended to a depth of only 5 km and the stress release was about 100 km cm^{-2} near to the fault, appears not to be explicable in terms of sliding friction. Many earthquakes occur at depths much greater than 5 km, in some areas down to 700 km, and in these cases, even allowing for a reasonable decrease in the coefficient of friction with depth, dry frictional sliding can have no relevance.

This argument is nullified, to some extent, if there is a fluid pore pressure in the rock. Assuming a constant coefficient of friction between rock surfaces, the shear stress to cause sliding is proportional to the normal stress between the surfaces. But hydrostatic pressure in the pores between them tends to hold the surfaces apart and so reduces this normal stress. Thus, allowing for pore pressure, the estimated depth to which sliding friction can occur is increased somewhat. Raleigh and Paterson (1965) have shown that pore pressure can actually be produced in a rock (serpentinite in their experiments) by dehydration, so that dehydration by heating could be responsible for initiating tectonic events. However, even allowing for this possibility, frictional sliding appears not to be relevant to tectonic processes deeper than about 25 km.

The inadmissibility of frictional sliding as an earthquake mechanism at depth within the Earth is avoided in the elastic rebound theory by appealing to a mechanism of unstable creep (Orowan, 1960). The necessary condition is that a zone of weakness be made selectively weaker by creep, thus concentrating deformation in the weak zone and accelerating the creep to a catastrophic rate. Normally experiments on the deformation of solids at room temperature are stabilized by work-hardening, which causes an initially weak spot to harden as it deforms, but at the high temperatures within the Earth work-hardening will be removed by spontaneous recovery (the condition for steady state creep; see Section 7.2). Various mechanisms of progressive local softening by creep have been suggested; the simplest is the temperature rise which results when a body is deformed adiabatically (Stacey, 1963b). The dimensions of a layer which may be selectively softened by a temperature rise depend upon the time scale of the deformation which causes the heating. This must be equated to the thermal time constant, which is of the order 4 days for 1 m thickness and 1 year for a 10-m thickness. Presumably the creep will be increasingly concentrated as it accelerates and the zone of final catastrophic failure will be much less than 1 m thick. Griggs and Baker (1968) have developed a more quantitative theory on this basis.* A sufficient fraction of earthquake energy may be released as heat to cause actual melting of a thin layer in the fault zone; for example, 10^{23} ergs on an area of 10^4 km^2 gives 25 cals cm^{-2}, sufficient to melt a layer of order 1 mm thick. Thus melting or partial melting may allow a complete release of local seismic stress by lubricating the fault (Griggs and Handin, 1960). In shallow shocks the lubrication could occur as a result of dehydration (Raleigh and Paterson, 1965). Byerlee and Brace (1968) have pointed out the probable relevance of laboratory observations of the stick-slip process, in which sliding rock surfaces under confining pressure exhibit earthquake-like stress drops. The phenomenon appears to be restricted to certain rock types. It is not clear whether the Griggs-Baker theory suffices to explain it.

The time scale of seismic stress buildup and/or release, which is a vital consideration in the unsolved problem of earthquake prediction, ought in principle to be derivable from a knowledge of the creep phenomenon. However, this is not yet possible and the best evidence available is from Japanese observations of local strain and tilt, which suggest that local strain in the region of an impending earthquake becomes "anomalous" during an interval of a few months before the shock (Figs. 4.35 and 4.36). Even more striking and quite unexplained are the tilts observed in a period of several hours before a shock (Fig. 4.37). There is, of course, no certainty that similar time-scales apply to other seismic areas, and it appears that the time scale increases with the magnitude of an earthquake. However, local magnetic changes observed by Breiner and Kovach (1967) to precede creep increments on the San Andreas fault, California, indicate that deep-seated stress changes occur some hours to weeks before surface movements.

Major earthquakes are commonly followed by sequences of aftershocks which eventually die away, and Benioff (1955) has shown that the integrated strain release generally follows one of two more or less regular patterns. Both are represented in Fig. 7.14, which shows separately the strain release patterns for aftershocks on opposite sides of the fault following an earthquake in Kern

---

* Griggs and Baker appealed to observations by Bridgman (1936). Relevant also are experiments by Basinski (1957, 1960), who recognized the phenomenon of thermally unstable creep.

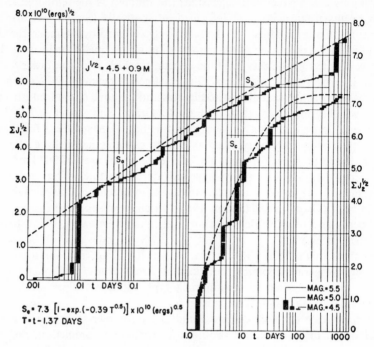

Figure 7.14: Integrated strain release for aftershock sequences on south-east (upper curve) and north-west (lower curve) of the fault following the Kern County earthquake of 1952. Reproduced, by permission, from Benioff (1955).

County, California. The strain release of each shock is taken to be proportional to the square root of its energy, as determined from the magnitude. It is difficult to draw fundamental conclusions from these curves. It would appear more significant to plot the integrated energy release, rather than the sum of the square roots of energy increments, since " strain " is not clearly identifiable with the square root of earthquake energy unless the focal volumes of all earthquakes are equal. Rather it may be supposed that the actual strain release is approximately the same for all shocks and the magnitudes are determined by the volumes from which the strain is released. However, curves of integrated energy release would be similar to those in Fig. 7.14. Aftershocks are attributed to delayed adjustment to redistributed stress in the focal region following a major shock and have been described in terms of simple mechanical models, such as Fig. 7.7*b*. If the time scale of aftershock sequences represents the time scale of seismic stresses generally, then they would appear to be in the range 1 month to 1 year or more for large shocks.

Chapter 8

RADIOACTIVITY AND THE
AGE OF THE EARTH

" In terms of observational facts alone the behaviour of the inner earth now appears even more fantastic than when we knew much less about it."

HOLMES, 1965, p. vii.

8.1 The Pre-radioactivity Age Problem

Until the discovery of radioactivity, the age of the Earth was a matter of hot dispute. Geologists required times of hundreds of millions of years for the accumulation of known sedimentary strata, but Kelvin (1899) argued that the heat flow through the crust was incompatible with an age greater than about 25 million years. It is of interest to reconsider Kelvin's argument, which is not based upon a limitation in the thermal capacity of the Earth but on the ineffectiveness of thermal conduction through the crust. The total heat capacity of the Earth, neglecting latent heat of solidification, is about 10^{27} cal deg^{-1}, which could maintain the present heat flux of 2.4×10^{20} cal year^{-1} for 4.5×10^9 years with a very modest average temperature drop of 1080°C. Thus if Kelvin had supposed that heat from the entire Earth was available to maintain the geothermal flux, he could have arrived at an age greater than the presently accepted 4.5×10^9 years. However, he considered a model of the Earth which was more plausible in terms of the physics known at the time, and assumed an initially molten crust which solidified progressively inward from the outside. The thermal gradient in the crust was determined by the fixed difference in temperature across it, i.e., melting point of rock minus surface temperature, and decreased as the crust thickened. Thus the heat flux to the surface and the consequent rate of solidification were determined by the thermal conductivity of the crust; the heat flux would have fallen to the present value after only 25 million years.

221

It is interesting to conjecture how Kelvin's model would have been modified to accord with geological evidence if the recognition of convection and radiative transfer of heat within the Earth (Section 9.3) had preceded the discovery of heat generation by radioactivity. However, an even greater difficulty arose in the case of the Sun, whose prodigious heat output can be estimated from the solar constant S (Section 1.4), the heat received per unit area at the radius of the Earth's orbit. In units convenient for the present calculation,

$$S = 1.05 \times 10^6 \text{ cal cm}^{-2} \text{ year} \qquad (8.1)$$

from which the total heat output of the Sun is found to be 3×10^{33} cal year^{-1}. The two sources of heat to which the pre-radioactivity physicists could appeal were gravitational energy, due to shrinking of the Sun, and the simple heat capacity of an initially very hot Sun. An approximate value of the gravitational energy E released by shrinking of mass M to the present size is obtained by assuming the Sun to be a sphere of uniform density and radius R, in which case

$$E_G = \frac{3}{5} \frac{GM^2}{R} = 5.5 \times 10^{40} \text{ cal} \qquad (8.2)$$

In fact, we can expect the mass to be more concentrated toward the center, so that E_G is somewhat higher. At the present rate of dissipation this energy would last for 20 million years, which is, by coincidence, in agreement with Kelvin's estimate of the age of the Earth and thus supports the estimate. Alternatively, we may assume the specific heat of the Sun to have the classical value, applied to atomic hydrogen,

$$C = 3R = 6 \text{ cal/mole deg} = 6 \text{ cal/gm deg} \qquad (8.3)$$

which gives a total heat capacity of 1.2×10^{34} cal deg^{-1}. On this basis the average, unmaintained temperature of the Sun would decrease by 0.25 deg year^{-1}. Temperatures of hundreds of millions of degrees were not seriously contemplated in Kelvin's time and since, from the fossil record, the Earth was apparently warmed by the Sun for hundreds of millions of years, the heat supply of the Sun would have remained a problem, even if Kelvin's thermal conductivity argument had been circumvented in estimating the age of the Earth. As is well known, the discovery of radioactivity offered a solution to both problems.

The uneven distribution of radioactive sources within the Earth was recognized in an early paper by Strutt (1906) (later Lord Rayleigh), who found that the concentration of radioactivity in igneous rocks greatly exceeded what was required in the Earth as a whole to produce the observed geothermal flux. He made the suggestion that the radioactivity of the Earth is confined to a crust a few tens of kilometers thick, which was, by that time, recognized seismologically as distinct from the deeper material or mantle. This shallow distribution of heat sources neatly removed the thermal conductivity problem which had led to Kelvin's erroneous conclusion.

Strutt (1906) also noted that basic igneous rocks (basalt, gabbro) were much less radioactive than the lighter acidic or granitic rocks. This observation is basic to the theory of the formation of the Earth's crust by differentiation from the mantle (Section 8.3) and to the thermal model of the Earth (Sections 9.1 and 9.3).

8.2 Radioactive Elements and the Principles of Radiometric Dating

Independently of the thermal problem, which is considered further in Chapter 9, radioactivity provides the means for precise dating of geological events. Faul (1966) gives very readable and up-to-date summary, and a monographic treatment with comprehensive references is by Hamilton (1965). The important isotopes, listed in Table 8.1, are those which have half-lives comparable to the age of the Earth and are widely distributed in sufficient quantities to be measurable in many rock types. A more complete list is given by Faul (1966). Potassium and rubidium decays are simple but the uranium and thorium decay schemes have intermediate daughter products with measured half-lives, in one case (U^{234}, granddaughter of U^{238}) exceeding 10^5 years, so that decay to lead is not immediate. However, the half-lives of all intermediate products are very small by comparison with those of the parent isotopes. Provided the ages to be measured are also much longer than 10^5 years, which is always the case with uranium-lead methods, then the estimation of elapsed time from the ratios of parent and final daughter isotopes requires only that the material under consideration be a closed system, i.e., there must be no fresh introduction to it or escape from it of any component, including all intermediate products. Fortunately, the gaseous intermediate product, radon, of which there are isotopes in all three decay series, is short-lived and appears not to diffuse out of rocks appreciably, although the steady outgassing of soil produces a measurable radon content of the atmosphere. The shorter-lived isotopes which are generated by cosmic rays in the upper atmosphere, or in interplanetary dust collected by the Earth, are not considered here, although they can be useful in dating sediments. The best known is C^{14}, which has become very important in the dating of archeological remains.

Almost all measurements of the ages of rocks and minerals use one or more of the four decay schemes of U^{238}, U^{235}, Rb^{87}, and K^{40}. Th^{232} is less useful than U^{238} and U^{235}, which, being chemically identical, exist in a ratio which is a function of time only and so, when considered together, allow an additional unknown factor in the history of a sample to be estimated or eliminated.

TABLE 8.1: IMPORTANT RADIOACTIVE ELEMENTS IN THE EARTH

Parent Isotope	Percent of Natural Element	Decay Mechanism	Stable Daughter	Decay Constant (Year^{-1})	Half Life (Years)
U^{238}	99.274	$(8\alpha + 6\beta)$ series decay	Pb^{206}	1.54×10^{-10}	4.51×10
		$4.5 \times 10^{-5}\%$ spontaneous fission	various	6.9×10^{-17}	
U^{235}	0.720	$(7\alpha + 4\beta)$ series decay	Pb^{207}	9.72×10^{-10}	7.13×10
		Neutron-induced fission	various	very small (Proportionl to neutron flux)	
Th^{232}	100	$(6\alpha + 4\beta)$ series decay	Pb^{208}	4.92×10^{-11}	1.41×10
Rb^{87}	27.85	β emission	Sr^{87}	1.39×10^{-11}	5.0×10^{10}
K^{40}	0.012	11% electron capture	A^{40}	0.585×10^{-10}	1.31×10^{9}
		89% β emission	Ca^{40}	4.72×10^{-10}	
Re^{187}	66	β emission	Os^{187}	1.7×10^{-11}	4×10^{10}

Radioactive decay is described in terms of the probability that a constituent particle in a nucleus will escape through the potential barrier binding it to the nucleus. The energies are so large (and the nuclear size is so small) that no ordinary physical conditions influence the probability that a particular nucleus will decay.* "Ordinary physical conditions" here include all pressures and temperatures in the interior of the Earth. It follows that the rate of decay of N nuclei of a particular species is directly proportional to N:

$$\frac{dN}{dt} = -\lambda N \tag{8.4}$$

where λ is the decay constant given in Table 8.1. Integrating from an initial number N_0 at time $t = 0$ we obtain the basic equation of all radioactive age work:

$$N = N_0 e^{-\lambda t} \tag{8.5}$$

* The decay of K^{40} to Ar^{40} may be slightly influenced by chemical bonding of the potassium and by pressure, because the decay in this case occurs by capture of an inner, orbital electron and thus depends upon the local electron density at the nucleus. A fully ionized, isolated K^{40} nucleus can decay only to Ca^{40}. However, under all conditions within the Earth, λ may be taken as constant for K^{40}.

The half-life of the isotope is obtained by substituting $N = N_0/2$ at $t = \tau_{1/2}$ in Eq. 8.5:

$$\tau_{1/2} = \frac{\ln 2}{\lambda} = \frac{0.69315}{\lambda} \tag{8.6}$$

If the initial concentration of an isotope is known, then the age of the body in which its present concentration is measured, can be found directly from Eq. 8.5. However, this is rarely the case.† Normally the concentration D^* of the daughter isotope is measured; then

$$D^* = N_0 - N = N_0(1 - e^{-\lambda t}) \tag{8.7}$$

so that, combining (8.5) and (8.7), we can eliminate the unknown N_0:

$$\frac{D^*}{N} = e^{\lambda t} - 1 \tag{8.8}$$

D^* is used to designate the number of *radiogenic* daughter nuclei, which are produced by decay, as, in general, the same isotope may occur independently of the decay and the non-radiogenic or *initial* contribution must be allowed for.

The simplest of the so-called accumulation clocks, which are based upon measurements of the concentrations of daughter isotopes, is K^{40}—Ar^{40}, because when an igneous rock is formed, its initial concentration of Ar^{40} is usually negligible. Correction for contamination by atmospheric argon may be necessary and can be made on the basis of the Ar^{36} and Ar^{38} contents, which are present in the atmosphere to the extent of 0.337% and 0.063%, but are not produced as decay products. This correction is very conveniently applied routinely as the measurement of Ar^{40} is made by driving the argon out of a sample by melting it in vacuum and mixing it in a mass spectrometer with a known quantity of isotopically separated Ar^{38} (the spike‡). A measurement is simply a determination of the ratios $Ar^{40}/Ar^{38}/Ar^{36}$. Potassium is commonly determined by a flame photometer comparison with a known standard and K^{40} is estimated by assuming that it is always a constant fraction (0.0119%) of the total potassium. A new method examined by Mitchell (1968) is to estimate potassium from the Ar^{39} which is produced from K^{39} by (n, p) reactions in a nuclear reactor. Since the measurement is a comparison of Ar^{39} and Ar^{40}, argon and potassium are measured in the same sample; the necessity for a spike is removed and extremely small samples can be examined. In the case of K^{40}—Ar^{40} decay, Eq. 8.8 must be modified to determine ages

---

† An important exception is C^{14} analysis.
‡ A spike is a precisely controlled addition of the element to be measured, but with a grossly different, accurately known isotope ratio.

because Ar^{40} is produced only by one of the two competing decay processes. However, the branching ratio* $\lambda_{Ar}/\lambda_{Ca}$ is well known, so that age t can be determined from the radiogenic (i.e., corrected) Ar^{40} and total K^{40} contents:

$$\left.\begin{array}{c} \dfrac{\lambda}{\lambda_{Ar}} \dfrac{Ar^{40}}{K^{40}} = e^{\lambda t} - 1 \\[1.5em] \text{where} \\[1em] \lambda = \lambda_{Ar} + \lambda_{Ca} \end{array}\right\} \qquad (8.9)$$

The most important limitation of the K—Ar method of dating is the loss of argon by diffusion, which has been studied as a function of temperature in a wide range of minerals (Fechtig and Kalbitzer, 1960). The argon loss is strongly dependent upon temperature, as expected for a rate process, but several activation energies are involved and argon appears to be held in minerals in several types of site, from some of which it diffuses out more readily than others. We can, for example, imagine that an argon atom occupying a lattice vacancy will be held much more tightly than an interstitial atom. Certain minerals, notably hornblende, resist argon diffusion better than others. The only test for diffusion losses in nature is to compare the apparent K—Ar ages of different minerals and also compare these with other age measurements. If agreement is found the ages are termed *concordant*. Lower K—Ar ages normally indicate argon loss, which may be due to metamorphic heating or to very slow initial cooling and consequent delay in the start of argon retention. Comparison between minerals also serves as a check on the possible assimilation by some minerals of original argon in magma which was incompletely outgassed when it solidified.

Another simple accumulation clock of a different kind is based upon the spontaneous fission of U^{238}. Although spontaneous fission is a very rare process, recoiling fission fragments are very energetic and cause intense radiation damage in mineral crystals. Individual tracks can be made visible for counting under a microscope by an etching process, developed by Fleischer et al. (1964, 1965, 1967). The number of fission tracks T can be treated as an accumulated daughter product of the spontaneous fission of U^{238}:

$$\frac{\lambda}{\lambda_F} \frac{T}{U^{238}} = e^{\lambda t} - 1 \qquad (8.10)$$

where λ_F is the decay constant for fission and λ is the total decay constant of U^{238}. It is not possible to count the total number of tracks in any sample

---

* The branching ratio, as here defined, gives the ratio of the two daughters which are produced by competing decay mechanisms.

because only one surface is etched, but the count is calibrated by comparison with the number of additional tracks produced by neutron-induced fission of U^{235} during a controlled neutron irradiation in a nuclear reactor.

The neutron-fission cross section of U^{235} is known ($\sigma_F = 582 \times 10^{-24}$ cm^2) and the present-day ratio $U^{238}/U^{235} = 237$ is a constant in all geological materials. Then the number of neutron-induced fission tracks is

$$T_N = \phi \sigma U^{235} \tag{8.11}$$

where ϕ is the total neutron flux, in particles per unit area, and, combining Eqs. 8.10 and 8.11,

$$\frac{T}{T_N} = (2) \frac{\lambda_F}{\lambda} \cdot \frac{U^{238}}{U^{235}} \frac{(e^{\lambda t} - 1)}{\phi \sigma} \tag{8.12}^*$$

Regarding the tracks as a daughter product, in the general sense, the method is an accumulation clock with the special merit of having no nonradiogenic daughter and depending only upon the measurement of a ratio, T/T_N, and not upon a count of an absolute number of tracks. The production of neutrons in the Earth is negligible, so that U^{235} fission in rocks can be neglected, but in work on meteorites the neutrons produced by cosmic rays lead to additional fission tracks, which must be allowed for. Fission tracks are annealed out at temperatures from about 50 to 600°C, depending upon the mineral, so that differences between fission track ages for different minerals in a rock may allow the dating of a mild metamorphism or possibly indicate uranium diffusion. Although still new, fission track dating has considerable promise, both in its own right and as a check on the isotope accumulation methods.

The other important clocks, based on Rb—Sr and U—Pb decays, are complicated by the occurrence of *original* as well as *radiogenic* daughter nuclides. In these cases the estimation of the proportion of a daughter isotope which is of radiogenic origin is made possible by two essential features in the occurrence of the elements concerned:

1. By virtue of their chemical differences the various minerals in a rock have quite different initial parent/daughter ratios.

2. At least one entirely nonradiogenic isotope of the daughter element is also present, and the initial isotopic ratio of the daughter element is homogeneous throughout the rock.

Considering the case of Rb—Sr, we can rewrite Eq. 8.7 with the addition of an original amount Sr^{87}_o of the daughter isotope Sr^{87}:

$$Sr^{87} = Sr^{87}_o + Sr^{87*} = Sr^{87}_o + Rb^{87}(e^{\lambda t} - 1) \tag{8.13}$$

---

* Equation 8.12 applies without the factor 2, if T_N is measured in a surface which is freshly cut after irradiation. If the same surface is used as for the count of T then the factor 2 is included because the surface was irradiated from both sides by the spontaneous fissions, but only from the one remaining side by neutron-induced fissions.

Measurements of relative abundances of isotopes are made by mass spectrometer comparisons using spiked samples although the equality of atomic masses of Rb^{87} and Sr^{87} necessitates a chemical separation of the two elements before measurement. All values are referred to the abundance of the nonradiogenic isotope Sr^{86}, so that

$$\frac{Sr^{87}}{Sr^{86}} = \frac{Sr^{87}_{o}}{Sr^{86}} + \frac{Rb^{87}}{Sr^{86}}(e^{\lambda t} - 1) \tag{8.14}$$

The unknowns in this equation are Sr^{87}_{o}/Sr^{86} and $(e^{\lambda t} - 1)$, but both factors are constant for all of the minerals in a rock with a simple igneous history, so that there is a direct linear relationship between the measured ratios Sr^{87}/Sr^{86} and Rb^{87}/Sr^{86} for the several minerals. From this relationship both Sr^{87}_{o}/Sr^{86} and $(e^{\lambda t} - 1)$ are determined. Since $\lambda t < 0.1$ even for $t = 5 \times 10^{9}$ yr $(e^{\lambda t} - 1) = \lambda t$ is normally a sufficient approximation. The determination of t dates the time when strontium was last isotopically homogeneous.

The Rb—Sr technique is useful in examining the metamorphic histories of rocks and several graphical methods of representing the data have been important to the development of the subject (Compston et al., 1960; Nicolaysen, 1961; Lanphere, et al., 1964). The principle is shown in Fig. 8.1. Suppose

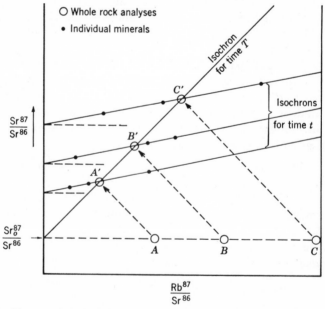

Figure 8.1: Rb-Sr evolution of three hypothetical rocks originating T years ago from a common source and undergoing simultaneous metamorphosis $t(\ll T)$ years ago. Original whole rock isotopic ratios are represented by A, B, C and present ratios by A', B', C'. Isochrons through individual mineral analyses for each rock date the metamorphic event and the isochron through the whole rock analyses dates the original magma differentiation.

three rocks A, B, C, to have differentiated from a common magma T years ago, with a common initial ratio Sr^{87}_o/Sr^{86}, but different ratios Rb^{87}/Sr^{86}, by virtue of their chemical differences. In the course of the time T, the Rb^{87} was depleted by decay and Sr^{87} was enriched correspondingly, so that the isotopic compositions of the rocks as a whole migrated from A, B, C to A', B', C' in Fig. 8.1. The line through A', B', C' is known as the T $isochron$; its gradient is, by Eq. 8.14, $(e^{\lambda T} - 1) \approx \lambda T$, from which T is determined, and the intercept gives the initial isotopic composition of Sr T years ago. However, the rocks were all reheated during a metamorphic event t years ago, much more recently than the original formation of the rocks, and during this event the minerals were changed somewhat and within each rock the strontium isotopes were rehomogenized. This gave a new, different starting point for strontium evolution in each of the rocks. The isotopes did not migrate appreciable distances so that the whole rock compositions and the determination of the T isochron were unaffected. However, isotopic measurements on individual minerals, indicated by the solid circles in Fig. 8.1, give isochrons, with gradients $(e^{\lambda t} - 1)$, for the time t when the rocks were individually rehomogenized with respect to strontium isotopes. Thus the times t and T and the original isotopic ratio of the parent magma, Sr^{87}_o/Sr^{86}, are all determined.

Lead-uranium evolution follows equations which are exact analogies of Eq. 8.14, except that there are two parallel decay schemes U^{238}—Pb^{206} and U^{235}—Pb^{207}, each of which is referred to the abundance of the nonradiogenic isotope Pb^{204}:

$$\frac{Pb^{206}}{Pb^{204}} = \frac{Pb^{206}_o}{Pb^{204}} + \frac{U^{238}}{Pb^{204}}(e^{\lambda_{238}t} - 1) \tag{8.15}$$

$$\frac{Pb^{207}}{Pb^{204}} = \frac{Pb^{207}_o}{Pb^{204}} + \frac{U^{235}}{Pb^{204}}(e^{\lambda_{235}t} - 1) \tag{8.16}$$

The simultaneous equations (8.15) and (8.16) give two semi-independent estimates of the age of a rock, in which lead and uranium isotopes can be measured for several minerals, or of a suite of related rocks. There is therefore a test for concordance solely from measurements on lead and uranium.

It is normally more convenient to make measurements of lead isotopes only; Eqs. 8.15 and 8.16 can be combined to give a relationship between Pb^{207}/Pb^{204} and Pb^{206}/Pb^{204}:

$$\frac{Pb^{207}}{Pb^{204}} = \left[\frac{U^{235}}{U^{238}} \cdot \frac{(e^{\lambda_{235}t} - 1)}{(e^{\lambda_{238}t} - 1)}\right] \cdot \frac{Pb^{206}}{Pb^{204}}$$

$$+ \left[\frac{Pb^{207}_o}{Pb^{204}} - \frac{U^{235}}{U^{238}}\frac{(e^{\lambda_{235}t} - 1)}{(e^{\lambda_{238}t} - 1)} \cdot \frac{Pb^{206}_o}{Pb^{204}}\right] \tag{8.17}$$

U^{235}/U^{238} is a constant (0.00725), so that the two bracketed quantities are constants for any set of samples which were isotopically homogeneous t years ago and have remained isolated (closed) systems since then. Equation 8.17 is therefore a linear relationship between Pb^{207}/Pb^{204} and Pb^{206}/Pb^{204} and is known as a lead-lead isochron; the age t may be calculated from its gradient.

Uranium-lead ages and the thorium-lead ages with which they can be compared are discordant sufficiently often to demonstrate that lead has had a complicated history in the Earth. However, the occurrence of the parallel decay schemes of U^{238} and U^{235}, with chemically identical parents and final products, provides a powerful tool for the study of this history and therefore of the chemical evolution of the Earth (Russell and Farquhar, 1960). The relevance to the age of the Earth as a whole is considered in Section 8.4.

8.3   Growth of the Continents and of Atmospheric Argon

Before the advent of dating methods based on the radioactive decay schemes, the ages of geological formations in different parts of the world were correlated using the fossil record, which is essentially similar everywhere. Sedimentary strata back to the early Cambrian period, about 600 million years ago, were placed in chronological order and an approximate time scale was obtained from the estimation of sedimentation rates (Hudson, 1964). Ages of igneous rocks were fitted into the time scale by determining which sedimentary strata were overlaid or cut by them and which were super-imposed. The whole scale was made quantitative by dating fixed points radiometrically. The validity of the geological time scale has been amply demonstrated; only minor discrepancies now remain between estimates of the ages of boundaries between geological periods. A table of dates is given in Appendix F.

The essential limitation of the fossil record is that, with the exception of a very limited range of microfossils, it extends back only 600 million years, little more than 10% of what we now understand to be the age of the Earth. The fundamental advance which radiometric methods have brought to the study of Earth history is to provide methods of dating pre-Cambrian events. However, an important difference between the two methods must be recognized. The methods of fossil dating use sedimentary rocks, and since erosion and sedimentation are continuous processes, by comparing sedimentary strata in different parts of the Earth we have an essentially continuous sedimentary record. Radiometric methods date events, igneous and metamorphic, which are discontinuous in both place and time.* Dates can be obtained only

---

* There is now a possibility of dating directly some fossils and sedimentary rocks; corals take up uranium selectively, relative to thorium, so that Th^{230}, a decay product of U^{238}, provides an accumulation clock. However, the mainstream of radiometric dating is concerned with igneous and metamorphic rocks.

for times and localities of igneous activity; if the present is typical of the past, then volcanism is quite limited in extent at any one time, and the basement rocks of the continents, the solid igneous rocks which underlie the veneer of sediments, etc., must have been produced by processes we can describe, in general terms, as igneous, although strictly they are probably better regarded as high-grade metamorphics. Thus a geographical plot of radiometric ages is a plot of the migration of zones of tectonic activity.

It has been known for a long time that the continents have "cores" of great ages, the pre-Cambrian "shield" areas, and that the surrounding basement rocks are younger. This observation has now developed into the theory of continental growth by lateral accretion; the continental cores, now the shields, were formed about 3×10^9 years ago, and the continents have grown, and are still growing, by progressive additions of younger material. This is illustrated in Fig. 8.2.

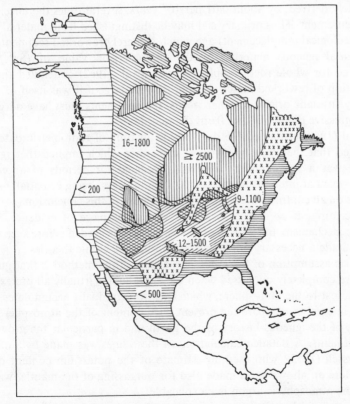

Figure 8.2: Age zones for basement rocks of North America. Figure drawn from data by Hurley et al. (1962) and Goldich et al. (1966). Numbers give ages in millions of years.

The idea that the continental material has originated by progressive differentiation from the mantle was given a quantitative basis by the work of Hurley et al. (1962) on Rb—Sr isotopes. They pointed out that the differentiation of sialic (continental basement) material from basic (low silica) sources gives an enrichment in Rb relative to Sr. They estimated that the weighted average Rb/Sr ratio in continental material was about 0.25 and that this was substantially greater than in the mantle. Using abundance ratios $Rb^{87}/Rb = 0.28$ and $Sr^{86}/Sr = 0.10$, the continental estimate corresponds to $Rb^{87}/Sr^{86} = 0.80$. Differentiating Eq. 8.14, and assuming $\lambda t \ll 1$, since λ is so small for Rb^{87}, we obtain as an average for continental material

$$\frac{d}{dt}\left(\frac{Sr^{87}}{Sr^{86}}\right) = \lambda\left(\frac{Rb^{87}}{Sr^{86}}\right) = 0.011 \times 10^{-9} \text{ year}^{-1} \qquad (8.18)$$

so that Hurley et al. (1962) suggested that the ratio Sr^{87}/Sr^{86} increased faster in sialic material than in the upper mantle, from which it was ultimately derived, by about 0.01 per 10^9 years, several times the uncertainty of measurement. Thus primary sial may be distinguished from material of the same geological (emplacement) age which is merely reworked sedimentary or continental igneous material, by the characteristic value of (Sr_o^{87}/Sr^{86}), measured for whole rock samples. Hurley et al. (1962) concluded that the proportion of reworked material could only be large if it was itself of recent primary subsialic origin; in other words, the continents must have developed by progressive differentiation from the mantle.

Gastil (1960) examined the distribution in time of radiometric dates and suggested that they were grouped in a manner which implied that igneous activity was a spasmodic process and occurred for periods of about 200 million years at intervals of 350 to 500 million years, being essentially simultaneous on all continents. The statistical evidence of this conclusion requires a close scrutiny; if correct, it is a very important piece of evidence of the tectonic mechanism, but a more extensive analysis is needed before such cycles can be made a necessary boundary condition on tectonic theories.

A basic assumption of the potassium-argon dating method is that igneous rocks are completely outgassed when they first cool. Virtually all pre-existing argon is lost to the atmosphere, which therefore gradually accumulates Ar^{40}. It is of interest to compare the present Ar^{40} content of the atmosphere with theories of the igneous history of the Earth and in particular the growth of the continents. A detailed calculation on these lines was made by Shillibeer and Russell (1955), who used an estimate of the potassium content of the crust. Here an allowance is made also for outgassing of the mantle, without which the atmospheric argon is inexplicable.

We first note that the atmosphere contains about 6.5×10^{19} gm of Ar^{40}, that the other argon isotopes are comparatively rare and that the other inert gases, notably neon and krypton, are also very much rarer, relative to argon, than would be expected from their cosmic abundances. There is little alternative but to suppose that the Earth has lost virtually all of its primeval atmosphere, if any, and that the present atmosphere, including practically all of the argon, has been derived from the crust and mantle. We therefore, have to account for 6.5×10^{19} gm of Ar^{40} in terms of the available K^{40}.

Geothermal considerations (Chapter 9) lead to an estimate of the total amounts of radioactive elements in the Earth and also indicate their distribution. Average potassium contents of common rock types are given in Table 9.3 and since K^{40} is 0.012 % of total potassium, the average fractions of K^{40} are 4.44×10^{-6} in granite and 0.72×10^{-6} in basalt. In Section 9.1 it is suggested that the continental crust is thermally equivalent to 10 km of granite (density 2.7 gm cm^{-3}) plus 30 km of basalt (2.9 gm cm^{-3}), which implies a K^{40} content of 18.25 gm cm^{-2}. The comparable figure for the oceanic crust, assumed to be 5 km of basalt, is 1.0 gm cm^{-2}, and the estimate of the total K^{40} content of crust, 2×10^{18} cm^2 of the continental type and 3×10^{18} cm^2 of the oceanic type, is 4×10^{19} gm. Since this is only a third of the estimate of the K^{40} content of the crust and upper mantle together (12.5×10^{19} gm), it is apparent that any estimate of Ar^{40} production depends critically upon the extent to which the mantle has been outgassed. The figures assumed here are, of course, uncertain, but the conclusion that mantle outgassing has been important appears to be unavoidable.

We assume that the Ar^{40} content of the Earth was zero at the time $\tau = 4.5 \times 10^9$ years ago, when the Earth is believed to have formed, and that Ar^{40} has accumulated in the mantle subsequently by decay of K^{40}, but that since some possibly later time t_o, parts of the upper mantle have been successively differentiated to produce the crust and that all of the argon present in any particular volume at the time of its differentiation was carried up and lost to the atmosphere. The total mass of K^{40} in the rock which has been outgassed since t_o is then

$$K_c^{64} = (4 + 8.5f) \times 10^{19} \text{ gm} \qquad (8.19)$$

where f is the fraction of the present upper mantle outgassed progressively in the past t_o years and the whole of the crustal material has been outgassed. The Ar^{40} liberated to the atmosphere is thus

$$Ar^{40} = K_c^{40} \frac{\lambda_{Ar}}{\lambda} \int_0^{t_o} (e^{\lambda\tau} - e^{\lambda t}) \frac{dt}{t_o}$$

$$= K_c^{40} \frac{\lambda_{Ar}}{\lambda} \left[e^{\lambda\tau} - \frac{1}{\lambda t_o}(e^{\lambda t_o} - 1) \right] \qquad (8.20)$$

All of the parameters of the model are specified except t_o and f, for which mutually consistent values can be found. Thus if $t_o = \tau = 4.5 \times 10^9$ years, we require $f = 0.56$, but if $t_o = 3 \times 10^9$ years, $f = 0.35$.

Although we can be reasonably sure that mantle outgassing has contributed to the atmospheric argon, there are too many unknowns in the histories of the mantle and atmosphere to use these estimates to assert that the upper mantle is only about 50% outgassed. In particular, argon may have been lost from the atmosphere, for example, during a violent event such as lunar capture (see Section 2.4), so that the present atmosphere gives an underestimate of outgassing. Alternatively, substantial mantle outgassing may have occurred early in the Earth's history, but not appreciably during the separation of crustal material. However, the conclusion that the mantle is only partly differentiated is not unreasonable; on that basis we can suppose that mantle differentiation will continue for several billion years more.

8.4   Age of the Earth and of Meteorites

We have, at least at present, no means of determining the age of the Earth independently of data on meteorites. The oldest well-dated rocks have ages slightly greater than 3×10^9 years and there is no geological record at all of the first billion years of the Earth's existence as a planet. This is not altogether surprising, when we realize that the internal heat generated by radioactive elements, especially U^{235}, was very much greater 4×10^9 years ago than now and that there had presumably been less differentiation of the crust from the mantle, so that the radioactivity was, on average, distributed more deeply. In these circumstances igneous activity would have been more intense, so that any very early rocks have been completely reworked.

A simple model of the Earth proposed independently by A. Holmes and F. G. Houtermans in 1946 has been widely used as a basis for discussions of its lead-uranium history. The supposition is that at or very soon after its accretion, the Earth differentiated into a number of subsystems, with different U/Pb ratios, and that each subsystem remained completely closed thereafter. The lead isotopes would then have been developed from a primordial lead, common to the whole system; with radiogenic lead distributed according to the U/Pb and Th/Pb ratios in the subsystems, the primary differentiation can be dated by the methods of Section 8.2. It is probable that the Earth's core and mantle are two such closed subsystems and perhaps also the lower mantle is a separate subsystem from the upper mantle. However, we have no means of sampling either the core or lower mantle and it has transpired that the upper mantle and crust have had too complex a history to allow us to find independent subsystems. Geological processes have led to partial rehomogenization of lead isotopes. Lead apparently diffuses quite readily in rocks, as is evident from the difficulty experienced in obtaining concordant lead ages.

A primary differentiation of the kind considered by Holmes and Houtermans evidently occurred in the material which formed the meteorites, many of which appear to have remained closed subsystems and can therefore be dated. Anders (1962) has reviewed the methods of dating meteorites. The three principal methods used for rocks are all available, but the method based on lead isotope ratios has received most attention, being, in principle, the most accurate. This is because the decay constants of uranium are the best known and because the range of variation in isotopic abundances is as great as 2:1. Most iron meteorites have closely similar lead isotope ratios, as measured by Patterson (1956), but negligible uranium and thorium contents, so that the lead ratios are taken to be those of unaltered " primordial" lead, i.e., the lead composition which is presumed to have been uniform throughout the material which accreted to form the meteorites and the planets. Stony meteorites have variable U/Pb ratios and their Pb^{207}/Pb^{204} and Pb^{206}/Pb^{204} ratios plot on a common isochron through the " primordial" lead point, from which Patterson (1956) obtained the widely quoted age estimate of 4.55×10^9 years.

The K—Ar method of dating is unsuitable for meteorites; although it is usable in principle and a large number of values has been obtained (Anders, 1962), two extraneous effects cause the estimates of age to be low. Argon diffusion appears to have caused appreciable losses in some cases and K^{40} is produced as a spallation product of cosmic radiation. It is therefore satisfactory that the age estimates range up to 4.6×10^9 years, in accord with the lead isotope value.

Gast (1962) has obtained a very satisfactory Rb—Sr isochron for stony meteorites. Members of the class known as achondrites are all very low in Rb, which very conveniently allows their strontium isotope ratio, $Sr^{87}/Sr^{86} = 0.701$, to be identified as that of " primordial" strontium. The chondrites are distributed along an isochron through the " primordial" point and having a gradient

$$\frac{d(Sr^{87}/Sr^{86})}{d(Rb^{87}/Sr^{86})} = 0.0664$$

The exact estimate of an age from this value has been subject to uncertainty in the decay constant of Rb^{87}. Using $\lambda = 1.39 \times 10^{-11}$ year^{-1}, as in Table 8.1, the calculated age by Eq. 8.14 is 4.6×10^9 years, in excellent agreement with the lead value. Shields et al., (1966) quoted preferred a Rb—Sr age of 4.45×10^9 years. Burnett and Wasserburg (1967a) have shown that silicate intrusions in iron meteorites give the same age. Although some meteorites appear to have undergone more recent processing (Compston et al., 1965; Burnett and Wasserburg, 1967b), the evidence that they are all products of a differentiation process which occurred about 4.6×10^9 years ago is convincing.

The evidence that the age of the Earth is the same as that of the meteorites is the coincidence of isotopic ratios of "average" terrestrial lead with the meteorite lead isochron, which was noted by Patterson (1956). The problem is to obtain a satisfactory "average" terrestrial lead. Chow and Patterson (1962) considered the lead deposited in marine sediments to represent the average of eroding continental rocks, but found the lead isotopes to be appreciably different in North Atlantic and North Pacific sediments. Their estimate of the average for the crust as a whole is given in Table 8.2, together with the isotopic ratios of "primordial" (iron meteorite) lead. Using subscripts c for the crustal average and m for iron meteorites, Eq. 8.17 gives

$$\frac{(Pb^{207}/Pb^{204})_c - (Pb^{207}/Pb^{204})_m}{(Pb^{206}/Pb^{204})_c - (Pb^{206}/Pb^{204})_m} \cdot \frac{U^{238}}{U^{235}} = \frac{e^{\lambda_{235}\tau} - 1}{e^{\lambda_{238}\tau} - 1} \tag{8.21}$$

where τ is the age of the Earth. With the decay constants in Table 8.1, this gives $\tau = 4.5_3 \times 10^9$ years.

The uncertainty in this estimate is much greater than is implied by the three-figure value; in particular, crustal lead, as represented by marine sediments, may not coincide with the average for the Earth as a whole, especially if uranium has been concentrated upward more than lead. Tilton and Steiger (1965) preferred an estimate based on the lead in very old rocks and arrived at an age of 4.75×10^9 years. The "age" of the Earth may therefore be quoted as $(4.6 \pm 0.2) \times 10^9$ years. In the case of the Earth we can be much less explicit about what happened 4.5×10^9 years ago than in the case of meteorites. In fact, all that we can say is that the Earth became an isolated chemical system at that time.

TABLE 8.2: LEAD ISOTOPE RATIOS IN THE EARTH'S CRUST (Estimated from marine sediments by Chow and Patterson, 1962, and in iron meteorites by V. R. Murthy and C. C. Patterson, quoted by Anders, 1962.)

	Pb^{206}/Pb^{204}	Pb^{207}/Pb^{204}	Pb^{208}/Pb^{204}
Earth's crust	18.5_8	15.7_7	38.8_7
Iron meteorites	9.56	10.42	29.71

8.5 Dating the Nuclear Synthesis

That radioactive species such as U^{235} with half-lives of 10^9 years or less have survived to the present time is evidence that the process or processes of nuclear synthesis that formed them must have occurred only a few billion years ago, i.e., not more than about 10^{10} years ago and probably less. If it is assumed that the uranium and lead isotopic ratios in the Earth's crust and in meteorites are representative of the solar system, then a more precise upper limit to the age of the elements can be imposed. The first approach was to adopt from theories of nuclear synthesis an original ratio of uranium isotopes $(U^{235}/U^{238})_o$. This is uncertain, as it depends upon the processes of nuclear synthesis; the heavier elements are presumed to have been built up by neutron bombardment, but this would rapidly destroy U^{235} by fission, so that U^{235} with which the solar system started was presumably itself produced by decay of now extinct heavier parents. However, we are unlikely to be far wrong by taking $(U^{235}/U^{238})_o = 0.8$, a value originally suggested by Rutherford. Using Eq. 8.5 for the U^{235} and U^{238} decays, the present and original isotopic ratios are related by the equation

$$\left(\frac{U^{235}}{U^{238}}\right) = \left(\frac{U^{235}}{U^{238}}\right)_o \exp[-(\lambda_{235} - \lambda_{238})\tau] \tag{8.22}$$

which gives $\tau = 5.7 \times 10^9$ yr as the time since nuclear synthesis.

A rather more satisfying method is to assume that all of the Pb^{207} has been derived by decay of U^{235}, which is unlikely to be the case and therefore gives a very clear upper limit. The relevant crustal ratio of U/Pb is itself obtained from the data of Section 8.4 as

$$\left(\frac{Pb^{207}}{Pb^{204}}\right)_c - \left(\frac{Pb^{207}}{Pb^{204}}\right)_m = \left(\frac{U^{235}}{Pb^{204}}\right)_c [\exp(\lambda_{235}\tau) - 1] \tag{8.23}$$

where subscripts c and m refer to values for the crust and iron meteorites, as in Table 8.2, and $\tau = 4.5_3 \times 10^9$ years is the age of the Earth, as determined from Eq. 8.21 with the data in Table 8.1. Then

$$\left(\frac{Pb^{207}}{Pb^{203}}\right)_c = \left(\frac{U^{235}}{Pb^{204}}\right)_c [\exp(\lambda_{235}\tau_o) - 1] \tag{8.24}$$

and from Eqs. 8.23 and 8.24 we obtain $\tau_o = 5.65 \times 10^9$ years. This result is subject to the doubt that we cannot allow for uranium/lead fractionation in the accretion of the Earth, but it indicates that the interval between nuclear synthesis and the accretion was no more than 10^9 years and could have been much shorter. We might therefore expect to find in the Earth, or in meteorites, evidence of orphaned daughter isotopes, whose short-lived parents no longer exist in measurable quantities.

Reynolds (1960) succeeded in identifying radiogenic Xe^{129} from extinct I^{129} in meteorites. There are many xenon isotopes and it is not only Xe^{129} which is variable with respect to the others, but the enrichment of Xe^{129} has been positively identified with iodine-bearing minerals,* so that the amount of enrichment allows the synthesis-accretion interval to be estimated in terms of the half-life of I^{129}, 1.64×10^7 years. By assuming that the non-radioactive isotope I^{127} and the extinct isotope I^{129} were produced with equal abundances in the nuclear synthesis, Reynolds (1960) estimated from the abundances of trapped Xe^{129} that the I^{129} abundance had fallen by a factor 10^5 between the synthesis and the accretion of the chondrites, giving an interval of 2.9×10^8 years. On the assumption of a very prolonged process of synthesis, the "initial" I^{129} is less abundant and the estimated interval is reduced to 1.2 to 1.4×10^8 years.

The interval which is estimated in this way is the time between cessation of nuclear synthesis and the solidification and cooling of the meteorite parent body to the point at which it could retain xenon. Anders (1962) noted that an iron meteorite with measurable radiogenic Xe^{129} (Sardis) gave an interval which was longer by 1.3 to 1.6×10^8 years than that estimated for chondrites This is consistent with the iron being part of the core of a parent body about 400 km across, in which the interior took 150 million years longer to cool than did the outer few kilometers. The gravitational energy of accretion of a body of this size is not significant, so that the internal heat must have been due to chemical reactions or else to strong but short-lived radioactivity. Presumably the same process operated in the Earth and dominated the early thermal history.

The state of the planetary material in the few hundred million years before accretion of the planets is not known. In particular, it is even possible that nuclear synthesis continued during the early stages of accretion. Certainly the interval between synthesis and accretion of the meteorites was very short compared with their ages, and there appears to be no possibility that the Earth is appreciably older than the meteorites.

* A clear and comprehensive discussion of this subject is given in Anders's (1962) review.

Chapter 9

THE EARTH'S INTERNAL HEAT

"... a direct proof that a particular hypothesis will account for particular data is not very strong confirmation of the hypothesis when both the data and the consequences of the hypothesis are known only vaguely. ..."

JEFFREYS, 1962, p. xii.

9.1 The Geothermal Flux

The outward flow of heat through the crust is the only directly measurable feature of the Earth's internal heat. Determinations of heat flow require measurements of temperature gradient in the crust and of the thermal conductivities of the actual rocks in which the gradient is measured. In principle these measurements are straightforward but in practice the necessary precautions, such as the avoidance of ventilation in mines and of rock formations in which flow of ground water could upset measurements, require the exercise of considerable care. Most of the useful data are very recent and since it is now easier to obtain data at sea, most of the values are of heat flow through the ocean floor. The whole subject is discussed in detail in a volume edited by Lee (1965) in which the available data have been tabulated and analyzed by Lee and Uyeda (1965). The conclusions of this analysis are summarized in Table 9.1.

Coverage of the Earth's surface by geothermal measurements is still quite uneven and large areas have not been examined at all, but the values listed in Table 9.1 represent a digest of over 1000 measurements, from which suspect observations have been excluded, and it seems unlikely that the general conclusions which can be drawn from these figures are seriously in error. The possibility of significant systematic errors due to penetration of the crust by climatic variations has also been discounted. Birch (1948) examined this problem for continental measurements; he found no significant evidence of climatic effects in vertical variations in the geothermal gradient and showed that in no case could the necessary correction amount to more than $3°C \text{ km}^{-1}$

239

TABLE 9.1: AVERAGE VALUES OF HEAT FLOW FROM EACH OF SEVERAL GEOLOGIC-ALLY DIFFERENT TYPES OF CRUST (Data from Lee and Uyeda, 1965. Values with standard deviations, are μcal cm^{-2} sec^{-1}. Numbers in parentheses give numbers of data points averaged.)

Continental areas		
Pre-Cambrian shields	0.92 ± 0.17	(26)
Post pre-Cambrian, nonorogenic areas	1.54 ± 0.38	(23)
Post pre-Cambrian, orogenic areas (excluding		
Cenozoic volcanic areas)	1.48 ± 0.56	(68))
Cenozoic volcanic areas	2.16 ± 0.46	(11)
Continental average (excluding geothermal areas)	1.43 ± 0.56	(128)
Continental " grid " average[a]	1.41 ± 0.52	(51)
Oceanic areas		
Ocean basins	1.28 ± 0.53	(273)
Ocean ridges	1.82 ± 1.56	(338)
Trenches	0.99 ± 0.61	(21)
Other (continental shelves, etc.)	1.71 ± 1.05	(281)
Ocean floor average	1.60 ± 1.18	(915)
Oceanic " grid " average[a]	1.42 ± 0.78	(338)
World average (all values)	1.58 ± 1.14	(1043)
World " grid " average[a]	1.43 ± 0.75	(389)

[a] Each of the values used to obtain a " grid " average is itself an average over a 300×300 nautical mile ($5° \times 5°$) square. This reduces the bias of the heavily sampled areas but may overemphasize isolated measurements. Additional data were used to obtain the world " grid " average, which accounts for the slight incompatibility of the three " grid " averages.

(about 15% of the average gradient). It is perhaps rather more important to be sure that climatic effects are absent in the oceanographic observations, because they are made with probes which penetrate the ocean floor sediment by only a few meters, and variations in temperature of the water of the bottom of the ocean could have a pronounced effect. Fortunately it appears that as long as the Earth has frozen polar caps, melt water flows along the ocean bottom to all deep parts, maintaining a constant low temperature. Coupled with the fact that there is evidently no flow of percolating water through marine sediments, this fact has enabled measurements of heat flow from the deep ocean floor to be made relatively more easily than the flow from continents. Von Herzen and Maxwell (1964) found that the temperature gradient down a deep ocean drill hole (preliminary Mohole) gave the same value of heat flow as simple probes nearby, so that it is now improbable that a significant fraction of the ocean floor heat flow is due to residual (stored) heat from a climatic warm period with no polar caps.

A general correlation of heat flow with geology is apparent in Table 9.1. The more recent the volcanic origin (orogeny) of an area, the higher its heat flow is likely to be. Although this generalization is presumed to be significant, it must be noted that heat flow can be locally very variable, especially over features such as the mid-Atlantic ridge, where there are apparently localized sources of heat within a few, or at least a few tens, of kilometres of the surface. Possibly these are merely equivalent to volcanic or thermal areas on land and do not need to be too highly regarded in the over-all pattern of heat flow, to which their contribution is small. Lee and MacDonald (1963) considered the large-scale variations of heat flow and produced spherical harmonic analyses with contoured maps of all available data and of area averages, and concluded that heat flow was negatively correlated with gravity, i.e. high heat flow corresponded to low gravity. An updated version with harmonics to third order, by Lee and Uyeda (1965), is reproduced as Fig. 9.1, for comparison with the satellite geoid (Fig. 3.1) and with a geoidal plot of harmonics limited to third order (Fig. 9.2). This analysis allows the conclusion that there may be a correlation between heat flow and gravity, but that the heat flow data are by no means adequate and that more extensive data will effect drastic revisions to the details of Fig. 9.1. (In particular the high heat flow from central Africa, where there are no measurements, is simply a product of the analysis and cannot be regarded as significant). An intriguing future development will be the revision of Fig. 9.1; will it converge to or diverge from similarity to the geoid? In any case it appears likely that the 3:1 variation in regional heat flow, apparent in Fig. 9.1, will be reduced in later analyses.

Spherical harmonic analysis is by no means entirely satisfactory as a method of representing such phenomena as heat flow, which have significant local variations on a scale too small to represent by a convenient number of harmonics. However, the orthogonality of spherical harmonic functions ensures that the validity of the correlation between Figs. 9.1 and 9.2 is limited only by adequacy of the primary data and not by the order of the harmonics taken. It is of interest to consider the physical significance of a correlation, even though, from the present data, we must doubt whether it is there. Areas of high heat flow and low geoid are both compatible with an underlying mantle which is hotter, and therefore less dense, than average; similarly, low heat flow and high geoid imply a cooler, denser mantle. This is, of course, not the only explanation, but, assuming for the moment that it is correct, we can note that continental and oceanic heat flows are equal within the uncertainties of the data; similarly, the features of the geoid are not apparently correlated with continents and oceans. On this basis we must conclude that the oceanic mantle is neither hotter nor colder, on average, than the mantle under the continents. We shall consider later a contrary argument.

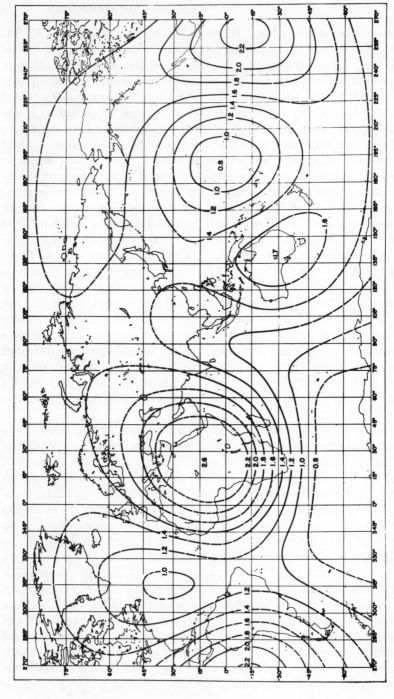

Figure 9.1: Contour representation of spherical harmonics in heat flow up to third order. Reproduced, by permission, from Lee and Uyeda (1965).

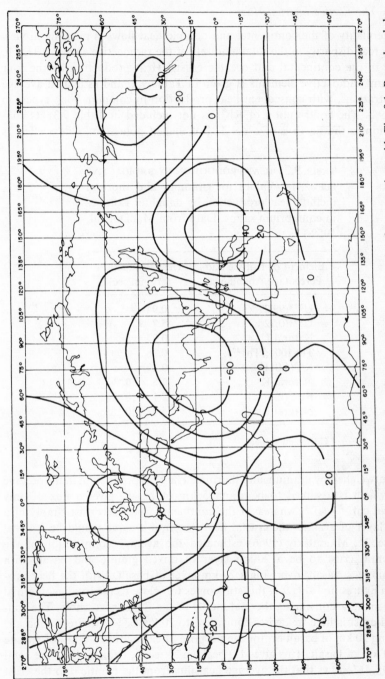

Figure 9.2: Contour representation of spherical harmonics in the geoid up to third order, for comparison with Fig. 9.1. Reproduced, by permission, from Lee and Uyeda (1965).

The equality of the continental and oceanic heat flows is puzzling in view of the great disparity in the total amounts of the radioactive elements U, Th, and K in the continental and oceanic crusts. Heat production by the four important radioactive elements is given in Table 9.2, and estimated average concentrations, with consequent heat production in common rock types, in Table 9.3. The contribution of Rb^{87} is not included in Table 9.3; its concentration in the Earth is of order 1% of that of potassium, whose effect is

TABLE 9.2: HEAT PRODUCTION BY RADIOACTIVE ELEMENTS (Values for uranium and thorium include heat production by daughter isotopes in equilibrium ratios. Data from MacDonald, 1959.)

Element or Isotope and Percentage Abundance	Heat Production (erg/gm sec)
$U^{238}(99.27\%)$	0.94
$U^{235}(0.72\%)$	5.7
Natural U	0.97
Th $(100\% \; Th^{232})$	0.26
$K^{40}(0.0119\%)$	0.29
Natural K	3.5×10^{-5}
$Rb^{87}(27.8\%)$	2.1×10^{-3}
Natural Rb	5.8×10^{-4} [a]

[a] Corrected value.

itself outweighed by uranium and thorium. The heat production in the oceanic crust, which is assumed to be about 5 km of basalt, gives a flux of only $0.05 \; \mu\text{cal cm}^{-2} \text{ sec}^{-1}$, whereas if the continents were to consist simply of a 20-km layer of granite, the crustal heat generation alone would be 1.4 μcal $\text{cm}^{-2} \text{ sec}^{-1}$. Magnitskiy (1966) estimated the average composition of the continental crust to be a 20-km layer of mixed granite and basalt with ($\frac{2}{3}$ granite $+ \frac{1}{3}$ basalt) overlying a 20-km layer of basalt. With the heat production figures in Table 9.3, this still comes too close to the total geothermal flux to make a convincing thermal model of the Earth as a whole. The continental crust is here treated as equivalent to 20 km of ($\frac{1}{2}$ granite $+ \frac{1}{2}$ basalt) overlying 20 km of basalt or, for the purpose of calculation, 10 km of granite plus 30 km of basalt. This gives 1.0 μcal $\text{cm}^{-2} \text{ sec}^{-1}$, two-thirds of the total heat flux. It follows that the average heat flux from the mantle into the crust must be about 1.45 μcal $\text{cm}^{-2} \text{ sec}^{-1}$ under the oceans but no more than

TABLE 9.3: APPROXIMATE AVERAGE CONCENTRATIONS OF RADIOACTIVE ELE-
MENTS AND HEAT PRODUCTION IN DIFFERENT GEOLOGICAL MATERIALS.
(Estimates by various authors have converged to substantial agreement.
The following figures are selected from reviews and recent data by
MacDonald, 1959, 1964; Tilton and Reed, 1963; Wasserburg et al., 1964;
Magnitskiy, 1966; Lovering and Morgan, 1963, 1964; Lambert and Heier,
1967; and Amiel et al., 1967. The heat production is to be compared with
the geothermal flux per gram of the Earth, 1.7 ergs/gm year.)

| Material | Concentration (parts per million by weight) | | | Heat production (ergs/gm year) |
	U	Th	K	
Crust { granite	5.0	20	37000	357
{ basalt, gabbro	0.8	2.7	6000	53
Upper { eclogite	0.052	0.22	500	3.9
mantle? { peridotite, dunite	0.006	0.02	10	0.35
Meteorites { chondrites	.013	.040	850	1.67
{ iron meteorites	—	—	—	$<10^{-4}$

0.5 μcal cm^{-2} sec^{-1} under the continents.* While these figures are subject to
some uncertainty, the essential disparity is too big to be in doubt.

The generally favored explanation for the equality of the heat flux is that
the upper mantle has undergone vertical differentiation to an appreciable
extent only under the continents† (see, for example, MacDonald, 1963b,
Birch, 1965, and Magnitskiy, 1966). Virtually all of the radioactive elements
in the continental mantle are presumed to have risen with the granitic dif-
ferentiates which formed the continents, whereas these elements are still
distributed through the oceanic upper mantle, down to perhaps 500 km, the
average integrated vertical concentration being the same for both continents
and oceans. A virtually inescapable conclusion of this hypothesis is that the
mantle is appreciably hotter under the oceans than under continents. Parkinson's (1964) geomagnetic observation that electrical conductivity appears to
be higher under the oceans (see Section 5.3) is taken to confirm this conclusion.
However, there is a serious difficulty. The hypothesis requires the oceanic

* Hyndman and Everett (1968) conclude that the low heat flows in shield areas indicate a
heat flux from the mantle of less than 0.4 μcal cm^{-2} sec^{-1}.
† The lower mantle is considered to have lost nearly all radioactive material at an early stage.
U, Th, K, and Rb have large ionic sizes and would have been expelled from the close-packed
atomic structures of the lower mantle. Similarly, the core is considered to be devoid of
radioactivity, as are the iron meteorites.

mantle to be less dense, both because it is hotter, and therefore thermally expanded, and because it is presumed still to contain the lighter constituents which have been lost by the continental mantle. Such a density difference is incompatible with the form of the geoid, which shows no distinction between continents and oceans. We must therefore consider an alternative suggestion by Elsasser (1967) that the equality of heat flux from continents and oceans is a result of a dynamic balance in the convecting upper mantle, although some differences between continental and oceanic structures evidently extend down to several hundred kilometres (see, for example, Dorman et al., 1960), and details of the proposed convective balance are unclear.

Another interesting coincidence, first noticed by Birch (1958) and apparent in the figures of Table 9.3, is the close equality between the measured geothermal flux and the radioactive heat generation in an Earth with the over-all composition of the chondrites. The agreement is still well within the uncertainties of observation, but it is now evident that the agreement is fortuitous. As Wasserburg et al. (1964) have pointed out, the concentration ratio of K/U in a wide range of rocks has a sensibly constant value of about 1×10^4, but in chondrites the ratio is consistently much higher, averaging (according to the revised figures in Table 9.3) 6×10^4. On the other hand the Th/U ratio of about 4:1 is similar in both terrestrial rocks and chondrites. Presumably the parent materials of the Earth and of the chondrites were similar and the more highly volatile potassium was lost during the formation of the Earth to a much greater extent than in the formation of the chondrites. If we assume the overall average concentrations of U and Th to be the same in both, then the reduced potassium in the Earth brings the estimated heat production down to 0.9 erg/gm year, in which case a substantial contribution from residual heat (plus, possibly, rotational energy) is required to produce the observed geothermal flux. The estimate is further reduced to about 0.6 erg/gm year if we suppose, as seems reasonable, a potassium-reduced chondritic composition for the mass of the mantle only. Therefore both the continental and oceanic mantles must have been depleted in radioactive elements to provide the observed radioactivity of the continents unless the Earth is enriched in uranium and thorium, perhaps by a factor as great as 2.5, as well as being depleted in potassium, relative to the chondrites. Such an enrichment would restore the calculated equality of geothermal flux and heat generation and allow the assumption that the Earth is almost in thermal balance.

9.2   Thermal Conduction in the Mantle

The distribution of heat sources within the Earth is sufficiently uncertain that calculations of the Earth's thermal history are based upon plausible guesses, or simple models, which are described in terms of manageable equations. There have been several quantitative calculations, in particular by Lubimova (1958, 1967) and MacDonald (1959), in which the Earth was assumed to be spherically symmetrical, so that thermal diffusion could be represented by the equation

$$\rho C \frac{\partial T}{\partial t} = \frac{1}{r^2} \frac{\partial}{\partial r} \left(K r^2 \frac{\partial T}{\partial r} \right) + H(r, t) \qquad (9.1)$$

where ρ = density, C = specific heat, K = thermal conductivity, and H = rate of heat generation per unit volume; these properties are, in general, functions of radius r, or temperature T, and H is also a function of time t, by virtue of the decay of radioactivity. For any assumed distribution of heat sources and initial temperatures, and if $K(T)$ is a known function (in particular, convection is neglected), Eq. 9.1 can be solved numerically to obtain T as a function of r at different stages in the Earth's history. The boundary conditions which can then be applied to limit the range of acceptable solutions are:

1. The present heat flux and temperature gradient at the surface must coincide with the observed values.

2. The mantle is solid to a depth of 2900 km, below which there is a liquid outer core and solid inner core.

3. There is a strong concentration of radioactive elements in the continental crust, but much less in the oceanic crust.

We strike an immediate difficulty—boundary condition 3 is not compatible with the assumed spherical symmetry. The distribution of radioactivity under oceans is evidently quite different from that in and beneath the continents. This difficulty has usually been avoided in the analyses by considering two alternative models, one representing an Earth entirely covered by continents, in which radioactivity is largely confined to a 40-km crust, and the other an oceanic Earth with the same total radioactivity distributed more deeply, but usually only within the upper mantle to a depth of about 500 km. These models lead to two related conclusions about the thermal conductivity of the Earth, which require investigation:

1. If $K \approx K_0$* throughout, then in a body the size of the Earth thermal diffusion is so slow that very little heat has escaped from depths greater than about 1000 km in the whole history of the Earth . . .

2. But for any reasonable distribution of radioactivity in the oceanic Earth model, the effective thermal conductivity at depth must be substantially greater than K_0 in order that the mantle not be molten.

---

* The average value of thermal conductivity of igneous rocks, measured at laboratory temperatures, is

$$K_0 = 2.5 \times 10^5 \text{ ergs sec}^{-1} \text{ cm}^{-1} \text{ deg}^{-1} = 6 \times 10^{-3} \text{ cal sec}^{-1} \text{ cm}^{-1} \text{ deg}^{-1}$$

Conclusion 1 can be expressed quantitatively in terms of Eq. 9.1. Neglecting *H*, thermal diffusion is seen to be governed by the parameter $K/\rho C$, known as thermal diffusivity, whose value for igneous rocks at 0°C is about 40 km² per 10^6 years. The thermal relaxation time for the crust alone, thermally isolated from the mantle, would be about 10^6 years, but, as follows from Eq. 9.1 and is implicit in the units of diffusivity, the relaxation time varies as the square of the dimensions of the body considered, being 10^8 years for 400 km, 10^{10} years for 4000 km, and so on. However, conclusion 2 can be satisfied only if the lattice conductivity K_0 is augmented by one or more additional heat transfer mechanisms in the deep interior, which drastically shorten the thermal relaxation time. Two mechanisms must be considered— radiative transfer and convection. In the thermal model of the Earth considered in Section 9.3, both radiative conduction and convection are necessary.

Radiative transfer of heat becomes important in materials of moderate transparency at elevated temperatures. Radiated energy of a particular wavelength is absorbed according to the simple exponential law of decay with distance:

$$E = E_0 e^{-\epsilon r} \qquad (9.2)$$

where ϵ is the extinction coefficient, for that particular wavelength, and is equal to the reciprocal of the mean free path of light quanta. If ϵ is known as a function of wavelength, then the radiative conductivity is calculable at any temperature and Clark (1957a, 1957b) has applied measured optical absorption spectra of minerals to calculate their conductivities. However, appropriate values of ϵ for the mantle are so far from being known accurately that it is sufficient for the present purpose to assume that ϵ is a constant, independent of wavelength*; the material is then known as a " grey body." By Stefan's law, as for a black body, the density of radiation energy E at temperature T is proportional to T^4 and the gradient of energy density is

$$\frac{dE}{dx} \propto 4T^3 \frac{dT}{dx} \qquad (9.3)$$

The radiative thermal conductivity K_R is thus proportional to the cube of absolute temperature. It is related to ϵ by the following equation, which is derived in Appendix C:

$$K_R = \frac{16}{3} \frac{n^2}{\epsilon} \sigma T^3 \qquad (9.4)$$

where n is the refractive index and σ is Stefan's constant.

Values of K_R for a grey body at various temperatures are given in Table 9.4,

---

* Fukao et al., (1968) have shown that the optical absorption spectrum of olivine becomes much more nearly wavelength-independent at high temperatures than at room temperature.

TABLE 9.4

Radiative thermal conductivity, in erg sec^{-1} cm^{-1} °C^{-1}, of a grey body (refractive index, $n = \sqrt{3}$) as a function of temperature for different values of extinction coefficient ϵ, compared with Clark's (1957b) values for olivine and diopside. Values below the heavy line are those which exceed the mean lattice conductivity of igneous rocks at 0°C (2.5×10^5 ergs sec^{-1} cm^{-1} °C^{-1}). Values of ϵ greater than 300 cm^{-1} are of little interest as radiative conductivity is then not competitive with lattice conductivity at any depth in the mantle.

T(°K) \ ϵ(cm^{-1})	300	100	30	10	3	1	0.3	Olivine (Clark, 1957b)	Diopside (Clark, 1957b)
500	3.8×10^2	1.1×10^3	3.8×10^3	1.1×10^4	3.8×10^4	1.1×10^5	3.8×10^5	4×10^{5a}	8×10^{4a}
1000	3.0×10^3	9.1×10^3	3.0×10^4	9.1×10^4	3.0×10^5	9.1×10^5	3.0×10^6	2.97×10^6	6.7×10^5
1500	1.0×10^4	3.1×10^4	1.0×10^5	3.1×10^5	1.0×10^6	3.1×10^6	1.0×10^7	8.6×10^6	2.38×10^6
2000	2.4×10^4	7.3×10^4	2.4×10^5	7.3×10^5	2.4×10^6	7.3×10^6	2.4×10^7	1.45×10^7	4.44×10^6
2500	4.7×10^4	1.4×10^5	4.7×10^5	1.4×10^6	4.7×10^6	1.4×10^7	4.7×10^7	2.02×10^7	7.25×10^6
3000	8.2×10^4	2.5×10^5	8.2×10^5	2.5×10^6	8.2×10^6	2.5×10^7	8.2×10^7		
3500	1.3×10^5	3.9×10^5	1.3×10^6	3.9×10^6	1.3×10^7	3.9×10^7	1.3×10^8		
4000	1.9×10^5	5.8×10^5	1.9×10^6	5.8×10^6	1.9×10^7	5.8×10^7	1.9×10^8		

^{a} Extrapolated values.

for the range of values of ϵ which is of interest. Also listed are the values calculated by Clark (1957b) from his measurements on optical absorption in good single crystals of two minerals; values for less perfect crystals are much lower, but appear to be very variable. From a survey of measurements on various basic and ultrabasic rocks, Clark (1957a) concluded that ϵ commonly fell in the range 10 cm^{-1} to 30 cm^{-1}.

Although the average value of ϵ appropriate to the mantle is not known, several arguments suggest that it may be quite low, at least in the lower mantle, which appears seismically to be very uniform and geochemically to have lost almost all atoms which do not fit into close-packed structures. In these circumstances we might expect the lower mantle to consist of large, good quality crystals. The greatest doubt that arises concerns the reduction in transparency by the electrical conductivity. Clark (1957a) attempted to allow for this quantitatively, but values of important parameters, such as electron mobility in the lower mantle, are not well known and only a very rough estimate is possible. The skin effect calculation of attenuation of an electromagnetic wave in a conductor, which is discussed in connection with geomagnetic variations in Section 5.4, refers only to waves whose frequencies are very low compared with the frequency of scattering of conduction electrons. Since, in this case, the field is virtually unchanged in the interval τ between scattering collisions, attenuation is calculable in terms of d.c. conductivity. However, at optical frequencies (10^{14}–10^{15} cps) the electromagnetic field oscillates more rapidly than the electrons are scattered, and the electrons behave as though almost free. Attenuation of the field is due only to the relatively infrequent scattering collisions by electrons. Quite a good electrical conductor may be optically transparent, as, for example, hydrochloric acid whose electrical conductivity is comparable to that of the lower mantle. According to classical theory, the effective conductivity σ_v at a high frequency v is related to the d.c. conductivity σ_0 by the relation* (Fan and Becker, 1951)

$$\sigma_v = \frac{\sigma_0}{1 + (2\pi v \tau)^2} \tag{9.5}$$

Measurements of electron mobility in semiconductors at room temperature give $\tau \approx 10^{-12}$ sec, so that at $v = 5 \times 10^{14}$ sec^{-1}; i.e., in the middle of the optical range,

$$\sigma_v \approx 10^{-7} \sigma_0 \tag{9.6}$$

Using this result, Eq. 5.18 can be adapted to the present discussion, to give the contribution ϵ_e of free electrons to the optical extinction coefficient, in terms

---

* Strictly, σ_v should be written as a complex quantity, but we are concerned here only with the real (resistive) part and not with the imaginary (dielectric) part.

of d.c. conductivity, σ_0:

$$\epsilon_e \approx 2\pi(v\sigma_0 \times 10^{-7})^{1/2} \qquad (9.7)$$

which, for $\sigma_0 \approx 3 \times 10^{-9}$ emu, as suggested for the lower mantle, gives $\epsilon_e \approx 2.4$ cm^{-1}. It appears likely that τ is appreciably larger in mantle silicates than in germanium and silicon but decreases with temperature, as electron scattering is largely due to phonons. The preceding estimate of ϵ_e is therefore subject to a wide range of uncertainty. As follows from calculations given in Section 9.3, the preferred range of total extinction coefficient ϵ is 10 to 30 cm^{-1}, in which case it appears that ϵ_e can reasonably be neglected; however, if ϵ_e has been underestimated by a factor greater than 10, then it may be the dominant term in ϵ in the lower mantle.

Now consider the effect of convection in the upper mantle. Whether it is strictly thermal convection or partly a continuing differentiation of the crust from the mantle is not important here. What matters is the reduction in thermal time constant of the Earth. Convection is presumed to be confined to the upper mantle, say to the top 700 km. Then if the rate of movement is 3 cm year^{-1}, as suggested in Chapter 7, convective overturn takes about 10^8 years, depending upon the lateral dimensions of the convective cells. This is very short compared with the thermal time-constant previously considered. We can also note that the core is in motion and therefore that an adiabatic gradient is presumably maintained in it. The thermal relaxation time for the Earth therefore appears to depend mainly on the conductivity of the non-convecting part of the mantle. If radiative transfer increases the mean effective conductivity of the lower mantle by a factor of 6, relative to lattice conductivity alone, as seems reasonable, then the mean effective thermal diffusivity becomes 100 km^2 10^{-6} years and the relaxation time for the whole Earth is then only 5×10^8 years. Since this is shorter than both the age of the Earth and the half-lives of the isotopes responsible for radioactive heating, which are, in any case, concentrated toward the surface, it implies that the temperature profile and thermal diffusion within the Earth are very nearly in equilibrium with the internal heat generation. Making this assumption greatly simplifies the solution of Eq. 9.1 because the time-dependence is neglected. Calculations given in Section 9.3 yield values of radiative conductivity high enough to justify this assumption.

The strong temperature dependence of radiative conductivity also has the effect of stabilizing internal temperatures, as may be seen by considering a very simple model. Suppose the mantle to be a plane slab, of thickness h, whose upper surface is maintained at a temperature T_0 and whose lower surface is at temperature $T_c \gg T_0$, and is in equilibrium with the heat flux from the core, q per unit area. Assume that there are no heat sources within the mantle itself and that thermal conductivity K is dominated by radiative transfer, so that we can write

$$K = AT^3 \tag{9.8}$$

A being a constant. Then the temperature gradient dT/dr is given by

$$q = -K\frac{dT}{dr} = -AT^3\frac{dT}{dr} \tag{9.9}$$

which integrates over the thickness h of the slab to give

$$qh = \frac{A}{4}(T_c^4 - T_0^4) \tag{9.10}$$

Since $T_c \gg T_0$ was assumed,

$$T_c \propto q^{1/4} \tag{9.11}$$

Thus a substantial change in heat generation within the Earth produces a much smaller fractional change in the equilibrium internal temperatures if thermal conductivity is determined largely by radiative transfer. The temperature profile then approximates more nearly to equilibrium with the heat generation, in spite of the time-dependence of the heat generation.

Thus we have strong reasons for supposing that the thermal conductivity of the mantle maintains quite closely a temperature profile in equilibrium with the internal heat generation. This is the basis of calculations of the temperature profile in the following section.

9.3 Temperatures in the Interior of the Earth

For any assumed distribution of heat sources in a nonconvecting, greybody mantle, we can calculate the equilibrium temperature profile by making either of two assumptions:

1. The opacity and hence the thermal conductivity are known at all temperatures.

2. The opacity has a constant, but initially unknown value and the temperature at a fixed point in the deep interior is known.

At present the second of these alternatives is the more reliable, because we have virtually no independent control on the estimate of opacity, but the presumed solid/liquid phase transition at the boundary of the inner and outer cores (boundary condition 2 on p. 247) allows the temperature at that point to be estimated in terms of the phase diagram of iron.

The first problem is therefore to estimate the melting point of iron at the pressure of the inner core-outer core boundary—3 megabars or 3×10^6 atm.

The variation of the melting point T_m of a pure substance with pressure p

is given by the Clausius-Clapeyron equation, which is derived in texts on thermodynamics and in a straightforward way by Joos (1934, pp. 503–507):

$$\frac{dT_m}{dp} = \frac{T_m}{L}(v_L - v_s), \qquad (9.12)$$

where L is the latent heat of melting and v_L and v_s are the specific volumes or reciprocal densities of the liquid and solid phases. This equation is used to plot the variation of melting point with depth in the mantle, but integration to the depth of the core with any accuracy demands a knowledge of the variations of L, v_L, and v_s at very high compressions. Kraut and Kennedy (1966) examined data on the melting of materials at pressures beyond the linear compression range and concluded that the variation in melting point of a solid, even at extreme compressions, was proportional to the negative dilation of the solid by virtue of the compression:

$$\frac{dT_m}{d(\Delta v_s/v_{s_0})} = \text{constant} \qquad (9.13)$$

where $\Delta v_s = (v_s - v_{s0})$ and v_{s0} is the specific volume of the solid at zero pressure. This allows the effect of high compressions to be estimated from the initial effect of pressure on T_m according to Eq. 9.12:

$$
\begin{aligned}
T_m(\Delta v_s) &= T_{m0} + K_0\left(\frac{dT_m}{dp}\right)_0 \cdot \frac{\Delta v_s}{v_{s0}} \\
&= T_{m0} + K_0\left(\frac{dT_m}{dp}\right)_0\left(1 - \frac{\rho_0}{\rho}\right)
\end{aligned}
\qquad (9.14)
$$

K being the bulk modulus and subscript o indicating initial or low-pressure values. Kennedy (1966) used shock wave data on the density of iron at 3 megabars and concluded that the melting point of iron at this pressure is 3725°C (4000°K). The principal uncertainty in this estimate of the temperature of the inner core-outer core boundary appears to arise from the unknown composition of the core; we must also expect the solubilities of the minor constituents to be different in the liquid and solid phases. Phase diagrams of iron alloys are given by Bozorth (1951). Nickel has no striking effect upon the melting point but 15% silicon reduces it by 300°C and the effect of dissolved MgO at high pressure is quite unknown. We must therefore suppose that Kennedy's estimate has an uncertainty of about 500°C and probably errs on the high side. However, in the absence of a better figure, Kennedy's value is assumed in the following calculation.

Since the outer core is fluid and, according to observations of geomagnetic secular variation, it has internal motions which are very rapid by comparison with other geophysical processes, we can assume its temperature gradient to be adiabatic and thus estimate the temperature of the core-mantle boundary. The variation of temperature T with pressure p during adiabatic compression is obtained from one of the Maxwell relations of thermodynamics:

$$\left(\frac{\partial T}{\partial p}\right)_S = \left(\frac{\partial V}{\partial S}\right)_p = \left(\frac{\partial V}{\partial T}\right)_p\left(\frac{\partial T}{\partial S}\right)_p \tag{9.15}$$

which applies to material of constant phase, V being the volume of the material in question and S is its entropy. This can be expressed in terms of ordinary physical properties. The volume expansion coefficient α is given by

$$\alpha = \frac{1}{V}\left(\frac{\partial V}{\partial T}\right)_p \tag{9.16}$$

and since entropy is defined by

$$dS = \frac{dQ}{T} = \frac{mC\,dT}{T} \tag{9.17}$$

where dQ is the quantity of heat supplied to a mass m and specific heat C to raise its temperature by dT, it follows that

$$\left(\frac{\partial T}{\partial S}\right)_p = \frac{T}{mC_p} \tag{9.18}$$

Substituting Eqs. 9.16 and 9.18 in 9.15, and writing $m/V = \rho$, we obtain

$$\left(\frac{\partial T}{\partial p}\right)_S = \frac{\alpha T}{\rho C_p} \tag{9.19}$$

This is the adiabatic increase in temperature with pressure, but we also know the increase in pressure with depth, or negative of radius, r:

$$\frac{dp}{dr} = -g\rho \tag{9.20}$$

where g is gravitational accleration, so that

$$-\left(\frac{dT}{dr}\right)_{\text{adiabatic}} = -\left(\frac{\partial T}{\partial p}\right)_S\frac{dp}{dr} = \frac{\alpha T g}{C_p} \tag{9.21}$$

The application of Eq. 9.21 to the outer core requires an estimate of α/C_p, since T and g are believed to be known with sufficient accuracy. C_p is probably not dramatically affected by pressure but α is reduced by an amount that is difficult to estimate reliably, but is almost certainly quite large. Applying to iron calculations by Birch (1952) and Jacobs (1953), we find that at average core pressures α may be reduced by a factor of about 10 relative to laboratory measurements. Therefore if we take $\alpha = 3.6 \times 10^{-6}\ °C^{-1}$ as representative of the outer core, we find

$$\overline{\left(\frac{dT}{dr}\right)}_{\text{adiabatic}} = -0.14\ °C\ km^{-1} \tag{9.22}$$

and the temperature difference between the inner and outer boundaries of the outer core is about 300°C. This makes the estimated temperature at the core-mantle boundary 3700°K, which is assumed in calculating the temperature gradient in the mantle.

We can now consider in detail a model of the Earth as represented in Fig. 9.3. This is a thermal model for a differentiated mantle and combines two different models, representing supposed continental and oceanic structures. Allowing for continental shelves we can consider the Earth to be 40% of the continental type and 60% of the oceanic type. The supposition upon which this model is based is that the mantle has undergone two stages of differentiation. At a very early stage in the Earth's history, and perhaps even during the late stages of accretion, the core separated out, taking virtually no radioactive material, and the mantle differentiated into upper and lower fractions with an enrichment of radioactive material in the upper mantle. Subsequently, sections of the upper mantle further differentiated to produce the continental

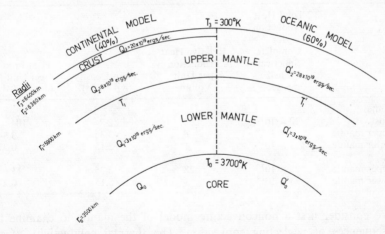

Figure 9.3: Thermal model of the Earth, based on the hypothesis of a differentiated mantle. Two spherically symmetrical models, representing continental and oceanic structures, are considered separately and then combined in the proportions of 40 : 60 to represent the Earth as a whole. The values of heat generation in each layer refer to the heat generation in a complete spherical shell with the composition of the layer in question. Since the model is only approximate, values have been rounded-off for numerical convenience.

crust, which is much more highly enriched in radioactive elements. This differentiation is presumed to have left the continental upper mantle relatively depleted in radioactive elements, but not to have affected the oceanic mantle, i.e., the differentiation occurred vertically. The existence of the geomagnetic field compels us to allow also for some heat flux from the core, although, as suggested in Chapter 5, this may originate as rotational energy rather than radioactivity. The essential features of this model appear in a number of calculations of the Earth's thermal history and have been devised to explain the equality of the continental and oceanic heat fluxes. However, as noted in Section 9.1, its validity is subject to considerable doubt, pending a satisfactory demonstration that it is compatible with the form of the geoid, which admits no distinction between continents and oceans. Meanwhile it is a usefully quantitative thermal model and will presumably become a basis for discussion of better theories.

TABLE 9.5: HEAT PRODUCTION IN DIFFERENT LAYERS OF THE MODEL EARTH REPRESENTED IN FIG. 9.3. EACH OF THE ASSUMED TWO STAGES OF DIFFERENTIATION PRODUCED APPROXIMATELY THE SAME UPWARD ENRICHMENT OF HEAT-PRODUCING ELEMENTS PER UNIT VOLUME (AT THE COMPRESSED DENSITIES). (1) Upper mantle (original oceanic type)/lower mantle = 28/1. (2) Crust/continental upper mantle = 25/1. Rounded values of total heat production were used, subject to the condition that the total for the Earth is 32×10^{19} ergs sec^{-1} (10^{28} ergs yr^{-1}), allowing for 1×10^{19} ergs sec^{-1} in the core

	Total Heat Production for Complete Spherical Shell (ergs sec^{-1})	Relative Heat Production per Unit Volume	Mean Density (gm cm^{-3})	Heat Production per Unit Mass (ergs gm^{-1} yr^{-1})
Continental				
Crust	20×10^{19}	200	2.7	104
Upper mantle	8×10^{19}	8	3.35	3.34
Lower mantle	3×10^{19}	1	5.15	0.27
Oceanic				
Upper mantle	28×10^{19}	28	3.35	11.6
Lower mantle	3×10^{19}	1	5.15	0.27

We consider first a nonconvecting model of the mantle to examine the consequences of neglecting convection. The thermal conductivity of the mantle is the same function of temperature for both continental and oceanic models, being due to constant (lattice) and temperature-dependent (radiative) conductivities:

$$K = K_L + AT^3 \tag{9.23}$$

where A is a constant, given by Eq. 9.4, but having a value which is initially an unknown parameter of the model. Then for either the continental or oceanic models we may equate the heat transfer, through a surface of radius r in the nth layer, to the total heat production within that surface*:

$$\sum_{l=0}^{n-1} Q_l + Q_n \frac{r^3 - r_{n-1}^3}{r_n^3 - r_{n-1}^3} = -(K_L + AT^3)4\pi r^2 \frac{dT}{dr} \tag{9.24}$$

Integrating from the inner boundary condition, $T = T_{n-1}$ at $r = r_{n-1}$, we obtain the temperature profile in the layer, referred to the temperature T_{n-1}, at its inner boundary:

$$\left(\sum_{l=0}^{n-1} Q_l - Q_n \frac{r_{n-1}^3}{r_n^3 - r_{n-1}^3}\right)\left(\frac{1}{r_{n-1}} - \frac{1}{r}\right) + \frac{Q_n}{2} \frac{r^2 - r_{n-1}^2}{r_n^3 - r_{n-1}^3}$$

$$= 4\pi \left[K_L(T_{n-1} - T) + \frac{A}{4}(T_{n-1}^4 - T^4)\right] \tag{9.25}$$

We can now substitute the upper boundary condition and add the equations for all layers, to obtain, for the continental model,

$$Q_0\left(\frac{1}{r_0} - \frac{1}{r_3}\right) - \frac{1}{r_3}(Q_1 + Q_2 + Q_3)$$

$$+ \frac{3}{2}\frac{Q_1}{r_0}\left[\frac{(r_1/r_0)^2 - 1}{(r_1/r_0)^3 - 1}\right] + \frac{3}{2}\frac{Q_2}{r_1}\left[\frac{(r_2/r_1)^2 - 1}{(r_2/r_1)^3 - 1}\right] + \frac{3}{2}\frac{Q_3}{r_2}\left[\frac{(r_3/r_2)^2 - 1}{(r_3/r_2)^3 - 1}\right]$$

$$= 4\pi\left[K_L(T_0 - T_3) + \frac{A}{4}(T_0^4 - T_3^4)\right] \tag{9.26}$$

and for the oceanic model,

$$Q_0'\left(\frac{1}{r_0} - \frac{1}{r_3}\right) - \frac{1}{r_3}(Q_1' + Q_2') + \frac{3}{2}\frac{Q_1'}{r_0}\left[\frac{(r_1/r_0)^2 - 1}{(r_1/r_0)^3 - 1}\right] + \frac{3}{2}\frac{Q_2'}{r_1}\left[\frac{(r_3/r_1)^2 - 1}{(r_3/r_1)^3 - 1}\right]$$

$$= 4\pi\left[K_L(T_0 - T_3) + \frac{A}{4}(T_0^4 - T_3^4)\right] \tag{9.27}$$

The right-hand sides of Eqs. 9.25 and 9.26 are identical and the only unspecified parameters on the left-hand sides are Q_0 and Q_0'. The model thus requires that Q_0 and Q_0' be unequal. This is expected since we suppose the temperature gradients at the bottom of the lower mantle, and therefore the heat flux from the core, to be different in the continental and oceanic models,

---

* This equation assumes a uniform concentration of heat sources within any layer. If composition were constant, compression would produce a slight increase in concentration downward. This effect is, however, negligible here.

even though the core temperature is the same for both. We can determine the values of Q_0 and Q'_0 from (9.26) and (9.27) by requiring, in addition, that the total heat flux from the core in the combined model (40% continental plus 60% oceanic) be 1×10^{19} ergs sec^{-1}:

$$0.4Q_0 + 0.6Q'_0 = 1 \times 10^{19} \text{ ergs sec}^{-1} \qquad (9.28)$$

Then we obtain, with other numerical values from Fig. 9.3,

$$Q_0 = 1.515 \times 10^{19} \text{ ergs sec}^{-1}$$
$$Q'_0 = 0.656_6 \times 10^{19} \text{ ergs sec}^{-1} \qquad (9.29)$$

Assuming a constant value for lattice conductivity, $K_L = 2.5 \times 10^5$ erg sec^{-1} cm^{-1} deg^{-1}, we obtain also the value of the radiative conductivity constant $A = 5.25_6 \times 10^{-5}$ erg sec^{-1} cm^{-1} deg^{-4}, and from Eq. 9.4 the corresponding extinction coefficient, $\epsilon = 17.26$ cm^{-1}.

After determining the numerical parameters Q_0, Q'_0, and A, the continental and oceanic temperature profiles of the model are formed by straightforward, if tedious numerical substitution of different radii into equations of the type (9.25), corresponding to each of the layers. The result is shown as the broken lines in Fig. 9.4.

It is important to compare the calculated temperature profile with the variation of melting point with depth in the mantle. The melting point curve shown in Fig. 9.4 was obtained from Eq. 9.14 by using as the initial gradient the variation of melting point with depth for "pyrolite," as given by Clark and Ringwood (1964). They suggest a zero pressure melting point of 1400°K and a gradient of 2.6°C/km. The density gradient in the upper mantle is about 8×10^{-4} gm cm^{-3} km^{-1} (Bullen, 1963) so that we have

$$T_m = 1400 + 1.0_{75} \times 10^4 \left(1 - \frac{\rho_0}{\rho}\right) °\text{K} \qquad (9.30)$$

With Bullen A model densities this gives the curve in Fig. 9.4, which is seen to have an inflection corresponding to the density increment in the transition zone between 400 and 1000 km.

The nonconvecting model is clearly deficient in that it gives temperatures exceeding the melting point in the upper mantle, whereas the mantle is, as far as we know, solid throughout. The difficulty could be avoided by concentrating the radioactivity even closer to the surface, but there is no other justification for such an adjustment to the model. The alternative is to allow for a substantial increase in the effective thermal conductivity of the upper mantle; the effect of convection must be allowed for. This is a very uncertain matter because convection is poorly understood. Before attempting a correction it is worth considering the acceptable features of the nonconvecting model.

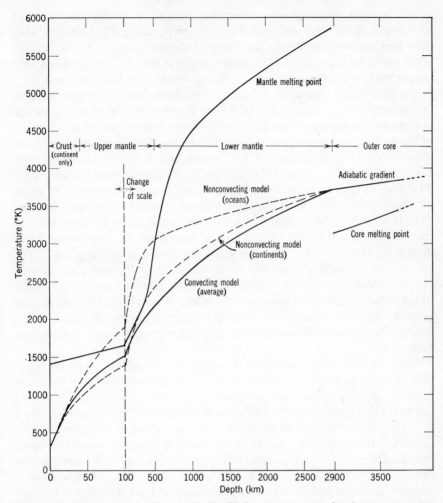

Figure 9.4: Temperatures in the mantle. Equilibrium temperatures for the nonconvecting model of the mantle, with heat sources distributed as in Fig. 9.3, are represented by broken lines. Since the heat flow in this model is assumed to be radial, the temperature difference between continents and oceans is exaggerated. Also shown as solid lines, are the mantle melting point from Eq. 9.30 and a temperature profile calculated with allowance for convection in the upper mantle.

A basic assumption of the nonconvecting model is the maintenance of equilibrium temperatures. Averaging the two profiles, the temperature exceeds $3000°K$ at depths greater than 800 km, so that in the major part of the mantle the radiative conductivity exceeds 1.5×10^6 erg sec^{-1} cm^{-1} deg^{-1}, in accord with the thermal time-constant requirement. The model is thus self-consistent in this respect, although this does not amount to a demonstration of its validity. It is suggested that the core temperature, $T_0 = 3700°K$, may have been overestimated by a few hundred degrees, but this has no serious effect upon the time-constant argument, because with a lower assumed core temperature, a smaller extinction coefficient would have been obtained.

The calculated inequality of continental and oceanic temperatures is incompatible with purely radial heat flow. Horizontal flow of heat from the oceanic mantle to the continental mantle would cause a partial equalization of the profiles. It is important to consider whether this is compatible with the observed equality of continental and oceanic heat flow. The computed temperature difference reaches 500°C and the scale of continental features is say 3000 km, so that in rough terms the effective horizontal gradient is about $\frac{1}{6}$°C km^{-1}, say one-third of the mean vertical gradient in the lower mantle. Thus this effect could be allowed for by transferring no more than a third of the heat generated in the oceanic lower mantle to the continental lower mantle, making $Q_1 = 4 \times 10^{19}$ ergs sec^{-1} and $Q'_1 = 2 \times 10^{19}$ ergs sec^{-1}. The consequent inequality of heat flux would be less than 7% and it will be very difficult indeed to determine whether such a small difference exists. Convection in the upper mantle must also have the effect of equalizing the temperature profiles in the lower mantle. The evidence from geomagnetic " deep sounding " will be vital in determining the reality of the continental-oceanic temperature difference.

Since the convective mechanism is poorly understood we can do no better than to adopt an extremely simple model for the thermal transfer in the upper mantle. The convective pattern is presumed to have a very large scale and not to break up into small " eddies," so that transfer of heat into and out of the convective cells must be by normal thermal diffusion, however effective the convection itself may be. Thus the effective conductivity of the upper mantle can be enhanced by a factor not greater than about 2. We can consider the conductivity to be increased only where the convective movement is vertical, i.e., in the middle of the convective zone and not at the top or bottom. The increase is proportional to the vertical component of convective motion. We can therefore make a simple, if crude correction to the conductivity of the convecting layer, by allowing it to be a function of depth z:

$$K_{\text{corrected}} = K\left[1 + \sin^2\left(\pi \frac{z}{z_0}\right)\right] \qquad (9.31)$$

where z_0 is the total depth of the layer and is taken to be 500 km in these calculations. The temperature curve for the "convecting model" in Fig. 9.4 was obtained by increasing the thermal conductivity according to Eq. 9.31. The weighted mean of the nonconvecting profiles, i.e., (0.6 × "oceanic" + 0.4 × "continental") temperatures, was used as a starting point and the upper mantle curve was adjusted by reducing the temperature gradient according to Eq. 9.31. This necessitated calculating a new profile for the lower mantle, which is somewhat less conducting than in the nonconvective model. The new value of the extinction coefficient ϵ, is 21 cm^{-1}.

An essential point which emerges from this thermal model of the mantle is that allowance for upper mantle convection is necessary if a plausible temperature profile is to be obtained. Radiative transfer alone does not sufficiently enhance the thermal conduction of the upper mantle. This conclusion can be avoided only by making drastic changes in the parameters of the model. No reasonable redistribution of radioactivity appears to suffice. The only serious possibility may be to increase very strongly the radiative conductivity in the upper mantle while reducing it in the lower mantle, perhaps by virtue of the electrical conductivity. However, it is more natural to allow for convection, of which there is independent evidence (see Chapter 7).

The closest approach of the mantle to its melting point is in the upper mantle; in Fig. 9.4 it occurs at a depth between 100 and 200 km. This is entirely consistent with the so-called weak layer in the upper mantle (Fig. 7.11). The creep strength or yield point of a material becomes relatively small near to its melting point. We can also suggest the weak layer as a source of volcanic magma in regions which are slightly hotter, for some reason, than average. The convecting model curve is thus compatible with what we know about the mantle. However, this must not be taken to imply that the curve is an accurate representation of the mantle temperatures, but merely a summary of our present very incomplete information.

The melting point curve of the core may also be obtained from Eq. 9.14. Using Bullen A model densities (Bullen, 1963), and assuming the uncompressed density of core material to be 7 gm cm^{-3}, we obtain

$$T_m(\text{core}) = 1800 + 4560\left(1 - \frac{\rho_0}{\rho}\right) \,^\circ\text{K} \tag{9.32}$$

which is shown as the core melting point curve on Fig. 9.4. This curve intersects the adiabatic curve, which is assumed to be the actual temperature profile, at the boundary of the inner core.

The melting point gradient is steeper than the adiabatic gradient in both mantle and core. Thus if the Earth ever had a molten stage, it would have had an adiabatic gradient in the lower mantle, and as it cooled the temperature profile would intersect the melting point curve at a radius which moved progessively outward from the bottom of the mantle. The mantle would thus have solidified from the inside outward and the upward chemical fractionation of elements such as uranium and potassium, which have large ionic sizes and do not fit easily into the close-packed crystal structures of the lower mantle, has a natural explanation.

9.4   Energy Source for the Geomagnetic Dynamo

Two possible energy sources for geomagnetic dynamo action are mentioned in Section 5.4. Verhoogen (1961) concluded that the latent heat of a progressively solidifying inner core could have maintained an adiabatic temperature gradient in the outer core for 3×10^9 years and therefore that the convective dynamos of W. M. Elsasser and E. C. Bullard were possible. Malkus (1963) re-examined the data and questioned this conclusion, preferring a dynamo mechanism driven by precessional torques. The choice between these alternatives is a vital one in the theory of the Earth's thermal history. If the core is convecting thermally, then it is cooling down and we can estimate the rate of cooling, which is decreasing, but is about $100°C/10^9$ years. On the other hand the precessional dynamo allows no direct conclusion for or against cooling of the deep interior, although the recession of the Moon implies diminution in the internal heat generated by precessional torques. The heat flux from the core need not be grossly different in the two cases since a stirred outer core implies the maintenance of an adiabatic temperature gradient against conduction in both.

The minimum possible rate of heat loss from the core is determined by the adiabatic temperature gradient and thermal conductivity and the actual rate will be greater than this by the amount of the convective heat transport. The adiabatic gradient is calculated from Eq. 9.21, being somewhat larger in the outer layers of the core than the core average, so that we may take $0.2°C \text{ km}^{-1}$ as an appropriate value. In metals heat transport is dominated by the electronic contribution K_e to thermal conductivity K and is linked with electrical conductivity σ_e or resistivity ρ_e by the Wiedemann-Franz relationship:

$$\frac{K_e}{\sigma_e T} = \frac{K_e \rho_e}{T} = L \approx \frac{\pi^2}{3}\left(\frac{k}{e}\right)^2 = 2.45 \times 10^8 \text{ erg-emu/deg}^2 \text{ sec}$$

$$= 2.45 \times 10^{-8} \text{ watt-ohm/deg}^2 \qquad (9.33)$$

This relationship is closely observed by all metals, including liquids, at all temperatures which are not low with respect to the characteristic Debye temperature. Applying it to the core, with the value of resistivity from Section 5.3 and temperature from Section 9.3, we obtain

$$K_e = 3 \times 10^5 \text{ erg cm}^{-1} \text{ sec}^{-1} \text{ deg}^{-1} \qquad (9.34)$$

This value is lower than that usually considered, for example, by Bullard (1950), because a lower electrical conductivity has been assumed for the core. In fact, it is so low that it does not mask completely the phonon contribution to thermal conduction, which by extrapolation from mercury and from glasses is estimated to be about $1 \times 10^5 \text{ erg cm}^{-1} \text{ sec}^{-1} \text{ deg}^{-1}$ for liquid iron. The total thermal conductivity of the core is thus estimated to be

$$K = 4 \times 10^5 \text{ erg cm}^{-1} \text{ sec}^{-1} \text{ deg}^{-1} = 0.01 \text{ cal cm}^{-1} \text{ sec}^{-1} \text{ deg}^{-1} \quad (9.35)$$

The corresponding heat flux is $0.8 \text{ erg sec}^{-1} \text{ cm}^{-2}$, or, for the whole core, $1.4 \times 10^{18} \text{ ergs sec}^{-1}$.

Most previous estimates of the conducted heat from the core have yielded much larger values, which have presented some difficulty for geomagnetic dynamo theories. With the revised electrical resistivity for the core, the conducted heat becomes relatively unimportant, being much less than 10^{19} ergs sec^{-1}, which was tentatively allowed for the dynamo in Section 5.4 and assumed in Section 9.3. If the heat flux from the core, whatever its source, exceeds the conducted heat due to the adiabatic gradient by a factor of 5 or 10, as is supposed here, then the core is necessarily convecting and the convection must play an important role in the geomagnetic dynamo action, even if the ultimate driving force is derived from the precession.

Now consider the magnitudes of the two possible energy sources. The possible availability of 10^{19} ergs sec^{-1} of rotational energy is noted in Sections 2.4 and 5.4. Alternatively, if the energy is derived from cooling and solidification of the core, then the total heat produced since the core was entirely liquid is

$$Q = M_c C_p \Delta T + M_{ic} L \quad (9.36)$$

where M_c and M_{ic} are the masses of the whole core and inner core, respectively, C_p is specific heat (at constant pressure), ΔT is the fall in temperature of the whole core, and L is the latent heat of solidification. The classical Dulong and Petit value appears appropriate for C_p, since the material is a compressed liquid at high temperature: $C_p = 0.11 \text{ cal gm}^{-1}$. L can be estimated from the Clausius-Clapeyron equation (9.12) if Kennedy's relation (9.13) is assumed, because Kennedy gives

$$\frac{dT_m}{dp} \cdot K = \text{constant} = \left(\frac{dT_m}{dp} \right)_0 K_0 \quad (9.37)$$

K being bulk modulus and subscript o indicating zero pressure values. Combining this with Eq. 9.12, we have:

$$L = \frac{K}{K_0} \frac{(v_L - v_s)}{(dT_m/dp)_0} \cdot T_m \quad (9.38)$$

or since we expect approximately

$$K(v_L - v_S) \approx K_0(v_L - v_S)_0 \quad (9.39)$$

we have

$$L = L_0\left(\frac{T_m}{T_{m0}}\right) \qquad (9.40)$$

With $L_0 = 65$ cal gm^{-1} and $T_m/T_{m0} = 2$, we obtain $L = 130$ cal gm^{-1}. The temperature drop ΔT is the difference in melting points between the center of the Earth and the inner core boundary minus the adiabatic temperature drop through the depth of the inner core, as in Fig. 9.5. From Eq. 9.14, the

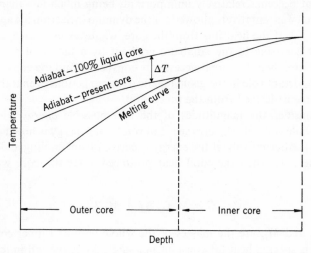

Figure 9.5: Fall in temperature of the core (ΔT) associated with the solidification of the inner core.

melting point difference is 400°C and integrating Eq. 9.21, the adiabatic drop is about 100°C. We therefore put $\Delta T = 300$°C. With these values Eq. 9.36 gives a total heat output of 7.6×10^{28} cal or 3.2×10^{36} ergs, the larger part being due to the specific heat, not the latent heat. If a convective dynamo has been operating for a large part of the Earth's history, then we must suppose, as did Verhoogen (1961), that the core has lost heat at a constant rate over the past 3×10^9 years or so, in which case the rate of heat loss is 3.4×10^{19} ergs sec^{-1}. Not all of this energy is directly available for dynamo action because convection does not produce a very efficient thermodynamic engine. As Bullard and Gellman (1954) point out, the ideal thermodynamic efficiency of an engine working on the temperature difference across the outer core is

$$\eta = \frac{T_i - T_o}{T_i} \approx 0.08 \qquad (9.41)$$

T_i and T_o being the temperatures of the inner and outer core boundaries. The convective energy available for dynamo action is therefore 2.7×10^{18} ergs sec^{-1}.

Thus we conclude that the energies available from the two possible driving mechanisms are similar. The convective engine appears to be capable of delivering only about a quarter of the power which is possible from the precessional engine but the uncertainties in this calculation are quite large and do not allow us to favor one mechanism rather than the other. However, we can examine the theories of Bullard and Gellman (1954) and Malkus (1963, 1968) in more detail in the light of the tenfold increase in estimated core resistivity, which is suggested in Section 5.4. Bullard and Gellman's dynamo gave an ohmic dissipation 0.9×10^{18} ergs sec^{-1}, which, as Malkus points out, probably errs on the low side because irregularities of the field were not taken into account. It leaves a factor of 3 in hand on the available energy of the convective engine, but upward revision of the resistivity estimate upsets the balance completely. It requires a general speeding up of the dynamo and increases dissipation by a factor 10^3 to the impossible figure of 9×10^{20} ergs sec^{-1}. The effectiveness of the convective dynamo is thus even more questionable than Malkus suggested. The precessional dynamo is much less strongly affected by the estimate of resistivity; the required dissipation appears to be about 2×10^{19} ergs sec^{-1}, but with a substantial uncertainty, so that the available 1×10^{19} ergs sec^{-1} may prove to be adequate.

As a tentative conclusion, it is suggested that precession of the Earth provides the basic driving mechanism for the geomagnetic dynamo, but that convection must occur and play at least some part. The maintenance of an adiabatic temperature gradient in the core is automatic as ohmic dissipation alone is more than sufficient.

Appendix A

SPHERICAL HARMONIC FUNCTIONS

The principal purpose of this summary is to avoid repetition. Spherical harmonic analysis is used or referred to in Chapters 2, 3, 4, and 5 and is basic to several problems in geophysics. As far as possible, a common notation for the harmonic coefficients is used in the sections on gravity, geomagnetism, and free oscillations, although independent treatments of these subjects use different notations. For a readable statement of the properties of spherical harmonic functions reference should be made to Chapter 3 of Sneddon (1961). The discussion by Chapman and Bartels (1940, Vol. 2) is also useful, particularly as a specification for procedure in harmonic analysis of the geomagnetic field.

The gravitational and magnetic fields of the Earth as a whole are analyzed in terms of the potentials V of the fields, which satisfy Laplace's equation at all points external to the Earth:

$$\nabla^2 V = \frac{\partial^2 V}{\partial x^2} + \frac{\partial^2 V}{\partial y^2} + \frac{\partial^2 V}{\partial z^2} = 0 \tag{A.1}$$

The natural boundaries which impose boundary conditions on solutions of Laplace's equation can be taken as spherical in most whole-Earth problems and it is therefore convenient to express the equation in spherical polar coordinates, (r, θ, λ):

$$\nabla^2 V = \frac{1}{r^2} \frac{\partial}{\partial r} \left(r^2 \frac{\partial V}{\partial r} \right) + \frac{1}{r^2 \sin \theta} \frac{\partial}{\partial \theta} \left(\sin \theta \frac{\partial V}{\partial \theta} \right) + \frac{1}{r^2 \sin^2 \theta} \frac{\partial^2 V}{\partial \lambda^2} = 0 \tag{A.2}$$

Here θ is the angle to the axis of the coordinate system and is thus the colatitude (90° minus latitude) if the axis is chosen to be the Earth's rotational axis (or geomagnetic latitude if the magnetic axis is chosen), and λ is longitude measured from a convenient reference. These are the commonly selected axes, but others may be used according to the symmetry of the problem.

The wave equation for propagation of seismic (or any other) waves has a similar geometrical form:

$$\frac{\partial^2 V}{\partial t^2} = c^2 \nabla^2 V \tag{A.3}$$

266

where c is wave velocity and V is the potential whose derivative in any direction gives the velocity in that direction. Thus the geometrical solutions of Eq. A.2 appear also in the dynamic problem of free oscillations of the Earth (Section 4.6).

As may be verified by differentiation and substitution, Eq. A.2 has solutions of the form

$$V = (r^l, r^{-(l+1)})(\cos m\lambda, \sin m\lambda)P_l{}^m(\cos \theta) \qquad (A.4)$$

where $(r^l, r^{-(l+1)})$ and $(\cos m\lambda, \sin m\lambda)$ represent alternative solutions, l and m are integers with $m \leqslant l$, and $P_l{}^m(\mu)$ satisfies the equation

$$(1 - \mu^2)\frac{d^2P}{d\mu^2} - 2\mu \frac{dP}{d\mu} + \left[l(l + 1) - \frac{m^2}{1 - \mu^2} \right]P = 0 \qquad (A.5)$$

This is Legendre's associated equation which reduces to Legendre's equation for $m = 0$. Considering first the special case $m = 0$, we can verify, again by differentiation and substitution, that

$$P_l{}^0(\mu) = \frac{1}{2^l l!} \frac{d^l}{d\mu^l} [(u^2 - 1)^l] \qquad (A.6)$$

As a solution of Eq. A.5, the constant factor $(2^l l!)^{-1}$ is arbitrary, but has been chosen so that $P_l^0(1) = 1$. The functions $P_l^0(\mu)$, normally written with the superscript 0 omitted, are the Legendre polynomials $P_l(\cos \theta)$ given in the $m = 0$ column of Table A.1. In some problems it is convenient to use latitude ϕ rather than colatitude θ by substituting $\sin \phi$ for $\cos \theta$.

The Legendre polynomials give solutions to Laplace's equation which have rotational symmetry, since by putting $m = 0$ in Eq. A.4, dependence of V upon longitude is excluded. These are the zonal harmonics, functions of latitude only. A potential which can be expressed in terms of zonal harmonics only can be written as a sum of powers of distance r from the origin of the coordinate system, with the Legendre polynomials appearing in the coefficients and giving the variation of potential with latitude. In geophysical problems it is convenient to make the coefficients dimensionally uniform by relating r to the Earth radius a:

$$V = \frac{1}{a}\sum_{l=0}^{\infty} \left[C_l\left(\frac{a}{r}\right)^{l+1} + C_l'\left(\frac{r}{a}\right)^l \right]P_l(\cos \theta) \qquad (A.7)$$

where C_l are constant coefficients arising from sources internal to the surface considered and C_l' are coefficients arising from external sources. Equation A.7 is simply a sum of terms of the form of Eq. A.4 with $m = 0$.

The form of Eq. A.7 is directly derivable by generalizing the method used in Section 2.1 to obtain MacCullagh's formula for gravitational potential.

If, instead of terminating the expansion of Eq. 2.9 at terms in $1/r^2$, it is continued to higher powers in $1/r$, the coefficients are higher-order Legendre polynomials:

$$\left[1 + \left(\frac{s}{r}\right)^2 - 2\frac{s}{r}\cos\psi\right]^{-1/2} = \sum_{l=0}^{\infty}\left(\frac{s}{r}\right)^l P_l(\cos\psi) \tag{A.8}$$

We can now consider the general case of Eqs. A.4 and A.5 in which $m \neq 0$. By writing

$$P_l^m(\mu) = (1 - \mu^2)^{\frac{1}{2}m}\frac{d^m}{d\mu^m}[P_l^0(\mu)]$$

or

$$P_l^m(\cos\theta) = \sin^m\theta\frac{d^m}{d(\cos\theta)^m}[P_l^0(\cos\theta)] \tag{A.9}$$

and carrying out the differentiation, we can verify that the function $P_l^m(\mu)$ is a solution of Eq. A.5. This therefore gives the form of the solutions (A.4) in which a longitude variation is allowed. By writing $P_l^0(\mu)$ in the form of Eq. A.6 we obtain

$$P_l^m(\mu) = \frac{(1 - \mu^2)^{\frac{1}{2}m}}{2^l l!}\frac{d^{l+m}}{d\mu^{l+m}}[(\mu^2 - 1)^l] \tag{A.10}$$

Since the highest power of μ in the expression to be differentiated is μ^{2l}, it follows that differentiation gives a polynomial in μ of order $(l - m)$, being zero if $m > l$. Thus the distribution on a spherical surface of a potential with the form of Eq. A.4 has $(l - m)$ nodal lines of latitude, arising from the $P_l^m(\cos\theta)$ term and $2m$ nodal lines of longitude, arising from the $(\cos m\lambda,\sin m\lambda)$ term. Examples are shown in Fig. A.1. The explicit expressions for $P_l^m(\cos\theta)$ up to $l = 4$ are given in Table A.1.

The general expression for a function V as a sum of spherical harmonics is thus

$$V = \frac{1}{a}\sum_{l=0}^{\infty}\sum_{m=0}^{l}\left\{\begin{array}{l}\left[C_l^m\left(\frac{a}{r}\right)^{l+1} + C_l'^m\left(\frac{r}{a}\right)^l\right]\cos m\lambda \\ + \left[S_l^m\left(\frac{a}{r}\right)^{l+1} + S_l'^m\left(\frac{r}{a}\right)^l\right]\sin m\lambda\end{array}\right\}P_l^m(\cos\theta) \tag{A.11}$$

In problems of the gravitational or magnetic potential external to the Earth the coefficients C' and S' vanish, being incompatible with finite V at $r \to \infty$. It must also be noted that the solutions considered here are restricted to those which remain finite everywhere on a spherical boundary.

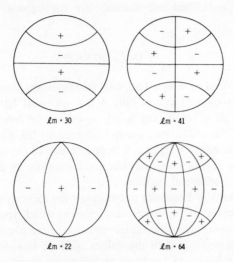

Figure A.1: Examples of spherical harmonics. $m = 0$ gives zonal harmonics, $m = l$ gives sectorial harmonics and the general cases, $m < l$, are known as tesseral harmonics. After Kaula (1965).

TABLE A.1: LEGENDRE POLYNOMIALS $P_l(\cos \theta)$ AND ASSOCIATED POLYNOMIALS $P_l^m(\cos \theta)$.

	$m = 0$			$m = 1$	
$l = 0$	1	$(1)^a$		—	
1	$\cos \theta$	(1)		$-\sin \theta$	(0.707)
2	$\frac{1}{2}(3 \cos^2 \theta - 1)$	(1)		$-3 \cos \theta \sin \theta$	(0.408)
3	$\frac{1}{2}(\cos^3 \theta - 3 \cos \theta)$	(1)		$-\frac{3}{2}(5 \cos^2 \theta - 1) \sin \theta$	(0.289)
4	$\frac{1}{8}(35 \cos^4 \theta - 30 \cos^2 \theta + 3)$	(1)		$-\frac{5}{2}(7 \cos^3 \theta - 3 \cos \theta) \sin \theta$	(0.224)

	$m = 2$		$m = 3$		$m = 4$	
$l = 0$	—		—		—	
1	—		—		—	
2	$3 \sin^2 \theta$	(0.204)	—		—	
3	$15 \cos \theta \sin^2 \theta$	(0.0913)	$-15 \sin^3 \theta$	(0.0373)	—	
4	$\frac{15}{2}(7 \cos^2 \theta - 1) \sin^2 \theta$	(0.0527)	$-105 \cos \theta \sin^3 \theta$	(0.0141)	$105 \sin^4 \theta$	(0.00498)

^{a} Numbers in brackets are factors which convert P_l^m to p_l^m by Eq. (A.13.)

Legendre and associated polynomials are orthogonal functions, which means that

$$\int_{-1}^{1} P_l^m(\mu)P_{l'}^{m'}(\mu)\,d\mu = 0 \tag{A.12}$$

unless both $l = l'$ and $m = m'$. This has an important consequence. If in a particular analysis of a function, only a few spherical harmonic terms are taken, the sum of these terms may poorly represent the function but nevetheless the values of the coefficients are not affected by taking additional terms in a more complete analysis. This is a property also of the terms in a Fourier series and we can regard spherical harmonic analysis as a development of Fourier analysis to spherical surfaces.

In some treatments of Legendre polynomials the subscript n is used in place of l. Here n is reserved for a further development which appears in the study of free oscillations—a harmonic radial variation. Free oscillations are classified in Section 4.6 according to the values of three integers; thus $_nS_l^m$ and $_nT_l^m$ denote spheroidal and torsional oscillations, respectively, where l, m represent variations on a spherical surface, as for $P_l^m(\cos\theta)\cos m\lambda$ and n is the number of internal spherical surfaces which are nodes of the motion.

The numerical factors in the associated polynomials defined by (A.9) and (A.10) increase rapidly with m; in order that the coefficients in a harmonic analysis relate more nearly to the physical significance of the terms they represent, various approximate normalizing factors are used with the harmonics. A commonly favored form is

$$p_l^m = \left[\frac{(l-m)!}{(l+m)!}\right]^{1/2} P_l^m \tag{A.13}$$

Values of this normalizing factor are given in parenthesis in Table A.1. The harmonic coefficients of the satellite geoid, as determined by Guier and Newton (1965) and listed in Table 3.1, refer to harmonics normalized by (A.13). Other normalizing factors which have been favored are $[(l-m)!/l!]$ (Jeffreys and Jeffreys, 1962), $[2(l-m)!/(l+m)!]^{1/2}$ (Chapman and Bartels, 1940; this alternative is widely used in geomagnetism), and the normalized form $[2(2l+1)(l-m)!/(l+m)!]^{1/2}$. A reversal of signs by the factor $(-1)^m$ is sometimes added. Of these, only (A.13) and the Jeffreys and Jeffreys factor satisfy the logical requirement that they are equal to unity for $m = 0$; the form used by Kozai (1966) and others provides a better normalization but is not applied to P_l^0. It is unfortunate that so many different conventions have been followed; care must be taken to specify the normalizing factors in any spherical harmonic analysis.

Appendix B

SOLUTION OF THE INTEGRAL EQUATION FOR THE VELOCITY PROFILE IN A SPHERICALLY LAYERED EARTH

Equation 4.45 is rewritten

$$\Delta = \int_{p}^{\eta_0} \frac{2p}{r(\eta^2 - p^2)^{1/2}} \frac{dr}{d\eta} \, d\eta \tag{B.1}$$

the limits being η_0 and $\eta' = p$. Now η_1 is the value of η corresponding to a radius r_1, between r_0 and r', and it must be assumed that η decreases monotonically with decreasing r so that $\eta_1 > \eta'$, the value at maximum penetration of the ray. Also Δ_1 is the value of Δ for the ray whose deepest point is at r_1. Multiply both sides of Eq. B.1 by $(p^2 - \eta_1^2)^{-1/2}$ and integrate with respect to p from η_1 to η_0:

$$\int_{\eta_1}^{\eta_0} \frac{\Delta \, dp}{(p^2 - \eta_1^2)^{1/2}} = \int_{\eta_1}^{\eta_0} \left\{ \int_{p}^{\eta} \frac{2p}{r[(p^2 - \eta_1^2)(\eta^2 - p^2)]^{1/2}} \frac{dr}{dp} \, d\eta \right\} dp \tag{B.2}$$

Now change the order of integration on the right-hand side, the limits being then η_1 to η for p and η_1 to η_0 for η:

$$\int_{\eta_1}^{\eta_0} \frac{\Delta \, dp}{(p^2 - \eta_1^2)^{1/2}} = \int_{\eta_1}^{\eta_0} \left\{ \int_{\eta_1}^{\eta} \frac{2p}{r[(p^2 - \eta_1^2)(\eta^2 - p^2)]^{1/2}} \frac{dr}{d\eta} \, dp \right\} d\eta \tag{B.3}$$

The left-hand side can be integrated by parts and the integration with respect to p on the right-hand side gives a considerable simplification (for $\eta > \eta_1$) because

$$\int_{\eta_1}^{\eta} \frac{p \, dp}{[(p^2 - \eta_1^2)(\eta^2 - p^2)]^{1/2}} = \frac{\pi}{2} \tag{B.4}$$

so that

$$\left[\Delta \cosh^{-1} \left(\frac{p}{\eta_1} \right) \right]_{\eta_1}^{\eta_0} - \int_{\eta_1}^{\eta_0} \frac{d\Delta}{dp} \cosh^{-1} \left(\frac{p}{\eta_1} \right) dp = \pi \int_{\eta_1}^{\eta_0} \frac{1}{r} \frac{dr}{d\eta} \, d\eta \tag{B.5}$$

271

The first term on the left-hand side is zero because $\Delta = 0$ at $p = \eta_0$ and $\cosh^{-1}(p/\eta_1) = 0$ for $p = \eta_1$. Thus

$$\int_0^{\Delta_1} \cosh^{-1}\left(\frac{p}{\eta_1}\right) d\Delta = \pi \ln\left(\frac{r_0}{r_1}\right) \tag{B.6}$$

Since η is the value of (r/v) at the point of deepest penetration of a ray which travels a distance Δ, it is convenient to put $\eta_1 = p_1$, as in Eq. 4.46. $p(\Delta)$ is tabulated from the observed travel times, $T(\Delta)$, since $p = dT/d\Delta$, and thus (r/v) is tabulated against (r/r_0) by Eq. 4.46.

Appendix C

RADIATIVE THERMAL CONDUCTIVITY OF A GREY BODY

We consider here a simple special case of radiative transfer: the refractive index of the medium and its extinction (or absorption) coefficient for radiation are assumed to be independent of wavelength (or frequency) and the medium is uniform and isotropic. This "grey body" is a generalization of the familiar "black body" of radiation laws. A more general treatment which does not assume constancy of the coefficients is given by Clark (1957a).

Stefan's law of thermal radiation gives the rate of energy emission per unit area of surface of an ideal black body at absolute temperature T:

$$\frac{dE}{dt} = \sigma T^4 \tag{C.1}$$

where $\sigma = 5.6686 \times 10^{-5}$ erg cm^{-2} sec^{-1} °C^{-4} is Stefan's constant, which is expressible in terms of the Planck and Boltzmann constants (see, for example, Joos, 1934, p. 587). This law gives the integral of energy over the solid angle 2π of the hemisphere about the surface. The intensity of the radiation at any angle ϕ to the normal to the radiating surface is reduced by the factor $\cos \phi$ relative to the normal radiation; i.e., it is proportional to the solid angle subtended by the surface element. It follows that the radiated energy per unit solid angle normal to the surface is twice the average over the hemisphere and the normal radiation into an elementary solid angle $d\Omega$ per unit area of radiating surface is thus

$$\frac{dE}{dt} = \sigma T^4 \frac{d\Omega}{\pi} \tag{C.2}$$

We now consider, instead of a surface element, a small volume element dV in a continuous medium of partly transparent material in thermal equilibrium. Attenuation of an incident beam of radiation with distance r in the material is given by

$$\frac{I}{I_0} = e^{-\epsilon r} \tag{C.3}$$

273

where ϵ is the extinction coefficient, which, as mentioned, is independent of wavelength. The dimensions of the elementary volume are small compared with ϵ^{-1}, so that the rate absorption of radiation from solid angle $d\Omega$ in a thickness dx of area (dV/dx) equals the emission by the volume into the same solid angle and is given by

$$d^3\left(\frac{dE}{dt}\right) = \sigma T^4 \frac{d\Omega}{\pi} \cdot n^2\left(\frac{dV}{dx}\right) \epsilon \, dx$$

$$= \sigma T^4 \frac{d\Omega}{\pi} \cdot n^2 \epsilon \, dV \tag{C.4}$$

Here d^3 is used to represent the differential with respect to the three coordinates of the volume element. n is the refractive index and the factor n^2 occurs because the radiation in the medium at temperature T is more intense by this factor than in a vacuum surrounded by surfaces at T. That this must be so can be seen by considering a cavity in the medium. Radiation entering the medium from the cavity is refracted into a smaller solid angle, isotropy of the radiation being maintained because radiation in the medium striking the boundary too obliquely for refraction is internally reflected.

The radiating volume element is now considered to be part of a spherical shell of radius r and thickness dr, centered on an elementary area ΔA which is normal to a temperature gradient (dT/dx), so that the net flux, through the area, of radiation originating in the sphere is directly calculable by integration. It is a requirement of the theory that the temperature gradient be small, that is the change in temperature with one quantum mean free path, ϵ^{-1}, is slight. By considering a volume element at θ to the normal to ΔA, as in Fig. C.1, the solid angle of emission of radiation crossing ΔA is

$$d\Omega = \frac{\Delta A}{r^2} \cos \theta \tag{C.5}$$

and the elementary volume is

$$dV = r \, d\theta \, dr \, dl \tag{C.6}$$

where dl is its length, measured around the ring in Fig. C.1. Since all elements in the ring are similarly situated with respect to ΔA, and $\int dl = 2\pi r \sin \theta$, this stage of integration gives, with substitution of (C.5) and (C.6) in (C.4) and multiplying the energy flux by $e^{-\epsilon r}$ to account for the absorption in distance r;

$$d^2\left(\frac{dE}{dt}\right) = 2n^2 \epsilon \sigma \Delta A T^4 e^{-\epsilon r} \cos \theta \sin \theta \, d\theta \, dr \tag{C.7}$$

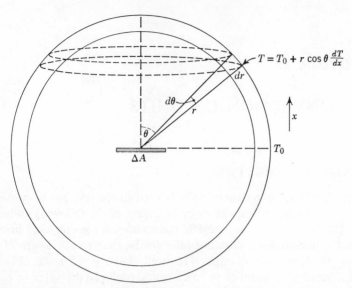

Figure C.1: Geometry for the integration of radiative flux through an elementary area ΔA normal to a temperature gradient dT/dx.

Since there is a gradient of temperature (dT/dx) normal to ΔA, T is a function of r, θ:

$$T = T_0 + \left(\frac{dT}{dx}\right) r \cos \theta \qquad \text{(C.8)}$$

and the gradient is assumed small so that

$$r\left(\frac{dT}{dx}\right) \ll T_0$$

and

$$T^4 \approx T_0{}^4 + 4T_0{}^3\left(\frac{dT}{dx}\right) r \cos \theta \qquad \text{(C.9)}$$

Equation C.9 is substituted into (C.7), which is then integrated with respect to $\theta(0$ to $\pi)$ and $r(0$ to $\infty)$ to give the net energy flux through ΔA due to all of the material surrounding it:

$$\frac{dE}{dt} = \frac{16}{3}\frac{n^2}{\epsilon}\sigma T_0{}^3 \Delta A \frac{dT}{dx} \qquad \text{(C.10)}$$

The radiative thermal conductivity is thus

$$K_R = \frac{16}{3}\frac{n^2}{\epsilon}\sigma T_0{}^3 \qquad \text{(C.11)}$$

Appendix D

CONVERSION OF UNITS

Electromagnetic and MKS Units

The electromagnetic system of units is used in the text but conversion to the MKS system may readily be made by means of the following conversion factors. This list gives the most useful conversions for geophysical problems but is not exhaustive and does not refer to the electrostatic sytem of units (esu), for which see, for example, Harnwell (1949, p. 655). To obtain the numerical value of a quantity in MKS units, multiply the value in electromagnetic units by the factor in the right-hand column. Thus the dipole moment of the earth is

$$8.09 \times 10^{25} \text{ emu} = 8 \times 10^{25} \times 4\pi \times 10^{-10} \text{ weber meters}$$
$$= 1.01_6 \times 10^{17} \text{ weber metres}$$
$$= 8 \times 10^{25} \times 10^{-3} \text{ ampere-turn meter}^2$$
$$= 8 \times 10^{22} \text{ ampere-turn meter}^2$$

	Electromagnetic Unit	Equivalent in MKS units
Mass	1 gram (gm)	10^{-3} kilogram (kg)
Length	1 centimeter (cm)	10^{-2} meter (m)
Time	1 second (sec)	1 sec
Force	1 dyne	10^{-5} newton (nt)
Energy	1 erg	10^{-7} joule
Current	1 electromagnetic unit (emu)	10 amperes (amp)
Potential difference (voltage)	1 emu	10^{-8} volt (v)
Electric field	1 emu	10^{-6} vm^{-1}
Magnetic flux	1 maxwell	10^{-8} weber
Magnetic intensity	1 gauss ($= 1$ maxwell cm^{-2})	10^{-4} weber m^{-2}

	Electromagnetic Unit	*Equivalent in MKS units*
Magnetic field strength	1 oersted (oe)	$10^3/4\pi$ amp-turns m^{-1}
Magnetic potential	1 gilbert	$10/4\pi$ amp-turn
Magnetic moment	1 emu	$4\pi \times 10^{-10}$ weber $m = 10^{-3}$ amp-turn m^2
Magnetization (magnetic moment/unit volume)	1 emu	$4\pi \times 10^{-4}$ weber m^{-2}
Inductance	1 emu	10^{-9} henry
Resistance	1 emu	10^{-9} ohm
Resistivity	1 emu	10^{-11} ohm m
Capacitance	1 emu	10^9 farads

Other Useful Conversions

1 statute mile	$= 1.609$ km	
1 nautical mile	$= 1.151$ statute miles	$= 1.852$ km
1 pound (lb)	$= 0.4536$ kg	
1 ton (U.S.A.)	$= 2000$ lb	$= 907.2$ kg
1 ton (imperial)	$= 2240$ lb	$= 1016$ kg
1 ton (metric)	$= 1000$ kg	
1 year	$= 3.156 \times 10^7$ sec	
1 kg-wt	$= 9.807$ newtons	$= 9.807 \times 10^5$ dynes
1 atmosphere (pressure) (atm)	$= 1.013$ bar (supports 76 cm Hg)	
1 bar	$= 10^6$ dynes cm^{-2}	$= 10^5$ newtons m^{-2}
1 kg cm^{-2}	$= 9.8 \times 10^5$ dynes cm^{-2}	$= 0.98$ bar $= 0.968$ atm
1 joule	$= 10^7$ erg	
1 calorie (cal)	$= 4.186$ joule	$= 4.186 \times 10^7$ erg
1 electron volt (ev)	$= 1.602 \times 10^{-12}$ erg	$= 1.602 \times 10^{-19}$ joule
1 radian (rad)	$= 57.30°$	$= 2.063 \times 10^5$ sec of arc
1 sec of arc	$= 4.848 \times 10^{-6}$ rad	
1 gamma	$= 10^{-5}$ oe ($=$gauss)	$= 10^{-9}$ weber m^{-2}
1 gal	$= 1$ cm sec^{-2}	$= 10^{-2}$ m sec^{-2}
$T°$C	$= (\frac{9}{5}T + 32)°$F	$= (T + 273.16)°$K

Appendix E

NUMERICAL DATA OF GEOPHYSICAL INTEREST

Physical Constants

Speed of light	2.997925×10^{10} cm $\sec^{-1}$	$= 2.998 \times 10^8$ m $\sec^{-1}$
Electronic charge	1.6021×10^{-20} emu	$= 1.6021 \times 10^{-19}$ coulomb (coul)
Electron rest mass	9.109×10^{-28} gm	$= 9.109 \times 10^{-31}$ kg
Proton mass	1.672×10^{-24} gm	$= 1.672 \times 10^{-27}$ kg
Mass of hydrogen atom	1.673×10^{-24} gm	$= 1.673 \times 10^{-27}$ kg
Planck's constant	6.625×10^{-27} erg sec	$= 6.626 \times 10^{-34}$ joule sec
Boltzmann's constant	1.3805×10^{-16} erg $\deg^{-1}$	$= 1.3805 \times 10^{-23}$ joule $\deg^{-1}$
Stefan-Boltzmann constant	5.670×10^{-5} erg cm^{-2} $\deg^{-4}$ $\sec^{-1}$	$= 5.67 \times 10^{-8}$ joule m^{-2} $\deg^{-4}$ $\sec^{-1}$
Gas constant	8.314×10^7 erg mole^{-1} $\deg^{-1} = 1.96$ cal mole^{-1} $\deg^{-1}$	$= 8.314$ joule mole^{-1} $\deg^{-1}$
Avogadro's number	6.0226×10^{23} mole^{-1}	
Gravitational constant	6.670×10^{-8} cm^3 gm^{-1} $\sec^{-2}$ (dyne cm^2 gm^{-2})	$= 6.670 \times 10^{-11}$ m^3 kg^{-1} $\sec^{-2}$) newton m^2 kg^{-2}

Dimensions and Properties of the Earth

Equatorial radius	$a = 6{,}378.16$ km
Polar radius	$c = 6{,}356.18$ km
Volume	$V = 1.083 \times 10^{27}$ cm^3
Radius of sphere of equal volume	$= 6{,}371$ km
Surface ellipticity	$\epsilon = \dfrac{a - c}{a} = \dfrac{1}{298.25} = 3.3529 \times 10^{-3}$
Surface area	$= 5.10 \times 10^{18}$ cm^2

278

Mass	$M = 5.976 \times 10^{27}$ gm
Product, gravitational constant × mass of Earth	$GM = 3.98603 \times 10^{20}$ cm^3 sec^{-2}
Mean density	$\bar{\rho} = 5.517$ gm cm^{-3}
Moments of inertia	
About polar axis	$C = 8.068 \times 10^{44}$ gm cm^2
About equatorial axis	$A = 8.042 \times 10^{44}$ gm cm^2
Dynamical ellipticity	$H = \dfrac{C - A}{C} = \dfrac{1}{305.51} = 3.2732 \times 10^{-3}$
Ellipticity coefficient	$J_2 = \dfrac{C - A}{Ma^2} = 1.0827 \times 10^{-3}$
Coefficient of moment of inertia	$\dfrac{J_2}{H} = \dfrac{C}{Ma^2} = 0.33078$
Solar day	$= 86{,}400$ sec
Sidereal day	$= 86{,}164$ sec
Angular velocity	$\omega = 7.29211_6 \times 10^{-5}$ sec^{-1}
Equatorial gravity (see Eq. 3.13 for latitude variation)	$g_e = 978.032$ cm sec^{-2}
Ratio $\dfrac{\text{centrifugal force}}{\text{equatorial gravity}}$	$m = \dfrac{\omega^2 a}{g_e} = 3.4678 \times 10^{-3}$
Mean orbital radius	$r_E = 1.4960 \times 10^8$ km $= 1$ astronomical unit (AU)
Orbital velocity	$= 29.77$ km sec^{-1}
Ratio $\dfrac{\text{mass of Sun}}{\text{mass of Earth}}$	$= 3.329 \times 10^5$
Solar constant	$S = 1.39 \times 10^6$ ergs/cm^2 sec
Mean Earth-Moon distance	$= 3.844 \times 10^5$ km
Ratio $\dfrac{\text{mass of Earth}}{\text{mass of Moon}}$	$\mu = 81.303$
Period of Chandler wobble	$\dfrac{2\pi}{\omega_c} = 431$ days $= 1.18$ years
Rate of precession of equinox	$\omega_p = 50.''25$ year^{-1}
Period of precession of equinox	$= 25{,}800$ years
Luni-solar precessional torque	$= 4.14 \times 10^{29}$ dyne cm
Lunar tidal torque	$= 3.9 \times 10^{23}$ dyne cm
Mass of atmosphere	$= 5.1 \times 10^{21}$ gm
Mass of oceans	$= 1.4 \times 10^{24}$ gm
Mass of crust	$= 2.4 \times 10^{25}$ gm
Mass of mantle	$= 4.1 \times 10^{27}$ gm
Mass of core	$= 1.9 \times 10^{27}$ gm

Total geothermal flux $= 1.0 \times 10^{28}$ erg yr^{-1} (1.5 μcal/cm^2 sec)

Magnetic dipole moment $= 8.09 \times 10^{25}$ emu

Masses of planets, see Table 1.1, p 2

Physical Properties of Geological Materials

Some values are only poorly known or are very variable between samples. For an extensive tabulation see Clark (1966).

Chemical Compositions of Rocks

	Percent SiO_2	Percent MgO	Percent Al_2O_3	Percent $(FeO + Fe_2O_3)$
Granite	71	1	15	3
Basalt	50	6	16	12
Eclogite	49	9	15	13
Dunite	40	43	1	14
Chondrites	47	30	3	15

Densities and Elastic Constants

	Density (gm cm^{-3})	Bulk Modulus (10^{11} dynes cm^{-2})	Rigidity (10^{11} dynes cm^{-2})
Sea water	1.025	0.2	0
Granite	2.7	5.5	2
Basalt	2.9	7.5	3.5
Eclogite	3.4	9	6
Dunite	3.3	10	6.5
Iron (20°C)	7.87	17	8
Liquid Iron (M.P.)	7.65	~ 8	0
10% Si-Iron	7.37	17	8

Thermal Properties

	Conductivity (cal cm^{-1} sec^{-1} deg^{-1})	Specific Heat (cal gm^{-1} deg^{-1})	Diffusivity (cm^2 sec^{-1})	Exp Coefficient (deg^{-1})
Sea water	1.4×10^{-3}	1.0	1.4×10^{-3}	—[a]
Average igneous rocks	6×10^{-3}	0.17	0.012	6×10^{-6}
Iron	0.16	0.113	0.18	12×10^{-6}

[a] Sea water has a maximum density at 2°C. Volume coefficient at 15°C is 1.5×10^{-4} °C^{-1}.

Electric and Magnetic Properties

	Electrical Resistivity (ohm cm)	Dielectric Constant	Magnetic Susceptibility (emu)
Sea water	23	80	10^{-6}
Granite (dry, 20°C)	3×10^{12} [a]	8[a]	2×10^{-4}
Basalt (dry, 20°C)	1.5×10^{11} [a]	20[a]	5×10^{-3}

[a] Values for mineral conduction with all water expelled.

TABLE E.1: MECHANICAL PARAMETERS OF EARTH MODEL HB1, BY BULLEN AND HADDON (1967a, 1967b) WITH AN ADJUSTMENT TO ALLOW FOR RIGIDITY IN THE INNER CORE

Depth (km)	Density (gm cm^{-3})	Pressure (10^{12} dyne cm^{-2})	Gravity (cm sec^{-2})	Bulk Modulus (10^{12} dyne cm^{-2})	Rigidity (10^{12} dyne cm^{-2})
0	2.840	0.000	982.2	0.650	0.358
15	2.840	0.004	983.2	0.650	0.358
15	3.313	0.004	983.2	1.017	0.709
60	3.332	0.019	984.7	1.093	0.713
60	3.332	0.019	984.7	1.144	0.675
100	3.348	0.032	986.1	1.213	0.678
200	3.387	0.065	989.6	1.394	0.686
300	3.424	0.099	993.4	1.599	0.693
350	3.441	0.116	995.5	1.706	0.697
350	3.700	0.116	995.5	1.834	0.749
400	3.775	0.135	996.4	1.886	0.840
413	3.795	0.140	996.6	1.900	0.865
500	3.925	0.174	997.5	2.267	1.041
600	4.075	0.213	998.6	2.581	1.270
650	4.150	0.234	998.7	2.698	1.396
650	4.200	0.234	998.7	2.665	1.462
800	4.380	0.298	997.8	3.100	1.656
984	4.529	0.380	996.0	3.471	1.827
1,000	4.538	0.387	995.8	3.490	1.837
1,200	4.655	0.478	994.3	3.756	1.967
1,400	4.768	0.572	993.7	4.069	2.090
1,600	4.877	0.668	994.5	4.387	2.209
1,800	4.983	0.766	997.1	4.725	2.324
2,000	5.087	0.867	1,002.1	5.074	2.439
2,200	5.188	0.970	1,010.2	5.407	2.555
2,400	5.288	1.076	1,022.3	5.741	2.678
2,600	5.387	1.187	1,039.3	6.073	2.804

Depth (km)	Density (gm cm^{-3})	Pressure (10^{12} dyne cm^{-2})	Gravity (cm sec^{-2})	Bulk Modulus (10^{12} dyne cm^{-2})	Rigidity (10^{12} dyne cm
2,800	5.487	1.301	1,062.6	6.294	2.934
2,878	5.527	1.347	1,073.8	6.349	2.954
2,878	9.927	1.347	1,073.8	6.522	0.000
3,000	10.121	1.476	1,046.7	6.883	0.000
3,200	10.421	1.686	999.6	7.522	0.000
3,400	10.697	1.892	949.5	8.266	0.000
3,600	10.948	2.092	896.7	8.984	0.000
3,800	11.176	2.284	841.4	9.680	0.000
4,000	11.383	2.467	783.9	10.340	0.000
4,200	11.570	2.640	724.4	10.956	0.000
4,400	11.737	2.802	663.0	11.521	0.000
4,600	11.887	2.951	600.0	12.115	0.000
4,800	12.017	3.087	535.6	12.710	0.000
4,982	12.121	3.198	475.9	13.282	0.000
5,000	12.130	3.208	469.9	12.949	0.000
5,121	12.197	3.275	429.6	10.842	0.000
5,121	12.197	3.275	429.6	13.627	1.20
5,200	12.229	3.315	403.1	13.724	1.22
5,400	12.301	3.405	335.4	13.882	1.25
5,600	12.360	3.480	267.1	14.006	1.28
5,800	12.405	3.537	198.2	14.113	1.30
6,000	12.437	3.579	129.0	14.177	1.32
6,200	12.455	3.602	59.5	14.221	1.33
6,371	12.460	3.608	0.0	14.244	1.33

Appendix F

THE GEOLOGICAL TIME SCALE

Ages of lower boundaries of geological periods in millions of years, according to three authors. For a finer subdivision see Harland et al. (1964).

Geological Period	Kulp (1961)	Holmes (1965)	Harland et al. (1964)
Cenozoic (Mammals)			
Quaternary			
Recent			
Pleistocene	1	2 to 3	1.5 to 2
Tertiary			
Pliocene	13	12 ± 1	~7
Miocene	25	25 ± 2	26
Oligocene	36	40 ± 2	37 to 38
Eocene	58	60 ± 2	53 to 54
Paleocene	63	70 ± 2	65
Mesozoic (Reptiles)			
Cretaceous	135	135 ± 5	136
Jurassic	181	180 ± 5	190 to 195
Triassic	230	225 ± 5	225
Paleozoic (Invertebrates)			
Permian	280	270 ± 5	280
Carboniferous	345	350 ± 10	345
Devonian	405	400 ± 10	395
Silurian	425	440 ± 10	430 to 440
Ordovician	500	500 ± 15	~500
Cambrian	600?	600 ± 20	570
Pre-Cambrian (No developed fossils)			

Bibliography

Aitken, M. J., and Weaver, G. H. (1965) Recent archaeomagnetic results in England. *J. Geomag. Geoelect.*, **17**, 391.

Akimoto, S., and Fujisawa, H. (1965) Demonstration of the electrical conductivity jump produced by the olivine-spinel transition. *J. Geophys. Res.*, **70**, 443.

Alder, B. J. (1966) Is the mantle soluble in the core? *J. Geophys. Res.*, **71**, 4973.

Alfvén, H. (1954) *The origin of the solar system.* Oxford: Clarendon Press.

Alfvén, H. (1965) Origin of the Moon. *Science*, **148**, 476.

Alfvén, H. (1967) Rotation of planets. In Runcorn (1967a), p. 213.

Allan, D. W. (1958) Reversals of the Earth's magnetic field. *Nature*, **182**, 469.

Alldredge, L. R. (1965) Analysis of long magnetic profiles. *J. Geomag. Geoelect.*, **17**, 173.

Alldredge, L. R., and Hurwitz, L. (1964) Radial dipoles as the sources of the Earth's main magnetic field. *J. Geophys. Res.*, **69**, 2631.

Alldredge, L. R., Van Voorhis, G. D., and Davis, T. M. (1963) A magnetic profile around the world. *J. Geophys. Res.*, **68**, 3679.

Allen, C. R. (1965) Transcurrent faults in continental areas. *Phil. Trans. Roy. Soc. A*, **258**, 82.

Alterman, Z., Jarosch, H., and Pekeris, C. L. (1959) Oscillations of the Earth. *Proc. Roy. Soc. A*, **252**, 80.

Altschuler, L. V., and Kormer, S. B. (1961) On the internal structure of the Earth. *Bull. Acad. Sci. U.S.S.R.*, 1961(1), 18.

Amiel, S., Gilat, J., and Heyman, D. (1967) Uranium content of chondrites by thermal neutron activation and delayed neutron counting. *Geochim. et Cosmochim. Acta*, **31**, 1499.

Anders, E. (1962) Meteorite ages. *Revs. Mod. Phys.*, **34**, 287.

Anders, E. (1963) Meteorite ages. In Middlehurst and Kuiper (1963), p. 402.

Anders, E. (1964) Origin, age and composition of meteorites. *Space Science Revs.*, **3**, 583.

Anderson, D. L. (1965). Recent evidence concerning the structure and composition of the Earth's mantle. *Phys. and Chem. of Earth*, **6**, 1.

Anderson, D. L. (1967a) The anelasticity of the mantle. *Geophys. J., R. Astr. Soc.*, **14**, 135.

Anderson, D. L. (1967b) Latest information from seismic observations. In Gaskell (1967), p. 355.

Anderson, D. L., and Archambeau, C. B. (1964) The anelasticity of the Earth. *J. Geophys. Res.*, **69**, 2071.

Anderson, O. L., Schreiber, E., Liebermann, R. C. and Soga, N. (1968) Some elastic constant data on minerals relevant to geophysics. *Revs. Geophys.*, **6**, 491.

As, J. A. (1967) Present and past changes in the Earth's magnetic field. In Hindmarsh et al. (1967), p. 29.

284

Ash, M. E., Shapiro, I. I., and Smith, W. B. (1967) Astronomical constants and planetary ephemerides deduced from radar and optical observations. *Astron. J.*, **72**, 338.

Babcock, H. W., and Babcock, H. D. (1955) The Sun's magnetic field, 1952-1954. *Astrophys. J.*, **121**, 349.

Baker, G. S. (1957) Internal friction in the presence of a static stress. *J. Appl. Phys.*, **28**, 734.

Balchan, A. S., and Cowan, G. R. (1966) Shock compression of two iron-silicon alloys to 2.7 megabars. *J. Geophys. Res.*, **71**, 3577.

Banks, R. J., and Bullard, E. C. (1966) The annual and 27 day magnetic variations. *Earth and Plan. Sci. Letters*, **1**, 118.

Barazangi, M., and Dorman, J. (1969) World seismicity map of ESSA Coast and Geodetic Survey epicentre data for 1961-1967. *Bull. Seism. Soc. Am.* **59**, 369.

Basinski, Z. S. (1957) The instability of plastic flow of metals at very low temperatures. *Proc. Roy. Soc. A*, **240**, 229.

Basinski, Z. S. (1960) The instability of plastic flow of metals at very low temperatures, II. *Austral. J. Phys.*, **13**, 354.

Bates, D. R. (ed.) (1964) *The planet Earth*. (2nd edition.) Oxford: Pergamon.

Båth, M. (1966a) Earthquake seismology. *Earth Sci. Revs.*, **1**, 69.

Båth, M. (1966b) Earthquake energy and magnitude. *Phys. and Chem. of Earth*, **7**, 115.

Båth, M., and Duda, S. J. (1964) Earthquake volume, fault plane area, seismic energy, strain, deformation and related quantities. *Annali di Geofisica*, **17**, 353.

Beck, A. E. (1961) Energy requirements of an expanding Earth. *J. Geophys. Res.*, **66**, 1485.

Bellomo, E., Colombo, G., and Shapiro, I. I. (1967) Theory of the axial rotations of Mercury and Venus. In Runcorn (1967a), p. 193.

Benioff, H. (1955) Seismic evidence for crustal structure and tectonic activity. *Geol. Soc. Am. Spec. Papers*, **62**, 61.

Benioff, H. (1959) Fused quartz extensometer for secular, tidal and seismic strains. *Bull. Seism. Soc. Am.*, **70**, 1019.

Benioff, H. (1962) Movements on major transcurrent faults. In Runcorn (1962), p. 103.

Benioff, H., Press, F., and Smith, S. (1961) Excitation of the free oscillations of the Earth by earthquakes. *J. Geophys. Res.*, **66**, 605.

Berg, E. (1966) Triggering of the Alaskan earthquake of March 28, 1964 and major aftershocks by low ocean tide loads. *Nature*, **210**, 893.

Birch, F. (1938) The effect of pressure upon the elastic parameters of isotropic solids, according to Murnaghan's theory of finite strain. *J. Appl. Phys.*, **9**, 279.

Birch, F. (1948) The effects of pleistocene climatic variations upon geothermal gradients. *Am. J. Sci.*, **246**, 729.

Birch, F. (1952) Elasticity and constitution of the Earth's interior. *J. Geophys. Res.*, **57**, 227.

Birch, F. (1958) Differentiation of the mantle. *Bull. Geol. Soc. Am.*, **69**, 483.

Birch, F. (1965) Speculations on the Earth's thermal history. *Bull. Geol. Soc. Am.*, **76**, 133.

Black, D. I. (1967) Cosmic ray effects and faunal extinctions at geomagnetic field reversals. *Earth and Plan. Sci. Letters*, **3**, 225.

Blackett, P. M. S. (1961) Comparison of ancient climates with ancient latitudes deduced from rock magnetic measurements. *Proc. Roy. Soc. A*, **263**, 1.

Blackett, P. M. S., Bullard, E. C., and Runcorn, S. K. (eds.) (1965) A symposium on continental drift. *Phil. Trans. Roy. Soc. A*, **258**, 1.

Blanco, V. M., and McCuskey, S. W. (1961) *Basic physics of the solar system.* Reading, Mass.: Addison-Wesley.

Bolt, B. A. (1964) Recent information on the Earth's interior from studies of mantle waves and eigenvibrations. *Phys. and Chem. of the Earth*, **5**, 55.

Bolt, B. A., and Nuttli, O. W. (1966) P Wave residuals as a function of azimuth 1: Observations. *J. Geophys. Res.*, **71**, 5977.

Bomford, G. (1962) *Geodesy.* (2nd edition) Oxford: Clarendon Press.

Bowen, R. (1966) *Palaeotemperature analysis.* Amsterdam: Elsevier.

Bozorth, R. M. (1951) *Ferromagnetism.* Princeton, N.J.: Van Nostrand.

Breiner, S. (1964) Piezomagnetic effect at the time of local earthquakes. *Nature*, **202**, 790.

Breiner, S., and Kovach, R. L. (1967) Local geomagnetic events associated with displacements on the San Andreas fault. *Science*, **158**, 116.

Brekhovskikh, L. M. (1960) *Waves in layered media.* New York: Academic Press.

Briden, J. C. (1966) Variation of the intensity of the geomagnetic field through geological time. *Nature*, **212**, 246.

Briden, J. C., and Irving, E. (1964) Palaeolatitude spectra of sedimentary palaeoclimatic indicators. In Nairn (1964).

Bridgman, P. W. (1936) Shearing phenomena at high pressure of possible importance for geology. *J. Geol.*, **44**, 653.

Bridgman, P. W. (1945) Polymorphic transitions and geological phenomena. *Am. J. Sci.*, **243A**, 90.

Brune, J. N., and Oliver, J. (1959) The seismic noise of the Earth's surface. *Bull. Seism. Soc. Am.*, **49**, 349.

Bucha, V. (1965) Results of archaeomagnetic research in Czechoslovakia for the epoch from 4400 B.C. to the present. *J. Geomag. Geoelect.*, **17**, 407.

Bullard, E. C. (1949) The magnetic field within the Earth. *Proc. Roy. Soc. A*, **197**, 433.

Bullard, E. C. (1950) The transfer of heat from the core of the Earth. *Geophys. Supp. Mon. Not. R. Astr. Soc.*, **6**, 36.

Bullard, E. C. (1956) Edmond Halley (1656-1742). *Endeavour*, **15**, 189.

Bullard, E. C. (1960) Response to award of Arthur L. Day medal. *Proc. Vol. for 1959, Geol. Soc. Am.*, p. 92.

Bullard, E. C. (1964) Continental drift. *Q. J. Geol. Soc. London*, **120**, 1.

Bullard, E. C. (1965) Concluding remarks, symposium on continental drift. *Phil. Trans. Roy. Soc. A*, **258**, 322.

Bullard, E. C., Freedman, C., Gellman, H., and Nixon, J. (1950) The westward drift of the Earth's magnetic field. *Phil. Trans. Roy. Soc. A*, **243**, 67.

Bullard, E. C., and Gellman, H. (1954) Homogeneous dynamos and terrestrial magnetism. *Phil. Trans. Roy. Soc., A*, **247**, 213.

Bullard, E. C. and Griggs, D. T. (1961) The nature of the Mohorovičic discontinuity. *Geophys. J., R. Astr. Soc.*, **6**, 118.

Bullen, K. E. (1954) *Seismology.* London: Methuen.

Bullen, K. E. (1963) *An introduction to the theory of seismology.* (3rd edition.) Cambridge: Cambridge University Press.

Bullen, K. E., and Haddon, R. A. W. (1967a) Earth oscillations and the Earth's interior. *Nature*, **213**, 574.

Bullen, K. E., and Haddon, R. A. W. (1967b) Derivation of an earth model from free oscillation data. *Proc. U.S. Nat. Acad. Sci.*, **58**, 846.

Burnett, D. S., and Wasserburg, G. J. (1967a) ^{87}Rb-^{87}Sr ages of silicate intrusions in iron meteorites. *Earth and Plan. Sci. Letters*, **2**, 397.

Burnett, D. S., and Wasserburg, G. J. (1967b) Evidence for the formation of an iron meteorite at 3.8×10^9 years. *Earth and Plan. Sci. Letters*, **2**, 137.

Byerlee, J. D., and Brace, W. F. (1968) Stick slip, stable sliding and earthquakes—effect of rock type, pressure, strain rate and stiffness. *J. Geophys. Res.*, **73**, 6031.

Cagniard, L. (1962) *Reflection and refraction of progressive seismic waves.* New York: McGraw-Hill.

Carder, D. S., Gordon, D. W., and Jordan, J. N. (1964) Travel times from central Pacific nuclear explosions and inferred mantle structure. *Bull. Seism. Soc. Am.*, **54**, 2271.

Carmichael, C. M. (1961) The magnetic properties of ilmenite-hematite crystals. *Proc. Roy. Soc. A*, **263**, 508.

Carmichael, C. M. (1967) An outline of the intensity of the paleomagnetic field of the Earth. *Earth and Plan. Sci. Letters*, **3**, 351.

Carpenter, E. W. (1966) A quantitative evaluation of teleseismic explosion records. *Proc. Roy. Soc. A*, **290**, 287.

Ceplecha, Z. (1961) Multiple fall of Pribram meteorites photographed. *Bull. Astr. Inst. Czech.*, **12**, 21.

Chapman, S., and Bartels, J. (1940) *Geomagnetism.* (2 volumes.) Oxford: Clarendon Press.

Chinnery, M. A. (1961) The deformation of the ground around surface faults. *Bull. Seism. Soc. Am.*, **51**, 355.

Chow, T. J., and Patterson, C. C. (1962) The occurrence and significance of lead isotopes in pelagic sediments. *Geochim. et Cosmochim. Acta*, **26**, 263.

Clark, S. P. (1957a) Radiative transfer in the Earth's mantle. *Trans. Am. Geophys. Un.*, **38**, 931.

Clark, S. P. (1957b) Absorption spectra of some silicates in the visible and near infra-red. *Am. Mineralogist*, **42**, 732.

Clark, S. P. (ed.) (1966) Handbook of physical constants, revised edition. *Memoir 97, Geol. Soc. Am.*

Clark, S. P., and Ringwood, A. E. (1964) Density distribution and constitution of the mantle. *Revs. Geophys.*, **2**, 35.

Cleary, J. R., and Hales, A. L. (1966) Azimuthal variation of U.S. station residuals. *Nature*, **210**, 619.

Collinson, D. W., Creer, K. M., and Runcorn, S. K. (eds.) (1967) *Methods in palaeomagnetism.* Amsterdam: Elsevier.

Compston, W., Jeffrey, P. M., and Riley, G. H. (1960) Age of emplacement of granites. *Nature*, **186**, 702.

Compston, W., Lovering, J. F., and Vernon, M. J. (1965) The rubidium-strontium age of the Bishopville aubrite and its component enstatite and feldspar. *Geochim. et Cosmochim. Acta*, **29**, 1085.

Cook, A. H. (1963a) The contribution of the observations of satellites to the determination of the Earth's gravitational potential. *Space Sci. Revs.*, **2**, 355.

Cook, A. H. (1963b) Sources of harmonics of low order in the external gravity field of the Earth. *Nature*, **198**, 1186.

Cook, K. L. (1962) The problem of the mantle-crust mix: lateral inhomogeneity in the uppermost part of the Earth's mantle. *Adv. in Geophys.*, **9**, 295.

Cooper, J. A., Richards, J. R., and Stacey, F. D. (1967) Possible new evidence bearing on the lunar capture hypothesis. *Nature*, **215**, 1256.

Cottrell, A. H. (1953) *Dislocations and plastic flow in crystals*. Oxford: Clarendon Press.

Cottrell, A. H. (1964) *The mechanical properties of matter*. New York: Wiley.

Cox, A. (1962) Analysis of the present geomagnetic field for comparison with paleomagnetic results. *J. Geomag. Geoelect.*, **13**, 101.

Cox, A., and Dalrymple, G. B. (1967a) Statistical analysis of geomagnetic reversal data and the precision of potassium-argon dating. *J. Geophys. Res.*, **72**, 2603.

Cox, A., and Dalrymple, G. B. (1967b) Geomagnetic polarity epochs: Nunivack Island, Alaska. *Earth and Plan. Sci. Letters*, **3**, 173.

Cox, A., Dalrymple, G. B., and Doell, R. R. (1967). Reversals of the Earth's magnetic field. *Scientific American*, **216**(2), 44.

Cox, A., and Doell, R. R. (1960) Review of paleomagnetism. *Bull. Geol. Soc. Am.*, **71**, 645.

Craig, H., Miller, S. L., and Wasserburg, G. J. (eds.) (1964) *Isotopic and cosmic chemistry*. Amsterdam: North Holland.

Creer, K. M. (1963) Geomagnetic and palaeomagnetic evidence of fossil axes of rotation of the Earth. *Nature*, **197**, 122.

Creer, K. M. (1965) Palaeomagnetic data from the gondwanic continents. *Phil. Trans. Roy. Soc. A*, **258**, 27.

Currie, R. G. (1968) Geomagnetic spectrum of internal origin and lower mantle conductivity. *J. Geophys. Res.*, **73**, 2779.

Dalrymple, G. B., Cox, A., Doell, R. R., and Gromme, C. S. (1967) Pliocene geomagnetic polarity epochs. *Earth and Plan. Sci. Letters*, **2**, 163.

Darbyshire, J. (1962) Microseisms. In Hill (1962), p. 700.

Darwin, G. H. (1962) *The tides and kindred phenomena in the solar system*. (Original edition 1898, reprinted 1962.) San Francisco: Freeman.

Deacon, G. E. R. (1947) Relations between sea waves and microseisms. *Nature*, **160**, 419.

Dietz, R. S. (1961) Continent and ocean basin evolution by spreading of the sea floor. *Nature*, **190**, 854.

Dobrin, M. B. (1960) *Introduction to geophysical prospecting*. (2nd edition.) New York: McGraw-Hill.

Dodd, R. T., and Teleky, L. S. (1967) Preferred orientation of olivine crystals in porphyritic chondrules. *Icarus*, **6**, 407.

Doell, R. R., and Cox, A. (1965) Paleomagnetism of Hawaiian lava flows. *J. Geophys. Res.*, **70**, 3377.

Dorman, J., Ewing, J., and Alsop, L. E. (1965) Oscillations of the Earth: New core-mantle boundary model based on low-order free vibrations. *Proc. Nat. Acad. Sci.*, **54**, 364.

Dorman, J., Ewing, M., and Oliver, J. (1960) Study of shear-velocity distribution in the upper mantle by Rayleigh waves. *Bull. Seism. Soc. Am.*, **50**, 87.

Du Fresne, E. R., and Anders, E. (1963) Chemical evolution of the carbonaceous chondrites. In Middlehurst and Kuiper (1963), p. 496.

Dyce, R. B., Pettengill, G. H., and Shapiro, I. I. (1967) Radar determinations of the rotations of Venus and Mercury. *Astron. J.*, **72**, 351.

Eaton, J. P., Richter, D. H., and Ault, W. U. (1961) The tsunami of May 23, 1960, on the island of Hawaii. *Bull. Seism. Soc. Am.*, **51**, 135.

Eirich, F. R. (ed.) (1958) *Rheology; theory and applications.* (3 vols.) New York: Academic Press.

Elsasser, W. M. (1950) The Earth's interior and geomagnetism. *Revs. Mod. Phys.*, **22**, 1.

Elsasser, W. M. (1956) Hydromagnetic dynamo theory. *Revs. Mod. Phys.*, **28**, 135.

Elsasser, W. M. (1963) Early history of the Earth. In Geiss and Goldberg (1963), p. 1.

Elsasser, W. M. (1966) Thermal structure of the upper mantle and convection. In Hurley (1966), p. 461.

Elsasser, W. M. (1967) Interpretation of heat flow equality. *J. Geophys. Res.*, **72**, 4768.

Emiliani, C. (1955) Pleistocene temperatures. *J. Geol.*, **63**, 538.

Engel, A. E. J., James, H. L., and Leonard, B. F. (eds.) (1962) *Petrological studies: A volume in honor of A. F. Buddington.* New York: Geological Society of America.

Evans, M. E., and McElhinny, M. W. (1966) The palaeomagnetism of the Modipe gabbro. *J. Geophys. Res.*, **71**, 6053.

Evison, F. F. (1963) Earthquakes and faults. *Bull. Seism. Soc. Am.*, **53**, 873.

Evison, F. F. (1967) On the occurrence of volume change at the earthquake source. *Bull. Seism. Soc. Am.*, **57**, 9.

Ewing, W. M., Jardetsky, W. S., and Press, F. (1957). *Elastic waves in layered media.* New York: McGraw-Hill.

Fan, H. Y., and Becker, M. (1951) Infrared optical properties of silicon and germanium. In Henisch (1951), p. 132.

Faul, H. (1966) *Ages of rocks, planets and stars.* New York: McGraw-Hill.

Fechtig, H., and Kalbitzer, S. (1966) The diffusion of argon in potassium-bearing solids. In Schaeffer and Zähringer (1966), p. 68.

Finch, H. F., and Leaton, B. R. (1957) The Earth's main magnetic field-epoch 1955.0. *Monthly Notices, R. Astr. Soc., Geophys. Suppl.*, **7**, 314.

Fisher, D. (1966) The origin of meteorites: space erosion and cosmic radiation ages. *J. Geophys. Res.*, **71**, 3251.

Fleischer, R. L., and Price, P. B. (1964) Techniques for geological dating of minerals by chemical etching of fission fragment tracks. *Geochim. et Cosmochim. Acta,* **28**, 1705.

Fleischer, R. L., Price, P. B., and Walker, R. M. (1965) Tracks of charged particles in solids. *Science,* **149**, 383.

Fleischer, R. L., Naesser, C. W., Price, P. B., and Walker, R. M. (1965) Cosmic ray exposure ages of tektites by the fission track technique. *J. Geophys. Res.,* **70**, 1491.

Fleischer, R. L., Price, P. B., Walker, R. M., and Maurette, M. (1967) Origins of fossil charged-particle tracks in meteorites. *J. Geophys. Res.,* **72**, 331.

Friedel, J. (1964) *Dislocations.* New York: Pergamon.

Fukao, Y., Hitoshi, M., and Uyeda, S. (1968) Optical absorption spectra at high temperatures and radiative thermal conductivity of olivines. *Phys. Earth Planet. Interiors.,* **1**, 57.

Gamow, G. (1963) *A planet called Earth.* New York: Viking Press.

Garland, G. D. (1965) *The Earth's shape and gravity.* Oxford: Pergamon.

Gaskell, T. F. (ed.) (1967) *The Earth's mantle.* London: Academic Press.

Gast, P. W. (1962) The isotopic composition of strontium and the age of stone meteorites. *Geochim. et Cosmochim. Acta,* **26**, 927.

Gastil, G. (1960) The distribution of mineral dates in time and space. *Amer. J. Sci.,* **258**, 1.

Geiss, J., and Goldberg, E. D. (eds.) (1963) *Earth science and meteoritics.* Amsterdam: North Holland.

Gerstenkorn, H. (1967a) The importance of tidal friction for the early history of the Moon. *Proc. Roy. Soc. A,* **296**, 293.

Gerstenkorn, H. (1967b) On the controversy over the effect of tidal friction upon the history of the Earth-Moon system. *Icarus,* **7**, 160.

Goldich, S. S., Muehlberger, W. R., Lidiak, E. G., and Hedge, C. E. (1966) Geochronology of the midcontinent region, United States 1. Scope, methods and principles. *J. Geophys. Res.,* **71**, 5375.

Goldreich, P. (1966) History of the lunar orbit. *Revs. Geophys.,* **4**, 411.

Goldreich, P., and Peale, S. (1966) Spin-orbit coupling in the solar system. *Astronomical Journal,* **71**, 425.

Goldreich, P., and Toomre, A. (1969) Some remarks on polar wandering. *J. Geophys. Res.,* **74**, 2555.

Goldstein, J. I., and Short, J. M. (1967) Cooling rates of 27 iron and stony iron meteorites. *Geochim. et Cosmochim. Acta,* **31**, 1001.

Gordon, R. B., and Nelson, C. W. (1966) Anelastic properties of the Earth, *Revs. Geophys.,* **4**, 457.

Graham, J. W. (1949) The stability and significance of magnetism in sedimentary rocks. *J. Geophys. Res.,* **54**, 131.

Grant, F. S., and West, G. F. (1965) *Interpretation theory in applied geophysics.* New York: McGraw-Hill.

Gray, A. (1959) *A treatise on gyrostatics and rotational motion.* New York: Dover (reprinted from 1918 edition).

Griffiths, D. H., and King, R. F. (1965) *Applied geophysics for engineers and geologists.* Oxford: Pergamon.

Griggs, D. T., and Baker, D. W. (1968) The origin of deep-focus earthquakes. In Mark and Fernbech (1968), p. 23.

Griggs, D., and Handin, J. (eds.) (1960a) Rock deformation. *Geol. Soc. Am. Memoir* **79**.

Griggs, D., and Handin, J. (1960b) Observations on fracture and a hypothesis of earthquakes. In Griggs and Handin (1960a), p. 347.

Griggs, D., Turner, F. J., and Heard, H. C. (1960). Deformation of rocks at 500° to 800°C. In Griggs and Handin (1960a), p. 39.

Grover, J. C. (1967) Forecasting earthquakes—correlation between deep foci and shallow events in Melanesia. *Nature*, **213**, 686.

Guier, W. H., and Newton, R. R. (1965) The Earth's gravity field as deduced from the Doppler tracking of five satellites. *J. Geophys. Res.*, **70**, 4613.

Gutenberg, B. (1958a) Velocity of seismic waves in the Earth's mantle. *Trans. Am. Geophys. Un.*, **39**, 486.

Gutenberg, B. (1958b) Rheological problems of the Earth's interior. In Eirich (1958), Vol. 2, p. 401.

Gutenberg, B. (1959) *Physics of the Earth's interior.* New York: Academic Press.

Gutenberg, B., and Richter, C. F. (1954) *Seismicity of the Earth and associated phenomena.* (2nd edition.) Princeton, N.J.: Princeton University Press.

Gutenberg, B., and Richter, C. F. (1956) Earthquake magnitude, intensity, energy and acceleration. *Bull. Seism. Soc. Am.*, **46**, 105.

Hagiwara, T., and Oliver, J. (eds.) (1964) *Proceedings of the United States-Japan conference on research related to earthquake prediction problems.* Washington: National Science Foundation; Tokyo: Japan Society for Promotion of Science.

Hales, A. L., and Doyle, H. A. (1967) P and S travel time anomalies and their interpretation. *Geophys. J., R. Astr. Soc.*, **13**, 403.

Hamilton, E. I. (1965) *Applied geochronology.* London: Academic Press.

Harland, W. B., Smith, A. G., and Wilcock, B. (eds.) (1964) Geological Society phanerozoic time scale 1964. *Quart J. Geol. Soc. Lond.*, **120S**, 260. (Supplement volume: The phanerozoic time scale.)

Harnwell, G. P. (1949) Principles of electricity and electromagnetism. (2nd edition.) New York: McGraw-Hill.

Harrison, C. G. A. (1968) Evolutionary processes and reversals of the Earth's magnetic field. *Nature*, **217**, 46.

Hartman, W. K., and Larson, S. M. (1967) Angular momenta of planetary bodies. *Icarus*, **7**, 257.

Hasselmann, K. (1963) A statistical analysis of the generation of microseisms. *Revs. Geophys.*, **1**, 177.

Haubrich, R. A., Munk, W. H., and Snodgrass, F. E. (1963) Comparative spectra of microseisms and swell. *Bull. Seism. Soc. Am.*, **53**, 27.

Hays, J. D., and Opdyke, N. D. (1967) Antarctic radiolaria, magnetic reversals and climatic change. *Science*, **158**, 1001.

Heard, H. C. (1963) Effect of large changes in strain rate in the experimental deformation of Yule marble. *J. Geol.*, **71**, 162.

Heard, H. C. (1968) Steady state flow in Yule marble at 500°-800°C. (Abstract) *Trans. Am. Geophys. Un.*, **49**, 312.

Heezen, B. C. (1962) The deep sea floor. In Runcorn (1962), p. 235.

Heiland, C. A. (1946) *Geophysical exploration.* Englewood Cliffs, N.J.: Prentice-Hall.

Heirtzler, J. R., Dickson, G. O., Herron, E. M., Pitman, W. C., and LePichon, X. (1968) Marine magnetic anomalies, geomagnetic field reversals and motions of the ocean floor and continents. *J. Geophys. Res.,* **73,** 2119.

Heiskanen, W. A., and Meinesz, F. A. V. (1958) *The Earth and its gravity field.* New York: McGraw-Hill.

Henisch, H. K. (ed.) (1951) *Semi-conducting materials* London: Butterworths.

Hess, H. H. (1962) History of the ocean basins. In Engel et al. (1968), p. 599.

Hess, W. N. (ed.) (1965) *Space science.* New York: Gordon and Breach.

Hide, R. (1966a) Planetary magnetic fields. *Planetary and Space Science,* **14,** 579.

Hide, R. (1966b) Free hydromagnetic oscillations of the Earth's core and the theory of the geomagnetic secular variation. *Phil. Trans. Roy. Soc.,* **A259,** 615.

Hide, R., and Roberts, P. H. (1961) The origin of the main geomagnetic field. *Phys. and Chem. of the Earth,* **4,** 27.

Hill, M. N. (ed.) (1962) *The sea.* Vol. 1: *Physical oceanography.* New York: Interscience.

Hindmarsh, W. R., Lowes, F. J., Roberts, P. H., and Runcorn, S. K. (eds.) (1967) *Magnetism and the cosmos.* Edinburgh: Oliver and Boyd.

Hoffman, R. B. (1968) *Geodimeter fault movement investigations in California.* Bulletin 116-6, California Department of Water Resources.

Holmes, A. (1965) *Principles of physical geology.* (Revised edition.) London: Nelson.

Howell, B. F. (1959) *Introduction to geophysics.* New York: McGraw-Hill.

Hoyle, F. (1960) On the origin of the solar nebula. *Quart. J. Roy. Astr. Soc.,* **1,** 28.

Hudson, J. D. (1964) Sedimentation rates in relation to the Phanerozoic time scale. *Quart. J. Geol. Soc. London,* **120S,** 37. (Review volume by Harland et al., 1964.)

Hurley, P. M. (ed.) (1966) *Advances in Earth science.* Cambridge, Mass.: M.I.T. Press.

Hurley, P. M., Hughes, H., Faure, G., Fairbairn, H. W., and Pinson, W. H. (1962) Radiogenic strontium-87 model of continent formation. *J. Geophys. Res.,* **67,** 5315.

Hyndman, R. D., and Everett, J. E. (1968) Heat flow measurements in a low radioactivity area of the Western Australian pre-Cambrian shield. *Geophys. J., R. Astr. Soc.,* **14,** 479.

Inglis, D. R. (1955) Theories of the Earth's magnetism. *Revs. Mod. Phys.,* **27,** 212.

International Union of Geodesy and Geophysics (1967) Resolution No. 1, XIV General Assembly. *Bulletin Geodesique,* **86,** 367.

Irving, E. (1964) *Paleomagnetism.* New York: Wiley.

Irving E. (1966) Paleomagnetism of some carboniferous rocks from New South Wales and its relation to geological events. *J. Geophys. Res.,* **71,** 6025.

Isacks, B., Oliver, J., and Sykes, L. R. (1968) Seismology and the new global tectonics. *J. Geophys. Res.,* **73,** 5855.

Ishikawa, Y., and Syono, Y. (1963) Order-disorder transformation and reverse thermoremanent magnetism in the $FeTiO_3$—Fe_2O_3 system. *J. Phys. Chem. Solids,* **24,** 517.

Jacchia, L. G. (1963) Meteors, meteorites and comets: Interrelations. In Middle-hurst and Kuiper (1963), p. 774.

Jacobs, J. A. (1953) Temperature-pressure hypothesis and the Earth's interior. *Canad. J. Phys.*, **31**, 370.

Jacobs, J. A., Russell, R. D, and Wilson, J. T. (1959) *Physics and geology.* New York: McGraw-Hill.

Jakosky, J. J. (ed.) (1950). *Exploration geophysics.* (2nd edition.) Los Angeles: Trija Publishing Co.

Jastrow, R., and Cameron, A. G. W. (eds.) (1963) *Origin of the solar system.* New York: Academic Press.

Jeffreys, H. (1962) *The Earth, its origin, history and physical constitution.* (4th edition revised.) Cambridge: Cambridge University Press.

Jeffreys, H. (1963) On the hydrostatic theory of the figure of the Earth. *Geophys. J., R. Astr. Soc.*, **8**, 196.

Jeffreys, H., and Bullen, K. E. (1940) *Seismological tables.* (Reprint issued 1958.) London: British Association for the Advancement of Science.

Jeffreys, H. and Jeffreys, B. S. (1962) *Methods of mathematical physics* (3rd edition.) Cambridge: Cambridge University Press.

Johnston, W. G., and Gilman, J. J. (1959) Dislocation velocities, dislocation densities and plastic flow in lithium fluoride crystals. *J. Appl. Phys.*, **30**, 129.

Joos, G. (1934) *Theoretical physics.* London: Blackie.

Kanamori, H. (1967) Spectrum of short period core phases in relation to the attenuation in the mantle. *J. Geophys. Res.*, **72**, 2181.

Karnik, V. (1961) Seismicity of Europe. Progress Report II. *International Union of Geodesy and Geophysics Monograph* **9**.

Kato, Y., and Utashiro, S. (1949) On the changes of terrestrial magnetic field accompanying the great Nankaido earthquake of 1946. *Sci. Repts. Tohoku Univ. Series* 5, *Geophysics*, **1**, 40.

Kaula, W. M. (1964) Tidal dissipation by solid friction and the resulting orbital evolution. *Revs. Geophys.*, **2**, 661.

Kaula, W. M. (1965) The shape of the Earth. In Hess (1965), p. 297.

Kaula, W. M. (1968) *An introduction to planetary physics; The terrestrial planets.* New York: Wiley.

Kelvin, Lord (1899) The age of the Earth as an abode fitted for life. *Phil. Mag.*, **47**, 66.

Kennedy, G. C. (1966) The effect of pressure on melting. (Abstract) *Trans. Am. Geophys. Un.*, **47**, 173.

Kern, J. W., and Vestine, E. H. (1963) Magnetic field of the Earth and planets. *Space Sci. Revs.*, **2**, 136.

Kertz, W. (1964) The conductivity anomaly in the upper mantle found in Europe. *J. Geomag. Geoelect.*, **15**, 185.

Kittel, C. (1949) Ferromagnetic domain theory. *Revs. Mod. Phys.*, **21**, 541.

Kittel, C. (1966) *Introduction to solid state physics.* (3rd edition) (New York; Wiley).

Knopoff, L. (1964a) "Q." *Revs. Geophys.*, **2**, 625.

Knopoff, L. (1964b) Earth tides as a triggering mechanism for earthquakes. *Bull. Seism. Soc. Am.*, **54**, 1865.

Knopoff, L. (1964c) The statistics of earthquakes in Southern California. *Bull. Seism. Soc. Am.*, **54**, 1871.

Kolsky, H. (1963) *Stress waves in solids.* New York: Dover.

Kozai, Y. (1966) The Earth gravitational potential derived from satellite motion. *Space Science Revs.*, **5**, 818.

Kraut, E. A., and Kennedy, G. C. (1966) New melting law at high pressure. *Phys. Rev.*, **151**, 668.

Krinov, E. L. (1960) *Principles of meteoritics.* Oxford: Pergamon.

Kulp, J. L. (1961) Geological time scale. *Science*, **133**, 1105.

Lahiri, B. N., and Price, A. T. (1939) Electromagnetic induction in non-uniform conductors, and the determination of the conductivity of the Earth from terrestrial magnetic variations. *Phil. Trans. Roy. Soc. A*, **237**, 509.

Lambert, I. B., and Heier, K. S. (1967) The vertical distribution of uranium, thorium and potassium in the continental crust. *Geochim. et Cosmochim. Acta.*, **31**, 377.

Lanphere, M. A., Wasserburg, G. J., Albee, A. L., and Tilton, G. R. (1964) Redistribution of strontium and rubidium isotopes during metamorphism, World Beater complex, Ponamint Range, California. In Craig et al. (1964), p. 269.

Lee, W. H. K. (ed.) (1965) *Terrestrial heat flow.* Geophysical Monograph No. 8. Washington: American Geophysical Union.

Lee, W. H. K., and MacDonald, G. J. F. (1963) The global variation of terrestrial heat flow. *J. Geophys. Res.*, **68**, 6481.

Lee, W. H. K., and Uyeda, S. (1965) Review of heat flow data. In Lee (1965), p. 87.

Legalley, D. P. (ed.) (1963) *Space science* New York: Wiley.

LePichon, X. (1968) Sea floor spreading and continental drift. *J. Geophys. Res.*, **73**, 3661.

Longuet-Higgins, M. S. (1950) A theory of the origin of microseisms. *Phil. Trans. Roy. Soc. A*, **243**, 1.

Longuet-Higgins, M. S., and Ursell, F. (1948) Sea Waves and microseisms. *Nature*, **162**, 700.

Lovell, A. C. B. (1954) *Meteor astronomy.* Oxford: Clarendon Press.

Lovering, J. F. (1962) Evolution of the meteorites—evidence for the coexistence of chondritic, achondritic and iron meteorites in a typical parent meteorite body. In Moore (1962).

Lovering, J. F., and Morgan, J. W. (1963) Uranium and thorium abundances in possible upper mantle materials. *Nature*, **197**, 138.

Lovering, J. F., and Morgan, J. W. (1964) Uranium and thorium abundances in stony meteorites. *J. Geophys. Res.*, **69**, 1979.

Lowes, F. J., and Runcorn, S. K. (1951) The analysis of the geomagnetic secular variation. *Phil. Trans. Roy. Soc. A.*, **243**, 525.

Lowes, F. J., and Wilkinson, I. (1963) Geomagnetic dynamo: a laboratory model. *Nature*, **198**, 1158.

Lowes, F. J., and Wilkinson, I. (1967) Laboratory self-exciting dynamo. In Hindmarsh et al. (1967), p. 121.

Lowes, F. J., and Wilkinson, I. (1968) Geomagnetic dynamo: an improved laboratory model. *Nature*, **219**, 717.

Lubimova, H. A. (1958) Thermal history of the Earth with consideration of the variable thermal conductivity of its mantle. *Geophys. J., R. Astr. Soc.*, **1**, 115.

Lubimova, E. A. (1967) Theory of thermal state of the Earth's mantle. In Gaskell (1967), p. 231.

MacDonald G. J. F. (1959) Calculations on the thermal history of the Earth. *J. Geophys. Res.* **64**, 1967.

MacDonald, J. G. F. (1963a) Internal constitutions of the inner planets and the Moon. *Space Sci. Revs.*, **2**, 473.

MacDonald, G. J. F. (1963b) The deep structure of continents. *Revs. Geophys.*, **1**, 587.

MacDonald, G. J. F. (1964a) Tidal friction. *Revs. Geophys.*, **2**, 467.

MacDonald, G. J. F. (1964b) Dependence of the surface heat flow on the radio-activity of the Earth. *J. Geophys. Res.*, **69**, 2933.

MacDonald, G. J. F. (1967) Evidence from the surface configuration of the Moon on its dynamical evolution. *Proc. Roy. Soc. A*, **296**, 298.

MacDonald, G. J. F., and Knopoff, L. (1958) On the chemical composition of the outer core. *Geophys. J., R. Astr. Soc.*, **1**, 284.

Magnitskiy, V. A. (1966) *The internal structure and physics of the Earth.* (English translation by National Aeronautics and Space Administration.) Springfield, V.: Clearing House for Technical and Scientific Information.

Major, M. W., Sutton, G. H., Oliver, J., and Metsger, R. (1964) On elastic strain of the Earth in the period range 5 seconds to 100 hours. *Bull. Seism. Soc. Am.*, **54**, 295.

Malkus, W. V. R. (1963) Precessional torques as the cause of geomagnetism. *J. Geophys. Res.*, **68**, 2871.

Malkus, W. V. R. (1968) Precession of the Earth as the cause of geomagnetism. *Science*, **160**, 259.

Malloy, R. J. (1964) Crustal uplift southwest of Montagu Island, Alaska. *Science*, **146**, 1048.

Mansinha, L., and Smylie, D. E. (1967) Effect of earthquakes on the Chandler wobble and the secular polar shift. *J. Geophys. Res.*, **72**, 4731.

Mansinha, L., and Smylie, D. E. (1968) Earthquakes and the Earth's wobble. *Science*, **161**, 1127.

Mark, H., and Fernbech, S. (1968) *Properties of matter under unusual conditions.* New York: Wiley-Interscience.

Markowitz, W., and Guinot, B. (1968) *Continental drift, secular motion of the pole and rotation of the Earth.* (International Astronomical Union Symposium 32, Stresa, March 1967.) Dortrecht: Reidel.

Markowitz, W., Stoyko, N., and Fedorov, E. P. (1964) Longitude and latitude. In Odishaw (1964), p. 149.

Mason, B. (1958) *Principles of geochemistry.* (2nd edition.) New York: Wiley.

Mason, B. (1962) *Meteorites.* New York: Wiley.

Mason, B. (1966) Composition of the Earth. *Nature*, **211**, 616.

Mason, R. G. (1958) A magnetic survey off the west coast of the United States between latitudes 32° and 36° N, longitudes 121° and 128° W. *Geophys. J., R. Astr. Soc.*, **1**, 320.

McDonald, K. L. (1957) Penetration of the geomagnetic secular field through a mantle with variable conductivity. *J. Geophys. Res.*, **62**, 117.

McElhinny, M. W. (In Press) The paleomagnetism of the southern continents—a survey and analysis. To appear in *Proc. Int. Union Geol. Sci.*, UNESCO symposium on continental drift, Montevideo, October 1967.

McElhinny, M. W., Briden, J. C., Jones, D. L., and Brock, A. (1968) Geological and geophysical implications of paleomagnetic results from Africa. *Revs. Geophys.*, 6, 201.

McKinley, D. W. R. (1961) *Meteor science and engineering*. New York: McGraw-Hill.

McQueen, R. G., and Marsh, S. P. (1966) Shock wave compression of iron-nickel alloys and the Earth's core. *J. Geophys. Res.*, **71**, 1751.

Meinesz, F. A. V. (1962) Thermal convection in the Earth's mantle. In Runcorn (1962), p. 145.

Melchior, P. J. (1957) Latitude variation. *Phys. and Chem. of the Earth*, **2**, 212.

Melchior, P. J. (1958) Earth tides. *Adv. Geophys.*, **4**, 391.

Melchior, P. (1964) Earth tides. In Odishaw (1964), p. 163.

Melchior, P. (1966) *The Earth tides*. Oxford: Pergamon.

Menard, H. W. (1965) The world wide oceanic rise-ridge system. *Phil. Trans. Roy. Soc. A.*, **258**, 1.

Merrihue, C. M. (1963) Excess Xenon-129 in chondrules from the Bruderheim meteorite. *J. Geophys. Res.*, **68**, 325.

Middlehurst, B. M., and Kuiper, G. P. (eds.) (1963) *The moon, meteorites and comets*. Vol. 4, *The solar system*. Chicago: University of Chicago Press.

Miller, G. R. (1966) The flux of tidal energy out of the deep oceans. *J. Geophys. Res.* **71**, 2485.

Misra, A. K. and Murrell, S. A. F. (1965) An experimental study of the effect of temperature and stress on the creep of rocks. *Geophys. J., R. Astr. Soc.*, **9**, 509.

Mitchell, J. G. (1968) The argon 40/argon 39 method for potassium-argon age determination. *Geochim. et Cosmochim. Acta*, **32**, 781.

Moore, C. B. (ed.) (1962) *Researches on meteorites*. New York: Wiley.

Morgan, W. J. (1968) Rises, trenches, great faults and crustal blocks. *J. Geophys. Res.*, **73**, 1959.

Munk, W. H., and Hassan, E. S. M. (1961) Atmospheric excitation of the Earth's wobble. *Geophys. J., R. Astr. Soc.*, **4**, 339.

Munk, W. H., and MacDonald, G. J. F. (1960a) Continentality and the gravitational field of the earth. *J. Geophys. Res.*, **65**, 2169.

Munk, W. H., and MacDonald, G. J. F. (1960b) *The Rotation of the Earth, a geophysical discussion*. Cambridge: Cambridge University Press.

Murnaghan, F. D. (1951) *Finite deformation of an elastic solid*. New York: Wiley.

Murrell, S. A. F., and Misra, A. K. (1962) Time dependent strain or "creep" in rocks and similar non-metallic materials. *Trans. Inst. Min. Met.*, **71**, 353.

Nabarro, F. R. N., Basinski, Z. S., and Holt, D. B. (1964) The plasticity of pure single crystals. *Adv. in Phys.*, **13**, 193.

Nagata, T. (1961) *Rock magnetism.* (2nd edition.) Tokyo: Maruzen. (The first edition (1953) is still useful.)

Nagata, T. (1965) Main characteristics of recent geomagnetic secular variation. *J. Geomag. Geoelect.*, **17**, 263.

Nairn, A. E. M., (ed.) (1961) *Descriptive palaeoclimatology.* New York: Wiley Interscience.

Nairn, A. E. M. (ed.) (1964) *Problems in palaeoclimatology.* New York: Wiley Interscience.

Néel, L. (1955) Some theoretical aspects of rock magnetism. *Adv. in Phys.*, **4**, 191

Ness, N. F., Behannon, K. W., Scearce, C. S., and Cantarano, S. C. (1967) Early results from the magnetic field experiment on Lunar Explorer 35. *J. Geophys. Res.*, **72**, 5769.

Nettleton, L. L. (1940) *Geophysical prospecting for oil.* New York: McGraw-Hill.

Newton, R. R. (1968) A satellite determination of tidal parameters and earth deceleration. *Geophys. J., R. Astr. Soc.*, **14**, 505.

Niblett, D. H., and Wilks, J. (1960) Dislocation damping in metals. *Adv. in Phys.*, **9**, 1.

Nicholls, G. D. (1955) The mineralogy of rock magnetism. *Adv. in Phys.*, **4**, 113.

Nicolaysen, L. O. (1961) Graphic interpretation of discordant age measurements on metamorphic rocks. *Annals of N.Y. Acad. Sci.*, **91**, 198.

Ninkovitch, D., Opdyke, N., Heezen, B. C., and Foster, J. H. (1966) Paleomagnetic stratigraphy, rates of deposition and tephrachronology in North Pacific deep sea sediments. *Earth and Plan. Sci. Letters*, **1**, 476.

Odishaw, H. (ed.) (1964) *Research in geophysics.* Vol. 2; *Solid Earth and interface phenomena.* Cambridge: Mass., M.I.T. Press.

O'Keefe, J. A. (1966) The origin of tektites. *Space Sci. Revs.*, **6**, 174.

Oliver, J. (1962) A summary of observed seismic surface wave dispersion. *Bull. Seism. Soc. Am.*, **52**, 81.

Oliver, J. (1966) Prospects for earthquake prediction. *Science Journal*, **2**, 44.

Opdyke, N. D., Glass, B., Hays, J. D. and Foster, J. H. (1966) Paleomagnetic study of Antarctic deep sea cores. *Science*, **154**, 349.

Öpik, E. J. (1961) Tidal deformations and the origin of the Moon. *Astronomical J.*, **66**, 60.

Öpik, E. J. (1966) The stray bodies in the solar system. Part 2. The cometory origin of meteorites. *Adv. in Astronomy and Astrophysics*, **4**, 301.

Öpik, E. J. (1967) Climatic changes. In Runcorn et al. (1967), p. 139.

Orowan, E. (1960) Mechanism of seismic faulting, In Griggs and Handin (1960a), p. 323.

Orowan, E. (1967a) Seismic damping and creep in the mantle. *Geophys. J., R. Astr. Soc.*, **14**,, 191.

Orowan, E. (1967b) Mechanical properties of crust and mantle. In Runcorn et al. (1967), p. 937.

Otsuka, M. (1966) Azimuth and slowness anomalies of seismic waves measured on the central California seismographic array. Part I. Observations. *Bull. Seism. Soc. Am.*, **56**, 223

Pannella, G., MacClintock, C., and Thompson, M. N. (1968) Paleontological evidence of variations in length of synodic month since late Cambrian. *Science*, **162**, 792.

Parasnis, D. S. (1962) *Principles of applied geophysics*. London: Methuen.

Parasnis, D. S. (1966) *Mining geophysics*. Amsterdam: Elsevier.

Parkinson, W. D. (1964) Conductivity anomalies in Australia and the ocean effect. *J. Geomag. Geoelect.*, **15**, 222.

Parry, L. G. (1965 Magnetic properties of dispersed magnetite powders. *Phil. Mag.*, **11**, 303.

Paterson, M. S. (1967) Effect of pressure on stress-strain properties of materials. *Geophys. J., R. Astr. Soc.*, **14**, 13.

Paterson, M. S. and Weiss, L. E. (1961) Symmetry concepts in the structural analysis of deformed rocks. *Bull. Geol. Soc. Am.*, **72**, 841.

Patterson, C. (1956) Age of meteorites and the Earth. *Geochim. et Cosmochim. Acta*, **10**, 230.

Pekeris, C. L., Alterman, Z., and Jarosch, H. (1961) Terrestrial spectroscopy. *Nature*, **190**, 498.

Plafker, G. (1965) Tectonic deformation associated with the 1964 Alaska earthquake. *Science*, **148**, 1675.

Powell, R. W. (1953) The electrical resistivity of liquid iron. *Phil. Mag.*, **44**, 772.

Press, F. (1965) Displacements, strains and tilts at teleseismic distances. *J. Geophys. Res.*, **70**, 2395.

Press, F. (1966) Free oscillations, aftershocks and Q. In Steinhart and Smith (1966), p. 498.

Press, F. (1968a) Density distribution in the Earth. *Science*, **160**, 1218..

Press, F. (1968b) Earth models obtained by Monte Carlo inversion. *J. Geophys. Res.*, **73**, 5223.

Press, F., Ben-Menahem, A., and Toksöz, M. N. (1961) Experimental determination of earthquake fault length and rupture velocity. *J. Geophys. Res.*, **66**, 3471.

Press, F., and Brace, W. F. (1966) Earthquake prediction. *Science*, **152**, 1575.

Press, F., and Jackson, D. (1965) Alaskan earthquake, 27 March, 1964: vertical extent of faulting and elastic strain release. *Science*, **147**, 867.

Proudman, J. (1953) *Dynamical oceanography* London: Methuen.

Radbruch, D. H., and others (1966) Tectonic creep in the Hayward fault zone, California. *U.S. Geol. Survey Circular*, **525**.

Raleigh, C. B., and Paterson, M. S. (1965) Experimental deformation of sepentinite and its tectonic implications. *J. Geophys. Res.*, **70**, 3965.

Reid, H. F. (1911) The elastic rebound theory of earthquakes. *Bull. Dept. Geology. Univ. California*, **6**, 413.

Reynolds, J. H. (1960) The age of the elements in the solar system. *Scientific American*, **203**(5), 171.

Richter, C. F. (1958) *Elementary seismology*. San Francisco: Freeman.

Rikitake, T. (1958) Oscillations of a system of disk dynamos. *Proc. Camb. Phil. Soc.*, **54**, 89.

Rikitake, T. (1966a) A five year plan for earthquake prediction research in Japan. *Tectonophysics*, **3**, 1.

Rikitake, T. (1966b) Westward drift of the equatorial component of the Earth's magnetic dipole. *J. Geomag. Geoelect.*, **18**, 383.

Rikitake, T. (1966c) *Electromagnetism and the Earth's interior.* Amsterdam: Elsevier.

Ringwood, A. E. (1959) On the chemical evolution and densities of the planets. *Geochim. et Cosmochim. Acta*, **15**, 157.

Ringwood, A. E. (1962) A model for the upper mantle. *J. Geophys. Res.*, **67**, 857 and 4473.

Ringwood, A. E. (1966a) Chemical evolution of the terrestrial planets. *Geochim. et Cosmochim. Acta*, **30**, 41.

Ringwood, A. E. (1966b) Genesis of chondritic meteorites. *Revs. Geophys.*, **4**, 113.

Ringwood, A. E. (1966c) The chemical composition and origin of the Earth. In Hurley (1966), p. 287.

Ringwood, A. E. (1967) New Light on the Earth's interior. *New Scientist*, **33**, 530.

Ringwood, A. E., and Green, D. H. (1966) An experimental investigation of the gabbro-ecolgite transformation and some geophysical implications. *Tectono-physics*, **3**, 383.

Ringwood, A. E., and Major, A. (1966a) Some high pressure transformations in olivines and pyroxenes. *J. Geophys. Res.*, **71**, 4448.

Ringwood, A. E., and Major, A. (1966b) Synthesis of $Mg_2 SiO_4$-$Fe_2 SiO_4$ spinel solid solutions. *Earth and Plan. Sci. Letters*, **1**, 241.

Roberts, P. H., and Scott, S. (1965) On analysis of the secular variation 1. A hydro-magnetic constraint: Theory. *J. Geomag. Geoelect.*, **17**, 137.

Rochester, M. G. (1960) Geomagnetic westward drift and irregularities in the Earth's rotation. *Phil. Trans. Roy. Soc. A*, **252**, 531.

Rochester, M. G., and Smylie, D. E. (1965) Geomagnetic core-mantle coupling and the Chandler wobble. *Geophys. J., R. Astr. Soc.*, **10**, 289.

Roden, R. B. (1963) Electromagnetic core-mantle coupling. *Geophys. J., R. Astr. Soc.*, **7**, 361.

Roy, A. E. (1967) Bode's law. In Runcorn et al. (1967), p. 146.

Runcorn, S. K. (1955) The electrical conductivity of the Earth's mantle. *Trans. Am. Geophys. Un.*, **36**, 191.

Runcorn, S. K. (ed.) (1962) *Continental drift.* New York: Academic Press.

Runcorn, S. K. (1964) Changes in the Earth's moment of inertia. *Nature*, **204**, 823.

Runcorn, S. K. (ed.) (1967a) *Mantles of the Earth and terrestrial planets.* New York: Wiley-Interscience.

Runcorn, S. K. (1967b) The problem of the figure of Mars. In Runcorn (1967a), p. 425.

Runcorn, S. K., et al. (eds.) (1967) *International Dictionary of Geophysics.* (2 vols.) Oxford: Pergamon.

Russell, R. D., and Farquhar, R. M. (1960) *Lead isotopes in geology.* New York: Interscience.

Sasajima, S., (1965) Geomagnetic secular variation revealed in the baked earths in West Japan (Part 2). Change of the field intensity. *J. Geomag. Geoelect.*, **17**, 413.

Sassa, K., and Nishimura, E. (1956) On phenomena forerunning earthquakes. *Bulletin of Disaster Prevention Research Institute, Kyoto University*, **13**, 1. (Also *Trans. Am. Geophys. Un.*, **32**, (1951), 1.

Savage, J. C., and Hastie, L. M. (1966) Surface deformation associated with dip slip faulting. *J. Geophys. Res.*, **71**, 4897.

Schaeffer, O. A., and Zähringer, J. (1966) *Potassium-argon dating.* New York: Springer.

Scheidegger, A. E. (1961) *Theoretical geomorphology.* Berlin: Springer.

Scheidegger, A. E. (1963) *Principles of geodynamics.* (2nd edition.) Berlin: Springer.

Schmucker, U. (1964) Anomalies of geomagnetic variations in the southwestern United States. *J. Geomag. Geoelect.*, **15**, 193.

Scholz, C. H. (1968) The frequency-magnitude relation of microfracturing in rock and its relation to earthquakes. *Bull. Seism. Soc. Am.*, **58**, 399.

Shields, R. M., Pinson, W. H., and Hurley, P. M. (1966) Rubidium-strontium analyses of the Bjurbole chondrite. *J. Geophys. Res.*, **71**, 2163.

Shillibeer, H. A., and Russell, R. D. (1955) The argon-40 content of the atmosphere and the age of the Earth. *Geochim. et Cosmochim. Acta*, **8**, 16.

Short, J. M., and Anderson, C. A. (1965) Electron microprobe analyses of the Widmanstätten structure of nine iron meteorites. *J. Geophys. Res.*, **70**, 3745.

Simpson, J. F. (1967) Earth tides as a triggering mechanism for earthquakes. *Earth and Plan. Sci. Letters*, **2**, 473.

Singer, S. F. (1968) The origin of the moon and geophysical consequences. *Geophys. J. R. Astr. Soc.*, **15**, 205.

Slichter, L. B. (1967) Free oscillations of the Earth. In Runcorn et al. (1967), p. 331.

Slichter, L. B., MacDonald, G. J. F., Caputo, M., and Hager, C. L. (1966) Comparison of spectra for spheroidal modes excited by the Chilean and Alaskan quakes. (Abstract) *Geophys. J., R. Astr. Soc.*, **11**, 256.

Small, J. B., and Parkin, E. J. (1967) Horizontal and vertical crustal movement in the Prince William Sound, Alaska, earthquake of 1964. Paper presented to the fifth United Nations Regional Cartographic Conference for Asia and the Far East, Canberra, March 1967. Washington: E.S.S.A., U.S. Department of Commerce.

Smith, E. J. (1967) A review of lunar and planetary magnetic field measurements using space probes. In Hindmarsh et al. (1967), p. 271.

Smith, P. J. (1967) The intensity of the ancient geomagnetic field: a review and analysis. *Geophys. J., R. Astr. Soc.*, **12**, 321.

Smith, S. W. (1967) Free vibrations of the Earth. In Runcorn et al. (1967), p. 344.

Sneddon, I. N. (1961) *Special functions of mathematical physics and chemistry.* Edinburgh: Oliver and Boyd.

Society of Exploration Geophysicists. (1967) *Mining geophysics.* Vols. I and II. Tulsa Okla: Society of Exploration Geophysicists.

Spencer Jones, H. (1956) The origin of the solar system. *Phys. and Chem. of the Earth*, **1**, 1.

Stacey, F. D. (1963a) The physical theory of rock magnetism. *Adv. in Phys.*, **12**, 45.

Stacey, F. D. (1963b) The theory of creep in rocks and the problem of convection in the Earth's mantle. *Icarus*, **1**, 304.

Stacey, F. D. (1964a) Could earthquakes be predicted? *New Scientist*, **21**, 70.

Stacey, F. D. (1964b) The seismomagnetic effect. *Pure. and Appl. Geophys.*, **58**, 5.

Stacey, F. D. (1967a) The Koenigsberger ratio and the nature of thermoremanence in igneous rocks. *Earth and Plan. Sci. Letters*, **2**, 67.

Stacey, F. D. (1967b) Consequences of energy dissipation by creep in the mantle. *Geophys. J., R. Astr. Soc.*, **14**, 433.

Stacey, F. D. (1967c) Convecting mantle as a thermodynamic engine. *Nature*, **214**, 476.

Stacey, F. D. (1967d) Electrical resistivity of the Earth's core. *Earth and Plan. Sci. Letters*, **3**, 204.

Stacey, F. D. (1967e) Palaeomagnetism of meteorites. In Runcorn et al. (1967), p. 1141.

Stacey, F. D. (1968) Energy balance of mantle convection. *Tectonophysics*, **5**, 441.

Stauder, W. (1962) The focal mechanism of earthquakes. *Advances in Geophysics*, **9**, 1.

Stauder, W., and Bollinger, G. A. (1966) The focal mechanism of the Alaska earthquake of March 28, 1964, and of its aftershock sequence. *J. Geophys. Res.*, **71**, 5283.

Steinbrugge, K. V., Zacher, E. G., Tocher, D., Whitten, C. A., and Claire, C. N. (1960) Creep on the San Andreas fault. *Bull Seism. Soc. Am.*, **50**, 389.

Steinhart, J. S., and Smith, T. J. (eds.) (1966) *The Earth beneath the continents.* (Geophysical Monograph 10). Washington: American Geophysical Union.

Stoneley, R. (1961) The oscillations of the Earth. *Phys. and Chem. of Earth*, **4**, 239.

Strutt, R. J. (1906) On the distribution of radium in the Earth's crust and on the Earth's internal heat. *Proc. Roy. Soc. A*, **77**, 472.

Sverdrup, H. U., Johnson, M. W., and Fleming, R. H. (1942) *The oceans—their physics, chemistry and general biology.* Englewood Cliffs, N.J.: Prentice-Hall.

Sykes, L. R. (1967) Mechanism of earthquakes and nature of faulting in the mid-ocean ridges. *J. Geophys. Res.*, **73**, 2131.

Sykes, L. R., Isacks, B. L., and Oliver, J. (1969) Spatial distribution of deep and shallow earthquakes of small magnitudes in the Fiji-Tonga region. *Bull. Seism. Soc. Am.* (in press).

Takeuchi, H. (1950) On the Earth tide of the compressible Earth of variable density and elasticity. *Trans. Am. Geophys. Un.*, **31**, 651.

Takeuchi, H. (1966) *Theory of the Earth's interior.* Waltham, Mass.: Blaisdell.

Tarling, D. H. (1965) The palaeomagnetism of some of the Hawaiian islands. *Geophys. J., R. Astr. Soc.*, **10**, 93.

TASS. (1967) Report, translated by C. C. Nikiforoff and others. Venus 4: an automatic interplanetary station. *Trans. Am. Geophys. Un.*, **48**, 931.

Taylor, C. E. (1968) New determination of the diameter of Neptune. *Nature*, **219**, 474.

Ter Haar, D., and Cameron, A. G. W. (1963) Historical review of theories of the origin of the solar system. In Jastrow and Cameron (1963), p. 1.

Thellier, E., and Thellier, O. (1959) Sur l'intensité du champ magnétique terrestre dans le passé historique e geologique. *Ann. de Geophys.*, **15**, 285.

Tilton, G. R., and Reed, G. W. (1963) Radioactive heat production in eclogite and some ultramafic rocks. In Geiss and Goldberg (1963), p. 31.

Tilton, G. R., and Steiger, R. H. (1965) Lead isotopes and the age of the Earth. *Science*, **150**, 1805.

Tocher, D. (1957) Anisotropy in rocks under simple compression. *Trans. Am. Geophys. Un.*, **38**, 89.

Tocher, D. (1960) Creep on the San Andreas fault: creep rate and related measurements at Vinyard, California. *Bull. Seism. Soc. Am.*, **50**, 396.

Toksöz, M. N., and Arkani-Hamed, J. (1967) Seismic delay times: correlation with other data. *Science*, **158**, 783.

Toksöz, M. N., Chinnery, M. A., and Anderson, D. L. (1967) Inhomogeneities in the Earth's mantle. *Geophys. J., R. Astr. Soc.*, **13**, 31.

Tomaschek, R. (1957) Tides of the solid Earth. *Handbuch der Physik*, **48** (*Geophysics* **2**), 775.

Torreson, O. W., Murphy, T., and Graham, J. W. (1949) Magnetic polarization of sedimentary rocks and the Earth's magnetic history. *J. Geophys. Res.*, **54**, 111.

Tozer, D. C. (1959) The electrical properties of the Earth's interior. *Phys. and Chem. of Earth*, **3**, 414.

Tozer, D. C. (ed.) (1967) The proceedings of the International Upper Mantle Committee symposium on non-elastic processes in the mantle, held at Newcastle-upon-Tyne 1966 February 21-25. *Geophys. J., R. Astr. Soc.*, **14**, 1-450.

Trendall, A. F. (1966) Carbon dioxide in the pre-Cambrian atmosphere. *Geochim. et Cosmochim. Acta*, **30**, 435.

Tsuboi, C. (1956) Earthquake energy, earthquake volume, aftershock area and strength of the Earth's crust. *J. Phys. of the Earth*, **4**, 63.

Tsuboi, C., Wadati, K., and Hagiwara, T. (1962) *Prediction of earthquakes—progress to date and plans for further development*. Tokyo: Earthquake Research Institute.

Uffen, R. J. (1963) Influence of the Earth's core on the origin and evolution of life. *Nature*, **198**, 143.

Urey, H. C. (1947) The thermodynamic properties of isotopic substances. *J. Chem. Soc.*, **1947 Part 1**, 562.

Urey, H. C. (1952) *The planets, their origin and development*. New Haven, Conn.: Yale University Press.

Urey, H. C. (1957) Boundary conditions for theories of the origin of the solar System. *Phys. and Chem. of the Earth*, **2**, 46.

Urey, H. C. (1963) The origin and evolution of the solar system. In Le Galley (1963), p. 123.

Vacquier, V. (1962) Magnetic evidence for horizontal displacements in the floor of the Pacific Ocean. In Runcorn (1962), p. 135.

Vacquier, V. (1965) Horizontal displacements in the Earth's crust: transcurrent faulting in the ocean floor. *Phil. Trans. Roy. Soc. A*, **258**, 77.

Van Allen, J. A., Krimigris, S. M., Frank, L. A., and Armstrong, T. P. (1967) Venus: an upper limit on intrinsic magnetic dipole moment based on absence of a radiation belt. *Science*, **158**, 1673.

Van den Heuvel, E. P. J. (1966) On the precession as a cause of Pleistocene variations of the Atlantic ocean water temperatures. *Geophys. J., R. Astr. Soc.*, **11**, 323.

Van Dorn, W. G. (1965) Tsunamis, *Advances in Hydroscience*, **2**, 1. (*New York: Academic Press*).

Van Zijl, J. S. V., Graham, K. W. T., and Hales, A. L. (1962) The palaeomagnetism of the Stormberg lavas of South Africa. *Geophys. J., R. Astr. Soc.*, **7**, 23 and 169.

Verhoogen, J. (1961) Heat balance of the Earth's core. *Geophys. J., R. Astr. Soc.*, **4**, 276.

Vestine, E. H. (1953) On variations of the geomagnetic field, fluid motions and the rate of the Earth's rotation. *J. Geophys. Res.*, **58**, 127.

Vestine, E. H., LaPorte, L., Lange, I., Cooper, C., and Hendrix, W. C. (1947) Description of the Earth's main magnetic field and its secular change, 1905-1945. Carnegie Institution of Washington: Publication **578**.

Vestine, E. H., Lange, I., LaPorte, L., and Scott, W. E. (1947) The geomagnetic field, its description and analysis. *Carnegie Institution of Wahington; Publication* **580**.

Vestine, E. H., Sibley, W. L., Kern, J. W., and Carlstedt, J. L. (1963) Integral and spherical-harmonic analysis of the geomagnetic field for 1955.0. *J. Geomag. Geoelect.*, **15**, 47 and 73.

Vine, F. J., and Mathews, D. H. (1963) Magnetic anomalies over ocean ridges. *Nature*, **199**, 947.

Volarovich, M. P., Balashov, D. B., Tomashevskaya, I. S., and Pavlogradskii, V. A., (1963) A study of the effect of uniaxial compression upon the velocity of elastic waves in rock samples under conditions of high hydrostatic pressure. *Bull. (Izv.) Acad. Sci. U.S.S.R.* (Geophys. Ser.), **1963**, 728.

Von Herzen, R. P., and Maxwell, A. E. (1964) Measurement of heat flow at the preliminary Mohole site off Mexico. *J. Geophys. Res.*, **69**, 741.

Waddington, C. J. (1967) Paleomagnetic field reversals and cosmic radiation. *Science* **158**, 913.

Wasserburg, G. J., MacDonald, G. J. F., Hoyle, F., and Fowler, W. A. (1964) The relative contributions of uranium, thorium and potassium to heat production in the Earth. *Science*, **143**, 465.

Watkins, N. D., and Goodell, H. G. (1967) Geomagnetic polarity change and faunal extinction in the Southern Ocean. *Science*, **156**, 1083.

Weertman, J. (1955) Theory of steady state creep based on a dislocation climb. *J. Appl. Phys.*, **26**, 1213.

Weertman, J. (1957) Steady state creep of crystals. *J. Appl. Phys.*, **28**, 1185.

Weizsäcker, C. F. von. (1943) Uber die Entstehung des Planetensystems. *Z. Astrophysik*, **22**, 319.

Wetherill, G. W. (1968) Stone meteorites: time of fall and origin. *Science*, **159**, 79.

Whitten, C. A. (1956) Crustal movement in California and Nevada. *Trans. Am. Geophys. Un.*, **37**, 393.

Wilkins, G. A. (1967) The determination of the mass and oblateness of Mars from the orbits of its satellites. In Runcorn (1967a), p. 77.

Wilson, J. T. (1965a) Convection currents and continental drift: evidence from ocean islands suggesting movement in the Earth. *Phil. Trans. Roy. Soc. A*, **258**, 145.

Wilson, J. T. (1965b) A new class of faults and their bearing on continental drift. *Nature*, **207**, 343.

Wilson, R. L., and Watkins, N. D. (1967) Correlation of petrology and natural magnetic polarity in Columbia Plateau basalts. *Geophys. J., R. Astr. Soc.*, **12**, 405.

Wood, J. A. (1963a) Physics and chemistry of meteorites. In Middlehurst and Kuiper (1963), p. 337.

Wood, J. A. (1963b) On the origin of chondrules and chondrites. *Icarus*, **2**, 152.

Yukutake, T. (1965) The solar cycle contribution to the secular change in the geomagnetic field. *J. Geomag. Geoelect.*, **17**, 287.

Name Index

Subject Index